Wastewater Treatment and Reuse Technologies

Wastewater Treatment and Reuse Technologies

Special Issue Editors

Faisal Ibney Hai
Kazuo Yamamoto
Jega Veeriah Jegatheesan

MDPI • Basel • Beijing • Wuhan • Barcelona • Belgrade

MDPI

Special Issue Editors

Faisal Ibney Hai
University of Wollongong
Australia

Kazuo Yamamoto
University of Tokyo
Japan

Jega Veeriah Jegatheesan
RMIT University
Australia

Editorial Office
MDPI
St. Alban-Anlage 66
Basel, Switzerland

This is a reprint of articles from the Special Issue published online in the open access journal *Applied Sciences* (ISSN 2076-3417) from 2017 to 2018 (available at: http://www.mdpi.com/journal/applsci/special_issues/reuse_technologies)

For citation purposes, cite each article independently as indicated on the article page online and as indicated below:

LastName, A.A.; LastName, B.B.; LastName, C.C. Article Title. *Journal Name* **Year**, *Article Number*, Page Range.

ISBN 978-3-03897-101-6 (Pbk)
ISBN 978-3-03897-102-3 (PDF)

Cover image courtesy of Jega Jegatheesan.

Contents

About the Special Issue Editors

Faisal Ibney Hai is the leader of the Strategic Water Infrastructure Laboratory at the University Of Wollongong, Australia. He has forged a strong collaboration with key industry partners (e.g., Sydney Water) and internationally leading researchers which has led to competitive grants and publications. A highly cited researcher, Prof. Hai has edited three recent books on the application of membrane technology in wastewater treatment, resource recovery, and biofuel production with distinguished overseas researchers as co-editors. He is the lead editor of one of these books (Membrane Biological Reactors, International Water Association (IWA) Publishing, UK, 2014), which is among the 5% best sellers of the IWA portfolio. Given his international research standing in membrane-based wastewater treatment processes, particularly in membrane bioreactor (MBR) technology, he has been appointed as an Associate Editor/ Editorial Board Member of *Water Science and Technology* (IWA, UK), *Journal of Water and Environment Technology* (Japan Society on Water Environment) and *Applied Sciences* (Environmental and Sustainable Science and Technology section), which are prime outlets for research communication to water professionals and researchers worldwide.

Kazuo Yamamoto Truly a leading authority in wastewater treatment and reuse, Prof. Yamamoto's revolutionary research in collaboration with international partners has provided the global water community with a better scientific framework to formulate policies and best practices. Professor Yamamoto's invention paved the way for the development of the membrane (MBR) technology itself and of the present-day membranes for water and wastewater treatment. With the increasing freshwater scarcity and the simultaneous drive to reuse wastewater, thanks to Professor Yamamoto's ongoing initiative, leadership, and dedication to the field, today MBR is considered by the industry as a technology providing consistent and high-quality product water at a reduced footprint, compared to the conventional wastewater treatment technologies.

Jega Veeriah Jegatheesan (Jega) has 20 years of experience in water research. His research focuses on sustainable catchment management through the application of novel treatment processes, resource recovery, and mathematical modelling. He has co-edited four books, was managing guest editor for 34 special issues in peer-reviewed journals, and has published over 120 journal articles. He is the chief editor of a book series entitled "Applied Environmental Science and Engineering for a Sustainable Future" published by Springer. He is an Editor for the journal *Sustainable Environment Research*, Associate Editor for the *Journal of Water Sustainability* and an Editorial Board Member of a number of journals. His core expertise includes Membrane system design, Aquaculture, Desalination, Forward osmosis, Resource recovery, and Water distribution maintenance management. Professor Jegatheesan is the co-founder and the chair of an international conference on the "Challenges in Environmental Science and Engineering (CESE)" which is held annually since 2008 around the world.

Preface to "Wastewater Treatment and Reuse Technologies"

Wastewater treatment allows for the safe disposal of municipal and industrial wastewater to protect public health and the ecosystem. Reclaimed or recycled water and adequately treated wastewater is reused for a variety of applications, including landscaping, irrigation, and recharging groundwater aquifers. In many parts of the world, the problem of water scarcity is being exacerbated by urban growth and increasingly erratic rainfall patterns due to climate change. This crisis has generated an ever-increasing drive for the use of alternative water sources, especially wastewater reclamation. However, water reuse practices raise concern due to the potential adverse health effects associated with wastewater-derived resistant pollutants. Conventional sewage treatment plants can effectively remove the total levels of organic carbon and nitrogen, as well as achieve some degree of disinfection. However, these plants have not been specifically designed to remove priority pollutants. Thus, the development of advanced wastewater treatment processes is necessary.

It is a great pleasure to present this edited volume on wastewater treatment and reuse. This is a collection of 12 publications from esteemed research groups around the globe. The articles belong to the following broad categories: biological treatment process parameters, sludge management and disinfection; removal of trace organic contaminants; removal of heavy metals; and synthesis and fouling control of membranes for wastewater treatment. We would like to thank the editorial team of MDPI, particularly managing editor Ryan Pei, for their great assistance in this project.

Faisal Hai dedicates this work to his late father Md. Abdul Hai, who was a great admirer of his work and a constant source of inspiration.

Faisal Ibney Hai, Kazuo Yamamoto , Jega Veeriah Jegatheesan
Special Issue Editors

applied
sciences

MDPI

Article

Dissolved Oxygen Control in Activated Sludge Process Using a Neural Network-Based Adaptive PID Algorithm

Xianjun Du [1,2,3,4], Junlu Wang [1,3,4], Veeriah Jegatheesan [2,*] and Guohua Shi [5]

[1] College of Electrical and Information Engineering, Lanzhou University of Technology, Lanzhou 730050, China; 27dxj@163.com (X.D.); wjllanzhou@126.com (J.W.)
[2] School of Engineering, Royal Melbourne Institute of Technology (RMIT) University, Melbourne 3000, Australia
[3] Key Laboratory of Gansu Advanced Control for Industrial Processes, Lanzhou University of Technology, Lanzhou 730050, China
[4] National Demonstration Center for Experimental Electrical and Control Engineering Education, Lanzhou University of Technology, Lanzhou 730050, China
[5] Department of Energy and Power Engineering, North China Electric Power University, Baoding 071003, China; ghuashi@outlook.com
* Correspondence: jega.jegatheesan@rmit.edu.au; Tel.: +61-3-9925-0810

Received: 20 December 2017; Accepted: 6 February 2018; Published: 9 February 2018

Featured Application: This work is currently undergoing field testing at Pingliang Wastewater Treatment Plant situated in Gansu province, China, especially for the control of dissolved oxygen concentration in the activated sludge process of the wastewater treatment. By implementing this control algorithm, we can achieve two goals, namely improving the efficiency of wastewater treatment and reducing the aeration energy. Meanwhile, the method proposed in this work can also be extended to other large- or medium-scale wastewater treatment plants in the future.

Abstract: The concentration of dissolved oxygen (DO) in the aeration tank(s) of an activated sludge system is one of the most important process control parameters. The DO concentration in the aeration tank(s) is maintained at a desired level by using a Proportional-Integral-Derivative (PID) controller. Since the traditional PID parameter adjustment is not adaptive, the unknown disturbances make it difficult to adjust the DO concentration rapidly and precisely to maintain at a desired level. A Radial Basis Function (RBF) neural network (NN)-based adaptive PID (RBFNNPID) algorithm is proposed and simulated in this paper for better control of DO in an activated sludge process-based wastewater treatment. The powerful learning and adaptive ability of the RBF neural network makes the adaptive adjustment of the PID parameters to be realized. Hence, when the wastewater quality and quantity fluctuate, adjustments to some parameters online can be made by RBFNNPID algorithm to improve the performance of the controller. The RBFNNPID algorithm is based on the gradient descent method. Simulation results comparing the performance of traditional PID and RBFNNPID in maintaining the DO concentration show that the RBFNNPID control algorithm can achieve better control performances. The RBFNNPID control algorithm has good tracking, anti-disturbance and strong robustness performances.

Keywords: dissolved oxygen concentration; radial basis function (RBF) neural network; adaptive PID; dynamic simulation

---

1. Introduction

Currently, the activated sludge process is the most widely used process in wastewater treatment plants to reduce the biochemical oxygen demand (BOD), nutrients and to some extent other micro-pollutants such as pharmaceuticals, personal care products and other household chemicals. The concentration of dissolved oxygen (DO) in the aeration tank(s) in an activated sludge process is an important process control parameter that has a great effect on the treatment efficiency, operational cost and system stability. As the DO drops, the quantity of these filamentous microorganisms increases, adversely affecting the settle-ability of the activated sludge. It is important to recognize these early warning signs and make corrections to dissolved-oxygen levels before the quality of the effluent deteriorates. If dissolved oxygen continues to drop, even low dissolved-oxygen filamentous microorganisms will not be present in the mixed liquor, and treatment efficiencies will be seriously affected. At this point, effluent turbidity will increase and treatment will deteriorate rapidly. Higher dissolved oxygen is often a target, but in reality, this is for the assurance of mixing. If dissolved oxygen is 5.0 or higher there is a good chance that dead zones are minimal since normal currents and mixing will transport the oxygenated mixed liquor throughout the reactor. However, if the dissolved oxygen is excessive then there could be problems in the settling of sludge due to shearing of flocs and re-suspension of inert materials. A high DO concentration also makes the denitrification less efficient. Both the above-mentioned factors will lead to waste of energy. On the other hand, a low DO level cannot supply enough oxygen to the microorganisms in the sludge, so the efficiency of organic matter degradation is reduced [1,2]. Therefore, the premise of how the wastewater treatment process can perform stably will depend on how effectively the concentration of DO is be maintained within a reasonable range [3]. Due to the complex nature of microbial activities that are present in an activated sludge process, even a small change introduced to the system (for example, change in flow rate, water quality of the influent, the temperature of the wastewater in the reactors and so on) can affect the concentration of DO. The air supplied to aeration tanks by blowers allows the oxygen to be transferred from the air to the liquid phase (wastewater). The oxygen transfer is a complex process characterized by large time-delays as well as strong nonlinearity, coupling and disturbance, which further increases the difficulty of controlling the concentration of DO [4,5]. A large number of studies have been carried out and achievements have been made by researchers all over the world to control the concentration of DO level; a series of control methods to control the concentration of DO have been put into practice and they have achieved some good effects.

Currently, the proportional–integral (PI) or proportional–integral–derivative (PID) control strategy is widely used in the process control of wastewater treatment plants. It is well known that the control effect might be affected by the unknown, unexpected disturbances and the great changes of operation conditions while using the PI or PID control strategy. In order to improve the dissolved oxygen control performance of the controller in the wastewater treatment process, various solutions are proposed, such as fuzzy adaptive PID, multivariable robust control and model predictive control (MPC) strategy [6–9]. MPC [2] is an effective way to control DO, not only maintaining the DO concentration at a set value, but also catching up with the real-time changes that occur in the process. Belchior et al. Proposed an adaptive fuzzy control (AFC) strategy for tracking the DO set-points applied to the Benchmark Simulation Model No. 1 (BSM1) [10] that was proposed by International Water Association (IWA) [11]. AFC is a supervised data-driven control method designed with a smooth switching scheme between supervisory and nonsupervisory modes. Results show that it can learn and improve control rules resulting in accurate DO control. Yu et al. simulated intelligent control method and traditional PID control method in combination. Based on their respective advantages, they achieved better control effect when they used the intelligent PID control algorithm into applications of control practice in Haicheng sewage treatment plant, China [12].

Scholars also introduced the neural network into the control of DO in wastewater treatment process, for example, back propagation (BP) neural network [13]. Furthermore, neural network is employed into some control strategies for the wastewater treatment process control. Macnab [14] and

Mirghasemi [15] proposed a robust adaptive neural network control strategy and used it to control the dissolved oxygen in activated sludge process application. The proposed method prevented weight drift and associated bursting, without sacrificing performance. They improved the control performance by using the algorithm, Cerebellar Model Arithmetic Computer (CMAC) to estimate the nonlinear behavior of the system. Results showed that it can effectively avoid state error. Ruan et al. proposed an on-line hybrid intelligent control system based on a genetic algorithm (GA) evolving fuzzy wavelet neural network software sensor to control dissolved oxygen (DO) in an anaerobic/anoxic/oxic (AAO) process for treating papermaking wastewater [16]. The results indicate that the reasonable forecasting and control performances were achieved with optimal DO, and the effluent quality was stable at and below the desired values in real time. It can be an effective control method, attaining not only adequate effluent quality but also minimizing the demand for energy, and is easily integrated into a global monitoring system for purposes of cost management [16]. Qiao Junfei et al. proposed a control method based on self-organizing T-S fuzzy neural network (SO-TSFNN), while using its powerful self-learning, fault-tolerant and adaptive abilities of the environment [17]. It realized the real-time control of dissolved oxygen of the BSM1 and achieved better control effect for DO concentration with good adaptability. Li Minghe et al. proposed a neural network predictive control method for dissolved oxygen based on Levenberg-Marquardt (LM) algorithm [18]. It overcomes the shortages of the BP neural network by combining with the LM algorithm to improve the prediction accuracy of neural network and the tracking performance of dissolved oxygen control. Xu et al. proposed a new control strategy of DO concentration based on fuzzy neural network (FNN). The minimum error of the gradient descent method is used to adjust the parameters of the neural network on-line. Simulation results show that the FNN controller is better than other compared methods [19]. Lin and Luo studied the design approach of a neural adaptive control method based on a disturbance observer. A RBF neural network is employed to approximate the uncertain dynamic model of the wastewater treatment process. The effectiveness of the controller is verified by simulation their study [20]. Han et al. proposed a self-organizing RBF neural network model predictive control (SORBF-MPC) method for controlling DO concentration in WWTP. The hidden nodes in RBF neural network can be added or deleted online on the basis of node activity and mutual information to achieve necessary dynamics of the network. The application results of DO concentration control show that SORBF-MPC can effectively control the process of dissolved oxygen [21]. Zhou Hongbiao proposed a self-organizing fuzzy neural network (SOFNN) control method based on. According to the activation strength and mutual information, the algorithm dynamically adds and reduces the number of neurons in the regular layer to meet the dynamic changes of the actual working conditions. At the same time, the gradient descent algorithm is used to optimize the center, width and output weight of the membership function online to ensure the convergence of SOFNN. Finally, experimental verification was carried out in the international benchmark simulation platform BSM1. Experimental results on the BSM1 show that, compared with control strategies of PID, fuzzy logic control (FLC) and FNN with fixed structure, SOFNN has a better performance on tracking accuracy, control stability and adaptive ability [22].

Although there are many studies on how to control the DO concentration in wastewater treatment system by using neural networks and predictive control methods with great outcomes, these kinds of methods have complicated structures and require large amount of computations. They are difficult to implement in practical engineering applications. Basically, most of the existing wastewater treatment plants (WWTPs) are still using PID, a simple and practical control strategy, to control the process. Unfortunately, since the parameters of the PID control algorithm are difficult to set up in advance which are strongly affected by the nonlinearity and large time-delay characters of the wastewater treatment process the control effect maybe unsatisfactory and the key problem is the parameters are not self-adjusted [23]. Therefore, combining intelligent algorithm with the PID algorithm becomes an effective way to realize simple structures and the control requirements of wastewater treatment process in actual WWTPs.

When we use intelligent algorithm into PID, the parameters can be adjusted real-time according to the control effect of current strategy (such as gradient descent method) to avoid the problem of difficult-to-adjust PID parameters. At the same time, they can be adaptively adjusted according to the change of operation environment and dynamic disturbances. There are two ways to improve the control accuracy: one is to improve the accuracy of the measurement equipment of dissolved oxygen concentration, and another is the selection of the center point and the node width of the neural network.

In this paper, a neural network-based adaptive PID control algorithm is proposed. The radial basis function (RBF) neural network is employed which has good generalization ability besides the strong self-learning and adaptive abilities and has a simple network structure. The proposed network already has research and application basis for the control of practical processes in some other areas [24–26]. Compared with the traditional PID control algorithm, the proposed RBF neural network-based adaptive PID (RBFNNPID) control algorithm comprises the advantages of these two methods. It is simple, easy to implement and has better control accuracy. More importantly, one does not need to set up the best parameters of PID in advance; that is to say, it can solve the problem of traditional PID controller that has difficulty in adjusting parameters online.

Considering the control problem of DO concentration level in the wastewater treatment process, in this paper, the Benchmark model of BSM1 is introduced and the implementation of the RBF neural network-based adaptive PID control algorithm is discussed. It can be seen from the comparison simulation results that RBFNNPID control algorithm can effectively improve the control accuracy of dissolved oxygen concentration under the Benchmark as opposed to traditional PID.

2. Materials and Methods

2.1. Activated Sludge Process (ASP) and Benchmark Simulation Model No. 1 (BSM1)

Activated sludge model No. 1 (ASM1) is a mathematical model that is widely accepted and applied in the research and application of activated sludge process (ASP) used in biological wastewater treatment systems. The typical ASP is shown in Figure 1, which includes two parts, the biological (more accurately biochemical) reaction tanks (or aeration tanks) and the secondary settler [27,28]. In the aeration tanks, the microorganisms are divided into active heterotrophic and autotrophic bacteria. The 13 reaction components and 8 reaction processes of the organic matter present in the influent are incorporated into the ASM1 [28–30]. In each process, all the organic substances and microorganisms have their own reaction rates and stoichiometry. Since the model has been published, researchers have been using the ASM1 model to verify their new proposed control algorithms of the DO concentration of ASP.

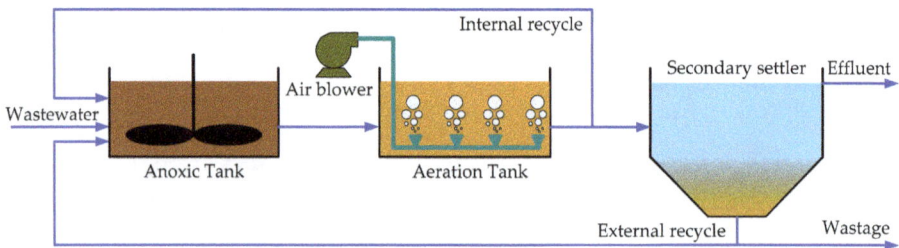

Figure 1. Typical biological (or biochemical) ASP to treat wastewater. ASP: Activated Sludge Process.

The activated sludge process aims to achieve, at minimum cost, sufficiently low concentrations of biodegradable matter and nutrients in the effluent together with minimal sludge production. In order to achieve this, the process has to be controlled [28]. However, it is difficult to predict the performance of the proposed or applied control strategy based on existing reference, process or location.

To enhance the acceptance of innovative control strategies the performance evaluation should be based on a rigorous methodology that includes a simulation model, plant layout, controllers, performance criteria and test procedures.

The first Benchmark Simulation Layout (BSM1), which was based on the ASM1, is relatively a simple layout and is shown in Figure 2. Similar to ASM1, the first part of BSM1 is also a biological (or biochemical) activated sludge reactor, which is comprised of five-compartments, two of them are anoxic tanks and the following three are aerobic tanks; the second part of BSM1 is a secondary settler. Reactors 1 and 2 are unaerated in open-loop, but fully mixed; reactors 3, 4 and 5 are aerated. For the open-loop case, the oxygen transfer coefficients ($K_L a$) are fixed; for reactors 3 and 4 the coefficient ($K_L a_3$ and $K_L a_4$) is set to a constant at 240 d^{-1} (10 h^{-1}), which means the air flow rate of the blower is constant; for reactor 5, the coefficient ($K_L a_5$) is selected as the control variable (or operational variable) in this paper to be manipulated for maintaining the DO concentration at a level of 2mg/L. Thus, the system can achieve biological nitrogen removal through nitrification in the aeration tanks and pre-denitrification in the anoxic tanks. The model equations to be implemented for the proposed layout, the procedure to test the implementation and the performance criteria to be used are described below along with the description of sensors and control handles [28]. For more information, it can be seen in literature [28,29].

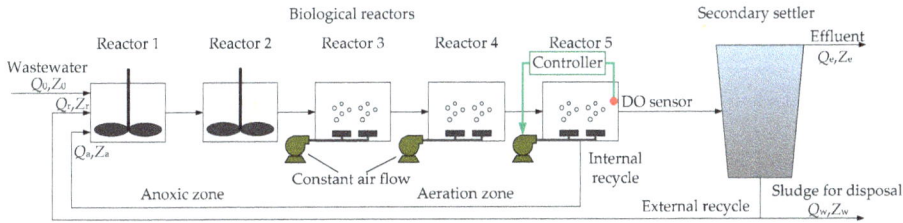

Figure 2. Schematic representation of Benchmark Simulation Model No. 1 (BSM1) model.

The ASM1 [27] has been selected to describe the biological phenomena taking place in the biological reactor and a double-exponential settling velocity function [31] has been selected to describe the secondary settler which is modeled as a 10 layers non-reactive unit (i.e., no biological reaction). In the activated sludge wastewater treatment system, the concentration of DO in the aeration tank is the most important parameter in the process of nitrogen removal [32]. Actually, the DO concentration has a direct impact on the effluent quality with respect to total nitrogen (N_{tot}), nitrate nitrogen (S_{NO}) and ammonia (S_{NH}). Therefore, the study of DO control has its important practical significance and prospect for application.

According to the mass balance of the system, the biochemical reactions that take place in each compartment (reactor) can be described as the follows.

Reactor 1

$$\frac{dZ_1}{dt} = \frac{1}{V_1}(Q_a Z_a + Q_r Z_r + Q_0 Z_0 + r_1 V_1 - Q_1 Z_1) \tag{1}$$

Reactors 2 through 5 ($k = 2$ to 5)

$$\frac{dZ_k}{dt} = \frac{1}{V_k}(Q_{k-1} Z_{k-1} + r_k V_k - Q_k Z_k) \tag{2}$$

Special case for oxygen ($S_{O,k}$)

$$\frac{dS_{O,k}}{dt} = \frac{1}{V_k}(Q_{k-1} S_{O,k-1} - Q_{k-1} S_{O,k})(K_L a)_k (S_O^* - S_{O,k}) + r_k \tag{3}$$

where, Q is the flow rate, Z is the mass concentration of either substrate or bacterial mass, V is the volume of the reactor, r is the reaction rate, $K_L a$ is the oxygen transfer coefficient, S_O is the dissolved oxygen concentration. S^* is the saturation concentration for oxygen ($S^* = 8$ g/m^3 at 15 °C); also $Q_1 = Q_a + Q_r + Q_0$; $Q_k = Q_{k-1}$.

2.2. A Neural Network Based Adaptive PID Algorithm

2.2.1. Radial Basis Function (RBF) Neural Network

Artificial neural network (ANN) is an artificial intelligence system to imitate biological neural networks (BNN). It uses nonlinear processing unit to simulate biological neurons for simulating the behavior of biological synapses among neurons by adjusting the variable weights between connected units. The specific topological structure of the network is organized from each processing unit in a certain connected form. Parallel processing ability and distributed storage are the main features of ANN. Furthermore, it has strong fault tolerance and nonlinear mapping ability with self-organization, self-learning and adaptive reasoning ability [33].

BP (backpropagation) network and RBF network are the most widely used forms of ANN. It is easily to be seen in the widely uses of pattern recognition, prediction, automatic control, etc. [34]. BP algorithm, a supervised learning algorithm, is based on gradient descent algorithm. The drawbacks of BP include an easy fall into local optimum, slow convergence speed, and disunity network structure. RBF network is a feedforward network based on the function approximation theory. It has strong global approximation ability, which can guarantee the network to approximation any kind of nonlinear function with arbitrary accuracy. It can fundamentally overcome the problem of local optimum occurs in BP network. The RBF network has the advantages of simple structure, fast convergence speed and strong generalization ability [35].

Radial basis function (RBF) neural network used in this paper is a three-layer forward network, which is a local approximation method of neural networks. The RBF neural network is composed of three layers, the input layer, the hidden layer and the output layer as shown in Figure 3. The mapping of the input layer to the output layer is nonlinear and the mapping of the space from the hidden layer to the output layer is linear. This kind of mapping configuration itself can speed up the learning rate and avoid the problem of local minima [18].

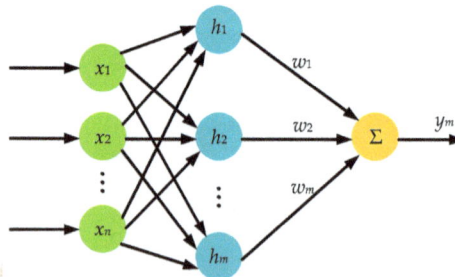

Figure 3. Topology of a radial basis function (RBF) neural network.

In Figure 3, the input vector of the input layer of the neural network is represented as:

$$X = [x_1, x_2, \cdots, x_s, \cdots, x_n]^T \tag{4}$$

where, $x_s = [u_s(k), y_s(k), y_s(k-1)]$, $s = 1, 2, \ldots, n$; $u(k)$ is the output of the controller; $y(k)$ is the present (measured) output of the system (or process), that is, the measured value of DO concentration; $y(k-1)$ is the last measured value of DO concentration output from the process.

The middle layer is the hidden layer. The activation function of the hidden layer is composed of radial basis functions. Each array of computing units of hidden layers is called node. The radial basis vector of the nodes in the RBF neural network is shown in Equation (5).

$$T = [h_1, h_2, \cdots, h_j, \cdots, h_m]^T \tag{5}$$

where, h_j is Gaussian function,

$$h_j = \exp(-\frac{\|X - C_j\|}{2b_j^2}) \tag{6}$$

where, $j = 1, 2, \ldots, m$. C_j is the central vector of the first j node of the hidden layer of the RBF neural network,

$$C_j = [c_{j1}, c_{j2}, \cdots, c_{ji}, \cdots, c_{jn}]^T \tag{7}$$

where, $i = 1, 2, \ldots, n$.

The basic width vector of the hidden layer node of the RBF neural network is

$$B = [b_1, b_2, \cdots, b_j, \cdots, b_m]^T \tag{8}$$

where, b_j is the parameter of the first j node and $j = 1, 2, \ldots, m$.

The weight vector of RBF neural network W is given by:

$$W = [w_1, w_2, \cdots, w_j, \cdots, w_m]^T \tag{9}$$

Then, the estimated output of the RBF network is defined as:

$$y_m = w_1 h_1 + w_2 h_2 + \cdots + w_m h_m \tag{10}$$

The performance index function of the RBF neural network is set as follows:

$$E_1 = \frac{1}{2}(y(k) - y_m(k))^2 \tag{11}$$

where, $y(k)$ is the system output and $y_m(k)$ is the estimated output of the RBF network.

From the above analysis, the three most important parameters C, W and B of a RBF neural network need to be obtained by the learning algorithm. In this paper, the gradient descent method is employed to obtain those three parameters of the nodes. The iterative algorithm used is as follows:

$$w_j(k) = w_j(k-1) + \eta(y(k) - y_m(k))h_j + \alpha(w_j(k-1) - w_j(k-2)) \tag{12}$$

$$\Delta b_j = (y(k) - y_m(k))w_j h_j \frac{\|X - C_j\|^2}{b_j^3} \tag{13}$$

$$b_j(k) = b_j(k-1) + \eta \Delta b_j + \alpha(b_j(k-1) - b_j(k-2)) \tag{14}$$

$$\Delta c_{ji} = (y(k) - y_m(k))w_j \frac{x_j - c_{ji}}{b_j^2} \tag{15}$$

$$c_{ji}(k) = c_{ji}(k-1) + \eta \Delta c_{ji} + \alpha(c_{ji}(k-1) - c_{ji}(k-2)) \tag{16}$$

and the Jacobian matrix:

$$\frac{\partial y(k)}{\partial u(k)} \approx \frac{\partial y_m(k)}{\partial u(k)} = \sum_{j=1}^{m} w_j h_j \frac{c_{ji} - x_1}{b_j^2} \tag{17}$$

in which, η is the learning rate, α is the momentum factor and $x_1 = \Delta u(k)$ is the control increment which is defined as the first input of the neural network.

2.2.2. Design of the RBF Neural Network Based Adaptive PID (RBFNNPID) Algorithm

In the past decades, Proportional-Integral-Derivative (PID) is the main control method for DO level [36,37]. However, owing to the WWTP's time-varying feature, strong nonlinearity, significant perturbations and large uncertainty, a fixed parameter linear controller is not able to maintain a satisfactory tracking performance under the full range of operating conditions [1,37].

The structure of the RBF neural network-based adaptive PID (RBFNNPID) algorithm is shown in Figure 4. The RBF neural network will adaptively calculate weighting coefficient and the parameter gradient information according to the operating state of the dissolved oxygen control system, by its own great learning ability. These results will be used to update the parameters of the PID controller in real time. Hence, such a repeated execution process realizes the adaptive adjustment of PID parameters and achieves the control of DO concentration.

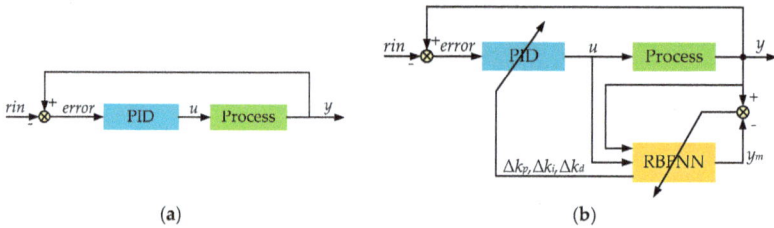

(a)   (b)

Figure 4. Block diagram comparing two controllers: (**a**) Block diagram of a traditional PID controller in a feedback loop; (**b**) Block diagram of proposed RBF neural network-based adaptive PID (RBFNNPID) controller. PID: Proportional-Integral-Derivative.

We have adopted the incremental PID controller and the control error is:

$$error(k) = rin(k) - y(k) \tag{18}$$

where, *rin* is the desired process value or setpoint of DO concentration; $y(k)$ is the measured process value of DO.

The input of the PID algorithm is three errors, which are defined as:

$$xc(1) = error(k) - error(k-1) \tag{19}$$

$$xc(2) = error(k) \tag{20}$$

$$xc(3) = error(k) - 2error(k-1) + error(k-2) \tag{21}$$

The output of the PID algorithm is:

$$u(k) = u(k-1) + \Delta u(k) \tag{22}$$

$$\Delta u(k) = k_p xc(1) + k_i xc(2) + k_d xc(3) \tag{23}$$

where, k_p, k_i and k_d are the three parameters of the PID controller, which represents the proportion, integration and differentiation. The performance function is defined as:

$$E(k) = \frac{1}{2}(error(k))^2 \tag{24}$$

According to the gradient descent method, the adjustment rules of three parameters are given as:

$$\Delta k_p = -\eta \frac{\partial E}{\partial k_p} = -\eta \frac{\partial E}{\partial y} \frac{\partial y}{\partial u} \frac{\partial u}{\partial k_p} = -\eta error(k) \frac{\partial y}{\partial u} xc(1) \tag{25}$$

$$\Delta k_i = -\eta \frac{\partial E}{\partial k_i} = -\eta \frac{\partial E}{\partial y} \frac{\partial y}{\partial u} \frac{\partial u}{\partial k_i} = -\eta \, error(k) \frac{\partial y}{\partial u} xc(2) \tag{26}$$

$$\Delta k_d = -\eta \frac{\partial E}{\partial k_d} = -\eta \frac{\partial E}{\partial y} \frac{\partial y}{\partial u} \frac{\partial u}{\partial k_d} = -\eta \, error(k) \frac{\partial y}{\partial u} xc(3) \tag{27}$$

in which, $\partial y / \partial u$ is the identification information for the Jacobian matrix of the controlled object and it can be obtained through the identification process of neural network. The Jacobian matrix reflects the sensitivity of the output of the controlled object to the change of the input of the control.

The steps of the proposed RBFNNPID control strategy are as follows:

Step 1: Initializing the network parameters, including the number of nodes in input layers and hidden layers, learning rate, inertia coefficient, the base width vector and the weight vector.

Step 2: Sampling to get input *rin* and output *y*, calculating error in terms of Equation (18).

Step 3: Calculating the output *u* of regulator according to Equation (22).

Step 4: Calculating network output y_m, adjusting center vector *C*, base width vector *B*, weight vector *W* and the Jacobian matrix in terms of Equations from (10) to (17) to obtain network identification information.

Step 5: Adjusting parameters of regulator in terms of Equations (25)–(27).

Step 6: Back to *Step 2* and repeat the subsequent steps until the end of the simulation time.

The DO control module and the main codes of the S-function module of RBFNNPID can be found in Appendixs A and B. Appendix C describes the stability and convergence analysis of the proposed RBFNNPID algorithm. An example to verify the convergence of the parameters of the neural network is shown in Appendix D.

3. Results

In order to verify the effectiveness and feasibility of the proposed neural network-based adaptive PID (RBFNNPID) algorithm for DO concentration control of the activated sludge wastewater treatment process, comparison simulation of RBFNNPID and traditional PID are designed in this section, including tracking performance and anti- disturbance performance.

We have selected the BSM1 as the simulation model and the dry weather wastewater data provided by IWA as the source data. The dry weather data contains two weeks long actual operational data of a wastewater treatment system, sampled at every 15 min. Figure 5 shows the dynamic influent data and the changes to the concentrations of some of the components between days 7 and 14.

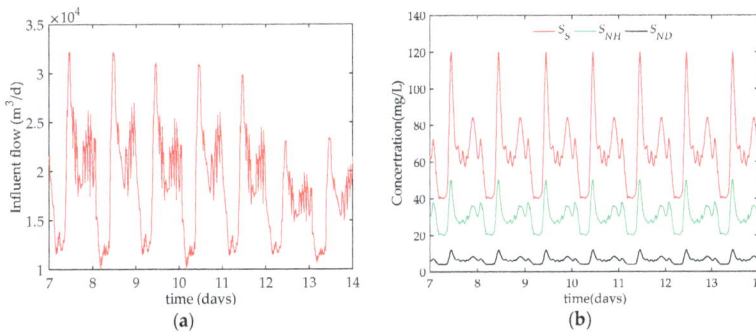

Figure 5. Dynamic data for experimental use: (**a**) Influent flow rate (days 7 to 14); (**b**) Concentration of some of the chemical species in the influent (days 7 to 14) (S_S—readily biodegradable substrate; S_{NH}—ammonium and ammonia nitrogen; S_{ND}—soluble biodegradable organic nitrogen).

In the simulation, the three parameters of both the traditional incremental PID algorithm and the proposed RBFNNPID algorithm are set as: $k_p = 5$, $k_i = 1$, $k_d = 0.5$; the learning rate, η of the three parameters of PID is 0.2; the momentum factor α is 0.05; the network sampling period is 0.001 s; the structure of the RBFNN is defined as "3-6-1", that is to say, the input layer has three nodes, the hidden layer has six nodes and the output layer has one node.

3.1. Tracking Performance Test 1

When the BSM1 wastewater treatment system is operating, due to the dynamic changes in the flow rate and composition of the influent, one need to adjust the oxygen transfer rate ($K_L a_5$) in the fifth tank in real-time to maintain the dissolved oxygen concentration in the appropriate range to ensure the effluent water quality meets the discharge standards. Therefore, how to control the DO concentration around the set point during the process is the aim of control algorithm that is being employed. The tracking performance is one of the deterministic criteria to evaluate whether an algorithm can be applied to an actual wastewater treatment system control.

Generally, the effluent water quality is the best when the DO concentration in the aeration tank is kept between 1~3 mg/L. Therefore, the DO concentration is setup to 2 mg/L in this simulation. The simulation results of last seven days are taken to evaluate the performance of the controller, which is shown in Figure 6.

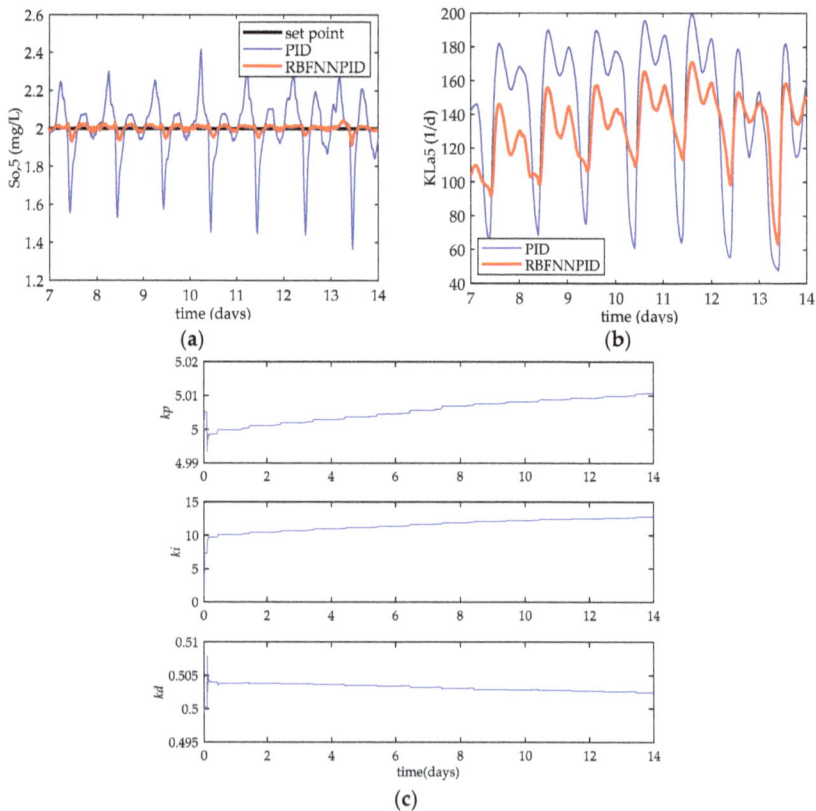

Figure 6. Comparison results of the tracking performance (dry weather): (a) DO concentration in the fifth tank; (b) Dynamic changes of the manipulated variable $K_L a_5$; (c) Dynamic adaptive adjustments of the three parameters *kp*, *ki*, *kd* of RBFNNPID algorithm.

3.2. Tracking Performance Test 2

The set point of DO concentration was changed on days 8, 10 and 12 to 2.5, 1.7 and back to 2 mg/L, respectively to verify the tracking performance of the new proposed RBFNNPID algorithm. Simulation results are shown in Figure 7.

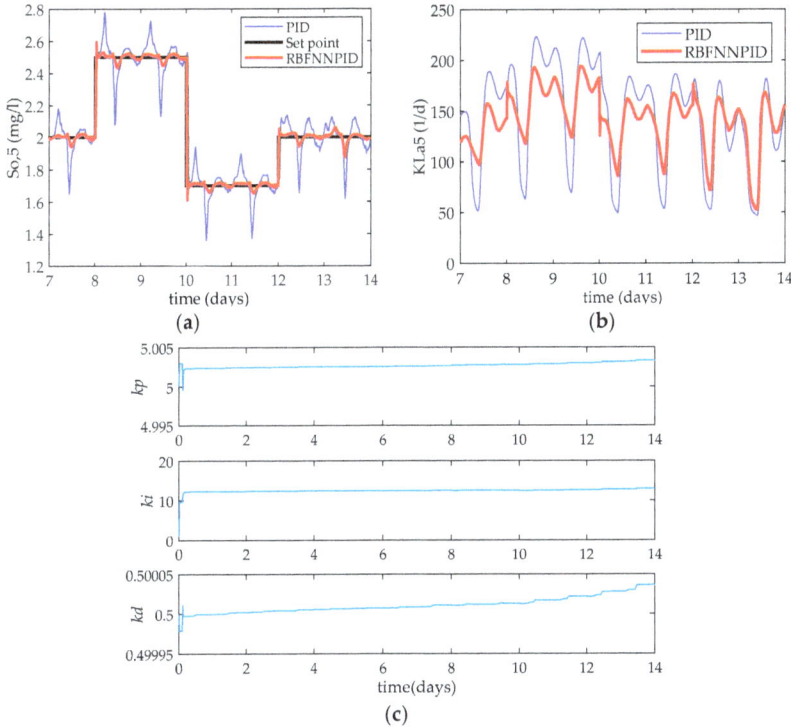

Figure 7. Comparison results of the tracking performance (changing set point of DO): (**a**) DO concentration in the fifth tank; (**b**) Dynamic changes of the manipulated variable K_La_5; (**c**) Dynamic adaptive adjustments of the three parameters kp, ki, kd of RBFNNPID algorithm.

3.3. Anti-Disturbance Performance Test

A good control algorithm should not only have a good tracking performance, but also have a strong anti-disturbance ability. By having these properties, it can be applied to control a complex system such as wastewater treatment process to achieve a precise control effect. To further verify the anti-disturbance ability of the RBFNNPID algorithm, we used the data collected in rain and storm weather to simulate the algorithm. Different weather condition can be looked as different disturbances in the influent. The results are shown in Figures 8 and 9.

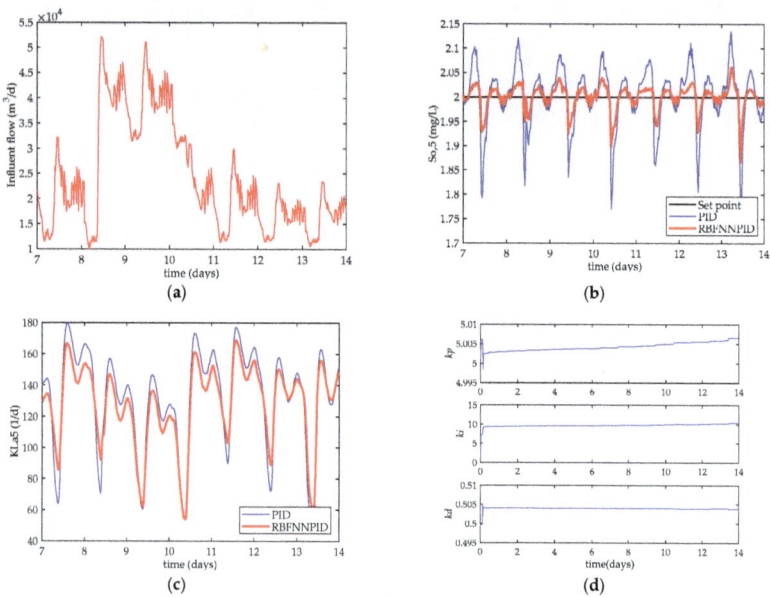

Figure 8. Comparison results of the anti-disturbance performance (rain weather): (**a**) Influent flow rate (days 7 to 14); (**b**) DO concentration in the fifth tank; (**c**) Dynamic changes of the manipulated variable K_La5; (**d**) Dynamic adaptive adjustments of the three parameters *kp*, *ki*, *kd* of RBFNNPID algorithm.

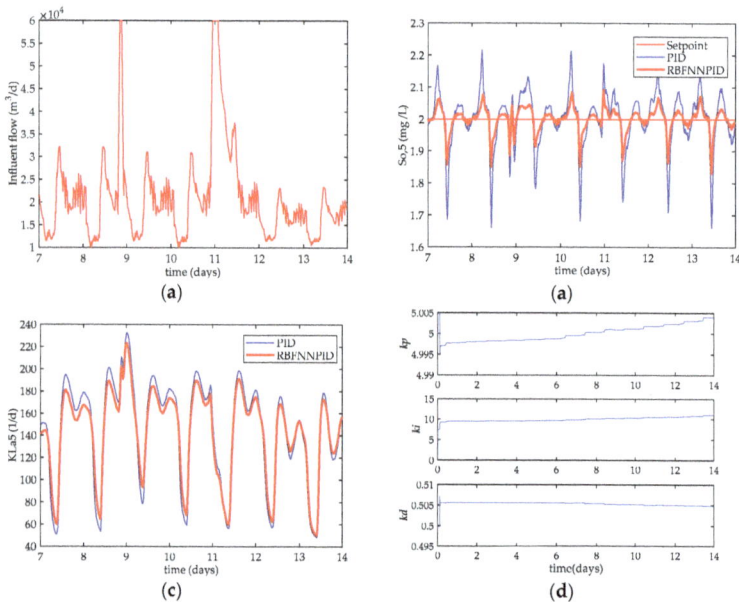

Figure 9. Comparison results of the anti-disturbance performance (storm weather): (**a**) Influent flow rate (days 7 to 14); (**b**) DO concentration in the fifth tank; (**c**) Dynamic changes of the manipulated variable K_La5; (**d**) Dynamic adaptive adjustments of the three parameters *kp*, *ki*, *kd* of RBFNNPID algorithm.

3.4. Controller Performance Evaluation Index

There are two main indices for evaluating the performance of the dissolved oxygen controller. One is the assessment of the underlying control strategy. Indices include the integral of absolute error (*IAE*), the integral of squared error (*ISE*), the maximal deviation from set point (*Dev*^{max}) and the variance of error (*Var$_e$*). The four indices are calculated by Equation (28) through to (31) as shown below.

$$IAE_i = \int_{t=7}^{t=14} |e_i| dt \tag{28}$$

$$ISE_i = \int_{t=7}^{t=14} e_i^2 dt \tag{29}$$

$$Dev_i^{max} = \max|e_i| \tag{30}$$

$$Var(e_i) = \frac{ISE_i}{T} - \left(\frac{IAE_i}{T}\right)^2 \tag{31}$$

The aeration cost can be calculated using aeration energy (*AE*), which will be the economic indicator. *AE* is mainly used in the last three units of the biochemical reaction tanks. *AE* can be obtained by using the oxygen transfer function (K_La) of the three units as shown in Equation (32)

$$AE = \frac{S_O^*}{T \cdot 1800} \int_{t=7}^{t=14} \sum_{i=1}^{i=5} V_i \cdot K_La(t) dt \tag{32}$$

where, S_O^* is the saturation value of dissolved oxygen, V_i is the volume of each unit, and T is the calculation period of *AE*, in this case $T = 7$ days.

Generally, the smaller the value of the above evaluation indices, the better the performance of the controller is. Results of the evaluation indices are shown in Tables 1 and 2. We can see that, under the different weather conditions, the RBFNNPID control strategy reduced the values of the above evaluation indices, compared with the traditional PID control strategy, indicating that the control performance of the system has been effectively improved, and the cost of aeration has also been reduced.

Table 1. Performance of two DO control methods.

Weather	Method	ISE	IAE	Dev^{max}	Var$_e$
Dry	RBFNNPID	1.64×10^{-2}	2.08×10^{-1}	1.89×10^{-1}	2.10×10^{-3}
	PID	4.44×10^{-2}	4.03×10^{-1}	3.43×10^{-1}	6.30×10^{-3}
Rain	RBFNNPID	2.50×10^{-3}	9.47×10^{-2}	6.94×10^{-2}	3.53×10^{-4}
	PID	3.59×10^{-2}	3.61×10^{-1}	2.95×10^{-1}	5.10×10^{-3}
Storm	RBFNNPID	5.70×10^{-3}	1.39×10^{-1}	1.46×10^{-1}	8.16×10^{-4}
	PID	1.38×10^{-2}	2.27×10^{-1}	1.97×10^{-1}	2.00×10^{-3}

Table 2. Aeration energy of two control methods.

Weather	PID (kWh/d)	RBFNNPID (kWh/d)
Dry	7149.9	7032.1
Rain	6955.8	6805.8
Storm	7199.6	6971.8

4. Discussion

PID may fail to achieve the control goal or effect of the process while using the traditional PID control algorithm due to unknown and unexpected disturbances as well as significant changes in operating conditions, such as a significant change in the (i) quality of influent; (ii) weather, etc. Therefore, in general, the parameters of the traditional PID controller need to be adjusted under different operating environments. However, a long period of accumulated experience and several tests are needed in order for the traditional PID to be adjusted to achieve satisfactory results under each operating environment. Clearly, it is not feasible in real time applications and increases the difficulty in applying it in different wastewater treatment plants. Our work will reduce the difficulty of parameter tuning of the traditional PID controller, which is essential to improve the adaptability of the PID control parameters in practice.

The simulation results in Figures 6a, 7a, 8b and 9b show that the DO concentration is difficult to maintain at set point under the control of the conventional incremental PID controller when the influent flow rate and quality changed greatly. On the contrary, RBFNNPID can effectively maintain the DO concentration around the set value with a relatively low error by adjusting the air flow (which can be seen in Figures 6b, 7b, 8c and 9c). It can be seen that the dynamic changes of the manipulated variable $K_L a_5$ is smooth under the control of RBFNNPID. This means we can get a more stable status by using less air supply to the aeration tank. Therefore, using RBFNNPID can reduce the aeration cost which is one of the major electrical costs of the wastewater treatment processes. It also can be verified from the results shown in Tables 1 and 2.

According to the results of the rain weather and storm weather, which can be considered as there has the disturbances of the influent, shown in Figures 8 and 9, compared with the conventional PID controller, the RBFNNPID controller can quickly and accurately track the desired output trajectory values, which means it not only has a good tracking performance, but also has a stronger anti-disturbance ability with the changes to the set points. Figures 6c, 7c, 8d and 9d show the curves of the PID parameters are being adjusted adaptively. Parameters adjusted rapidly at the start of the simulation and small adjustments took place as the simulation goes on.

Applying precise control of the concentration of dissolved oxygen can not only avoid the occurrence of sludge bulking, but also reduce the aeration energy in a wastewater treatment plant. Intermittent aeration can be successfully implemented in a small-scale wastewater treatment plant to reduce the aeration energy while ensuring good effluent water quality [38]; however the same cannot be said in a large-scale wastewater treatment plant such as Pingliang Wastewater Treatment Plant that is situated in Gansu Province, China. In a large-scale wastewater treatment plant, intermittent aeration can reduce only a fraction of aeration energy. However, continuous aeration can effectively reduce the emissions of volatile organic compounds (VOCs) from a wastewater treatment plant, which has been proven as a factor for the increase of haze in some Chinese cities [39].

The characteristics of strong coupling, nonlinearity and large time delay of dissolved oxygen control system in activated sludge wastewater treatment and a control algorithm called RBFNNPID algorithm are discussed in this paper. However, this algorithm has not been applied directly into practice so far as it has certain complexity. We are currently undertaking the following two studies to verify its feasibility, validity, and superiority: (i) Simplifying the algorithm for practical use; (ii) Trialing the algorithm at Pingliang Wastewater Treatment Plant in Gansu Province, China.

5. Conclusions

In this paper, an adaptive PID control algorithm based on RBF neural network is proposed. The RBFNNPID algorithm combines the good learning and adaptive ability of neural networks and the practical advantages of PID algorithm. The gradient descent method is used to adaptively adjust the increment of the three parameters of the PID controller to achieve an optimal control effect on the control of DO concentration. The simulation results show that the RBFNNPID algorithm not only has a better performance of tracking and anti-jamming, but also has a great improvement to the robustness compared to that of the traditional PID. Thus, it can reduce the aeration costs of a wastewater treatment plant employing ASP.

Acknowledgments: China Scholarship Council (CSC) supported the first author to visit RMIT University as visiting scholar while undertaking this work. This work is also supported by the other funding include the National Natural Science Foundation of China (No. 61563032), the Natural Science Foundation of Gansu province (No. 1506RJZA104), University Scientific Research Project of Gansu province (No. 2015B-030), and the Excellent Young Teacher Project of Lanzhou University of Technology (No. Q201408).

Author Contributions: Xianjun Du, Guohua Shi and Junlu Wang conceived and designed the experiments; Junlu Wang performed the experiments; Xianjun Du, Junlu Wang and Veeriah Jegatheesan analyzed the data; Xianjun Du and Veeriah Jegatheesan wrote the paper.

Conflicts of Interest: The authors declare no conflict of interest.

Appendix A

We designed a DO control module in the Matlab Simulink environment which is shown in Figure A1. The output one "kla_out" is used to adjust the air flow of the blower, equipped for reactor 5, to maintain the DO concentration at a level of desired value. The scope is used to monitor the dynamic changes of the three parameters of the controller, which have been shown in Figures 6c, 7c, 8d and 9d. Detail of the inner structure of the controller is shown in Figure A2.

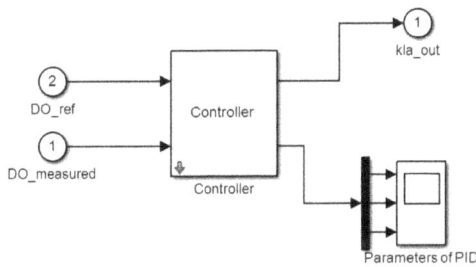

Figure A1. Overview of the DO control module.

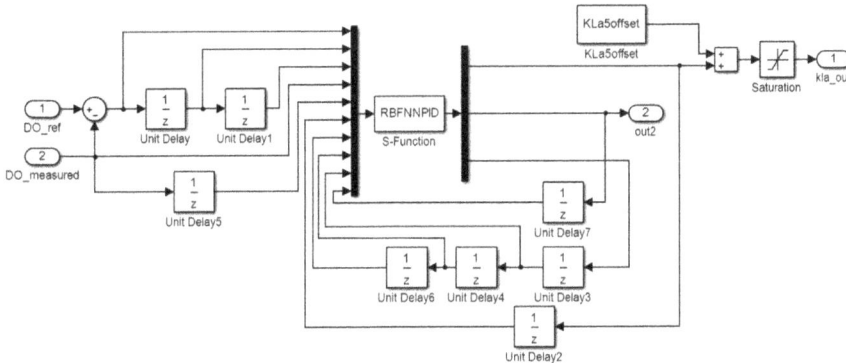

Figure A2. Detailed structure of the RBFNNPID controller.

Appendix B

The main codes of the S-function module, shown in Figure A2, are given below:

```matlab
function [sys, x0, str, ts] = nnrbf_pid(t,x,u,flag,T,nn,K_pid, eta_pid, xite, alfa, beta0, w0)
    switch flag,
        case 0, [sys, x0, str, ts] = mdlInitializeSizes(T,nn);
        case 2, sys = mdlUpdates(u);
        case 3, sys = mdlOutputs(t, x, u, T,nn, K_pid,eta_pid, xite, alfa, beta0, w0);
        case {1, 4, 9}, sys = [];
        otherwise, error (['Unhandled flag = ' , num2str(flag)]);
    end
function [sys,x0,str,ts] = mdlInitializeSizes(T, nn)
    sizes = simsizes;
    sizes. NumContStates = 0;
    sizes.NumDiscStates = 3;
    sizes. NumOutputs = 4 + 5* nn;
    sizes.NumInputs = 9 + 15* nn;
    sizes. DirFeedthrough = 1;
    sizes. NumSampleTimes =1;
    sys = simsizes(sizes) ;
    x0 = zeros(3, 1);
    str = [];
    ts = [T0];
function sys = mdlUpdates(u)
    sys = [ u(1) − u(2); u(1); u(1) + u(3) − 2* u(2)];
function sys = mdlOutputs(t, x, u,T, nn, K_pid, eta_pid, xite, alfa, beta0, w0)
    % Initialization of the radial basis centers
    ci3 = reshape(u(7: 6 + 3* nn), 3, nn);
    ci2 = reshape(u(7 + 5* nn: 6 + 8* nn), 3, nn);
    ci1 = reshape(u(7 + 10* nn: 6 + 13* nn), 3, nn);
    % Initialization of the radial basis width
    bi3 = u(7 + 3* nn: 6 + 4* nn);
    bi2 = u(7 + 8*nn: 6 + 9* nn);
    bi1 = u(7 + 13* nn: 6 + 14* nn);
    % Initialization of the weights
    w3 = u(7 + 4* nn: 6+ 5* nn) ;
    w2 = u(7 + 9* nn: 6+ 10* nn) ;
    w1 = u(7 + 14* nn: 6+ 15* nn) ;
    xx = u([6; 4; 5]);
    if t = 0
        % Initialize the PID parameters
        ci1 = w0(1) * ones(3, nn);
        bi1 = w0(2) *ones(nn, 1);
        w1 = w0(3) * ones(nn, 1);
        K_pid0 = K_pid;
    else
        % Update the PID parameters
        K_pid0 = u(end-2: end);
    end
    for j = 1: nn
        % Gaussian
        h(j, 1) = exp(−norm(xx−ci1( : , j))^2/(2* bi1(j) * bi1(j)));
    end
    % Dynamic of gradient descent method
    dym = u(4) − w1'* h;
    W = w1 + xite* dym* h + alfa* (w1 − w2) + beta0*(w2 − w3) ;
    for j = 1: nn
        dbi(j,1) = xite* dym* w1(j) * h(j) * (bi1(j) ^(−3)) * norm(xx − ci1(:,j))^2;
        dci( : ,j) = xite*dym* w1(j)* h(j) * (xx − ci1(:,j)) * (bi1(j)^(−2));
    end
    bi = bi1 + dbi + alfa* (bi1 − bi2) + beta0*(bi2 − bi3) ;
    ci = ci1 + dci + alfa* (ci1 − ci2) + beta0*(ci2 − ci3) ;
    % Jacobian
    dJac = sum(w.*h.*(−xx (1) + ci (1,:)') ./bi.^2);
    % adjustments of the PID parameters
    KK(1) = K_pid0(1) + u(1) * dJac* eta_pid(1)* x(1);
    KK(2) = K_pid0(2) + u(1) * dJac* eta_pid(2)* x(2);
    KK(3) = K_pid0(3) + u(1) * dJac* eta_pid(3)* x(3);
    sys= [ u(6) + KK* x; KK'; ci( : ) ; bi( : ) ; w( : ) ] ;
```

Appendix C

This section describes the stability and convergence analysis of the proposed RBFNNPID algorithm. The basic knowledge of the stability and convergence analysis is Lyapunov theorem, also known as Lyapunov stability.

$V(x_1, x_2, K, x_N)$ is an arbitrary function defined in the neighborhood of the origin Ω, where Ω is a state of equilibrium and x_1, x_2, K, x_N are variables, then

$$|x_i| \leq H, i = 1, 2, K, N \tag{A1}$$

where, H is a positive constant.

Assuming that V is a continuous differentiable function in Ω and $V(0, 0, K, 0) = 0$. Such that

(i) $V(x) > 0$ is positive definite or $V(x) < 0$ is negative definite, $x \in \Omega$ and $x \neq 0$;
(ii) $V(x) > 0$ is positive semi-definite or $V(x) < 0$ is negative semidefinite, $x \in \Omega$;

Consider an autonomous nonlinear dynamical system

$$\dot{x} = f(x) \tag{A2}$$

where, $f(0) = 0$.

Assuming $x_i = x_i(t)$, $(i = 1, 2, K, N)$ is the solution of the system (A2). We can obtain the derivation

$$\frac{dV}{dt} = \frac{\partial V}{\partial x_1} \frac{\partial x_1}{\partial t} + \frac{\partial V}{\partial x_2} \frac{\partial x_2}{\partial t} + K + \frac{\partial V}{\partial x_N} \frac{\partial x_N}{\partial t} \tag{A3}$$

Introducing the gradient vector (Equation (A4)) into Equation (A3)

$$\nabla V(x) = \left[\frac{\partial V}{\partial x_1}, \frac{\partial V}{\partial x_2}, L, \frac{\partial V}{\partial x_N} \right]^T \tag{A4}$$

We will arrive at the final equation as below:

$$V = [\nabla V(x)]^T f(x) = w(x) \tag{A5}$$

The following conclusions can be made from the above analysis:

(i) If $V(x)$ is positive (or negative) definite, and if derivation $V = w(x)$ is negative (or positive) semi-definite, the system is said to be Lyapunov stable at the equilibrium of the origin;
(ii) If $V(x)$ is positive (or negative) definite, and if derivation $V = w(x)$ is negative (or positive) definite, the system is said to be exponentially stable at the equilibrium of the origin;
(iii) If $V(x)$ is positive (or negative) definite, and if derivation $V = w(x)$ is also positive (or negative) definite, the system is said to be unstable at the equilibrium of the origin;

For the adjustment of the weights of the neural network, we need a parameter called learning rate η. If η is too large, NN will be unstable; but if η is too small, the convergence rate will be too slow. Therefore, the selection of the value of learning rate is crucial to the stability and convergence of the system.

Assuming the indicator function of the RBFNNPID controller

$$J(k) = \frac{1}{2}[y(k) - y_m(k)]^2 = \frac{1}{2}e^2(k) \tag{A6}$$

where, $e(k)$ is the learning error of the network.

In order to ensure the adjustment of the weight coefficient is carried out in the direction of the negative gradient relative to $w(k)$, there must be

$$w(k+1) = w(k) - \eta \frac{\partial J(k)}{\partial w(k)}$$

(A7)

From Equations (A6) and (A7), we can get

$$\Delta w(k) = \frac{\partial J(k)}{\partial w(k)} = e(k) \frac{\partial e(k)}{\partial w(k)} = e(k) \frac{\partial e(k)}{\partial \Delta u(k)} \frac{\partial \Delta u(k)}{\partial w(k)}$$

(A8)

Defining a Lyapunov function of a discrete-time systems as

$$v(k) = \frac{1}{2} e^2(k)$$

(A9)

As we introduced in the paper, the gradient descent method is used as the change of the network-learning algorithm.

$$\Delta v(k) = v(k+1) - v(k) = \frac{1}{2} e^2(k+1) - \frac{1}{2} e^2(k) = \frac{1}{2} \Delta e(k)[2e(k) + \Delta e(k)]$$

(A10)

where, $e(0) = 0$, and,

$$\Delta e(k) = \frac{\partial J(k)}{\partial e(k)} = e(k)$$

(A11)

It can be obtained from Equation (A8), that

$$\Delta e(k) = e(k) = (\frac{\partial e(k)}{\partial w(k)})^T \Delta w(k)$$

(A12)

and

$$\Delta w(k) = -\eta \frac{\partial J(k)}{\partial w(k)} = -\eta e(k) \frac{\partial e(k)}{\partial \Delta u(k)} \frac{\partial \Delta u(k)}{\partial w(k)}$$

(A13)

By substituting Equation (A13) into Equation (A12)

$$\Delta e(k) = -\eta \| \frac{\partial J(k)}{\partial w(k)} \|^2 e(k)$$

(A14)

Then, substituting Equation (A14) into Equation (A10)

$$\Delta v(k) = -\frac{1}{2} \eta \| \frac{\partial J(k)}{\partial w(k)} \|^2 e(k) \left[2e(k) - \eta \| \frac{\partial J(k)}{\partial w(k)} \|^2 e(k) \right] = -\frac{1}{2} \eta \| \frac{\partial J(k)}{\partial w(k)} \|^2 e^2(k) \left[2 - \eta \| \frac{\partial J(k)}{\partial w(k)} \|^2 \right]$$

(A15)

Knowing from the Lyapunov stability theory, the system is stable when $\Delta v < 0$. In addition, because of $\eta > 0$,

$$2 - \eta \| \frac{\partial J(k)}{\partial w(k)} \|^2 > 0$$

(A16)

That is,

$$0 < \eta < \frac{2}{\| \frac{\partial J(k)}{\partial w(k)} \|^2}$$

(A17)

Therefore, the system is stable.
When $\Delta v < 0$,

$$\frac{1}{2} e^2(k+1) < \frac{1}{2} e^2(k)$$

(A18)

$$\lim_{k \to \infty} e(k) = 0 \tag{A19}$$

It means, with the increase of k, $e(k)$ gradually reaches to zero, and it guarantees the convergence of the learning algorithm. Based on the above analysis, if the value of η is according to Equation (A17), the system is stable and the learning algorithm will converge.

Appendix D

To verify the convergence of the parameters of the neural network, the dynamic changes of the centers (Figure A3) as well as weights and widths (Figure A4) of the neural network were computed, taking the first neuron of the hidden layer neurons as the example. From those figures, it can be seen that the parameters are converging with adaptive changes while the control process is going on.

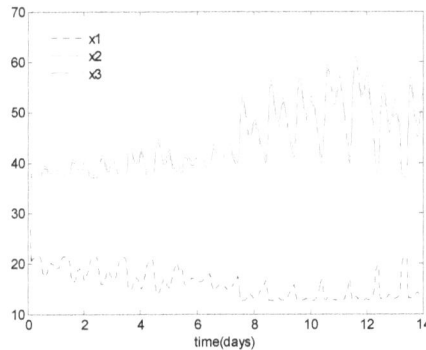

Figure A3. Dynamic changes of the centers of the first neuron (of the hidden layer neurons) of the neural network. (Where, $\times 1$, $\times 2$ and $\times 3$ represent the three centers of the first neuron).

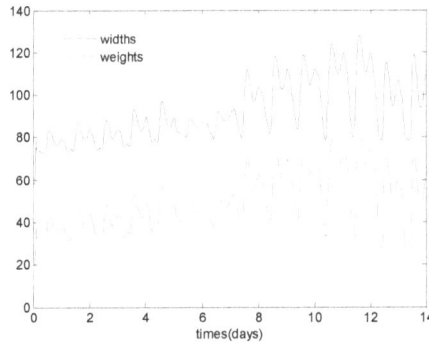

Figure A4. Dynamic changes of the weights and widths of the first neuron (of the hidden layer neurons) of the neural network.

References

1. Bo, Y.C.; Zhang, X. Online adaptive dynamic programming based on echo state networks for dissolved oxygen control. *Appl. Soft Comput.* **2017**. [CrossRef]
2. Holenda, B.; Domokos, E.; Rédey, Á.; Fazakas, J. Dissolved oxygen control of the activated sludge wastewater treatment process using model predictive control. *Comput. Chem. Eng.* **2008**, *32*, 1270–1278. [CrossRef]
3. Zhang, P.; Yuan, M.; Wang, H. Study on Dissolved Oxygen Control Method Based on International Evaluation Benchmark. *Inf. Control* **2007**, *36*, 199–203.

4. Yu, K.; Muhetaer, A.; Wei, L. Evaluation indexes of sewage stabilization from municipal wastewater treatment plant. *China Water Wastewater* **2016**, *5*, 93–97.

5. Chen, C.S. Robust self-organizing neural-fuzzy control with uncertainty observer for MIMO nonlinear systems. *IEEE Trans. Fuzzy Syst.* **2011**, *19*, 694–706. [CrossRef]

6. Nascu, I.; Vlad, G.; Folea, S.; Buzdugan, T. Development and application of a PID auto-tuning method to a wastewater treatment process. In Proceedings of the IEEE International Conference on Automation, Quality and Testing, Robotics (AQTR 2008), Cluj-Napoca, Romania, 22–25 May 2008.

7. Ye, H.T.; Li, Z.Q.; Luo, W.G. Dissolved oxygen control of the activated sludge wastewater treatment process using adaptive fuzzy PID control. In Proceedings of the 32nd Chinese Control Conference (CCC2013), Xi'an, China, 26–28 July 2013.

8. Goldar, A.; Revollar, S.; Lamanna, R.; Vega, P. Neural-MPC for N-removal in activated-sludge plants. In Proceedings of the IEEE European Control Conference (ECC 2014), Strasbourg, France, 24–27 June 2014.

9. Goldar, A.; Revollar, S.; Lamanna, R.; Vega, P. Neural NLMPC schemes for the control of the activated sludge process. In Proceedings of the 11th IFAC Symposium on Dynamics and Control of Process Systems Including Biosystems, Trondheim, Norway, 6–8 June 2016.

10. Belchior, C.A.C.; Araújo, R.A.M.; Landeck, J.A.C. Dissolved oxygen control of the activated sludge wastewater treatment process using stable adaptive fuzzy control. *Comput. Chem. Eng.* **2012**, *37*, 152–162. [CrossRef]

11. Jeppsson, U.; Rosen, C.; Alex, J.; Copp, J.; Gernaey, K.V.; Pons, M.N.; Vanrolleghem, P.A. Towards a benchmark simulation model for plant-wide control strategy performance evaluation of WWTPs. *Water Sci. Technol.* **2006**, *53*, 287–295. [CrossRef] [PubMed]

12. Yu, G.; Zhang, P.; Wei, S.; Fan, M.; Wang, H. Human-simulation intelligent PID control theory and its application for dissolved oxygen in wastewater treatment. *Microcomput. Inf.* **2006**, *22*, 13–15.

13. Syu, M.J.; Chen, B.C. Back-propagation neural network adaptive control of a continuous wastewater treatment process. *Ind. Eng. Chem. Res.* **1998**, *37*, 3625–3630. [CrossRef]

14. Macnab, C.J.B. Stable Neural-Adaptive Control of Activated Sludge Bioreactors. In Proceedings of the 2014 American Control Conference (ACC2014), Portland, OR, USA, 4–6 June 2014.

15. Mirghasemi, S.; Macnab, C.J.B.; Chu, A. Dissolved oxygen control of activated sludge bioreactors using neural-adaptive control. In Proceedings of the IEEE Symposium on Computational Intelligence in Control and Automation (CICA 2014), Orlando, FL, USA, 9–12 December 2014.

16. Ruan, J.; Zhang, C.; Li, Y.; Li, P.; Yang, Z.; Chen, X.; Huang, M.; Zhang, T. Improving the efficiency of dissolved oxygen control using an on-line control system based on a genetic algorithm evolving FWNN software sensor. *J. Environ. Manag.* **2017**, *187*, 550–559. [CrossRef] [PubMed]

17. Qiao, J.; Fu, W.; Han, H. Dissolved oxygen control method based on self-organizing T-S fuzzy neural network. *CIESC J.* **2016**, *67*, 960–966.

18. Li, M.; Zhou, L.; Wang, J. Neural network predictive control for dissolved oxygen based on Levenberg-Marquardt algorithm. *Trans. Chin. Soc. Agric. Mach.* **2016**, *47*, 297–302.

19. Xu, J.; Yang, C.; Qiao, J. A novel dissolve oxygen control method based on fuzzy neural network. In Proceedings of the 36th Chinese Control Conference (CCC2017), Dalian, China, 26–28 July 2017.

20. Lin, M.J.; Luo, F. Adaptive neural control of the dissolved oxygen concentration in WWTPs based on disturbance observer. *Neurocomputing* **2016**, *185*, 133–141. [CrossRef]

21. Han, H.G.; Qiao, J.F.; Chen, Q.L. Model predictive control of dissolved oxygen concentration based on a self-organizing RBF neural network. *Control Eng. Pract.* **2012**, *20*, 465–476. [CrossRef]

22. Zhou, H. Dissolved oxygen control of wastewater treatment process using self-organizing fuzzy neural network. *CIESC J.* **2017**, *68*, 1516–1524.

23. Huang, M.Z.; Han, W.; Wan, J.Q.; Chen, X. Multi-objective optimization for design and operation of anaerobic digestion using GA-ANN and NSGA-II. *J. Chem. Technol. Biotechnol.* **2016**, *91*, 226–233. [CrossRef]

24. Li, Y.; Li, T.; Jiang, Y.; Fan, J.-L. Adaptive PID control of quadrotor based on RBF neural network. *Control Eng. China* **2016**, *23*, 378–382.

25. Chamsai, T.; Jirawattana, P.; Radpukdee, T. Robust adaptive PID controller for a class of uncertain nonlinear systems: An application for speed tracking control of an SI engine. *Math. Probl. Eng.* **2015**, *17*, 1–12. [CrossRef]

26. Lin, C.M.; Chung, C.C. Fuzzy brain emotional learning control system design for nonlinear systems. *Int. J. Fuzzy Syst.* **2015**, *17*, 117–128. [CrossRef]

27. Henze, M.; Grady, C.P.L.; Gujer, W.; Matsuo, T. *Activated Sludge Model No. 1*; IAWPRC Publishing: London, UK, 1987.

28. Gernaey, K.V.; Jeppsson, U.; Vanrolleghem, P.A.; Copp, J.B. *Benchmarking of Control Strategies for Wastewater Treatment Plants*; IWA Publishing: London, UK, 2014.

29. Du, X.J.; Hao, X.H.; Li, H.J.; Ma, Y.W. Study on modelling and simulation of wastewater biochemical treatment activated sludge process. *Asian J. Chem.* **2011**, *23*, 4457–4460.

30. Simsek, H. Mathematical modeling of wastewater-derived biodegradable dissolved organic nitrogen. *Environ. Technol.* **2016**, *37*, 2879–2889. [CrossRef] [PubMed]

31. Takács, I.; Patry, G.G.; Nolasco, D. A dynamic model of the clarification thickening process. *Water Res.* **1991**, *25*, 1263–1271. [CrossRef]

32. Olsson, G.; Newell, B. *Wastewater Treatment Systems: Modelling, Diagnosis and Control*, 1st ed.; IWA Publishing: London, UK, 1999.

33. Maier, H.R.; Dandy, G.C. Neural networks for the prediction and forecasting of water resources variables: A review of modelling issues and applications. *Environ. Model. Softw.* **2000**, *15*, 101–124. [CrossRef]

34. Wang, J.; Shi, P.; Jiang, P.; Xiao, P. Application of BP Neural Network Algorithm in Traditional Hydrological Model for Flood Forecasting. *Water* **2017**, *9*, 48. [CrossRef]

35. Zebardast, B.; Maleki, I. A New Radial Basis Function Artificial Neural Network based Recognition for Kurdish Manuscript. *Int. J. Appl. Evolut. Comput.* **2013**, *4*, 72–87. [CrossRef]

36. Wahab, N.A.; Katebi, R.; Balderud, J. Multivariable PID control design for activated sludge process with nitrification and denitrification. *Biochem. Eng. J.* **2009**, *45*, 239–248. [CrossRef]

37. Luo, F.; Hoang, B.L.; Tien, D.N.; Nguyen, P.H. Hybrid PI controller design and hedge algebras for control problem of dissolved oxygen in the wastewater treatment system using activated sludge method. *Int. Res. J. Eng. Technol.* **2015**, *2*, 733–738.

38. Sánchezmonedero, M.A.; Aguilar, M.I.; Fenoll, R.; Roig, A. Effect of the aeration system on the levels of airborne microorganisms generated at wastewater treatment plants. *Water Res.* **2008**, *42*, 3739–3744. [CrossRef] [PubMed]

39. Chen, W.H. Influences of Aeration and Biological Treatment on the Fates of Aromatic VOCs in Wastewater Treatment Processes. *Aerosol Air Qual. Res.* **2013**, *13*, 225–236. [CrossRef]

*applied
sciences*

MDPI

Article

Effect of Seasonal Temperature on the Performance and on the Microbial Community of a Novel AWFR for Decentralized Domestic Wastewater Pretreatment

Juanhong Li and Xiwu Lu *

School of Energy and Environment, Southeast University, Nanjing 210096, China; li_juanhong1@163.com
* Correspondence: xiwulu@seu.edu.cn; Tel.: +86-25-8379-2614

Academic Editors: Faisal Ibney Hai, Kazuo Yamamoto and Jega Veeriah Jegatheesan
Received: 27 February 2017; Accepted: 5 June 2017; Published: 11 June 2017

Abstract: Due to environmental burden and human health risks in developing countries, the treatment of decentralized domestic wastewater has been a matter of great concern in recent years. A novel pilot-scale three-stage anaerobic wool-felt filter reactor (AWFR) was designed to treat real decentralized domestic wastewater at seasonal temperature variations of 8 to 35 °C for 364 days. The results showed that the average chemical oxygen demand (COD) removal efficiencies of AWFR in summer and winter were 76 ± 7.2% and 52 ± 5.9% at one day and three days Hydraulic Retention Time (HRT), respectively. COD mass balance analysis demonstrated that even though COD removal was lower in winter, approximately 43.5% of influent COD was still converted to methane. High-throughput MiSeq sequencing analyses indicated that *Methanosaeta*, *Methanobacterium*, and *Methanolinea* were the predominant methanogens, whereas the genus *Bacillus* probably played important roles in fermentation processes throughout the whole operation period. The performance and microbial community composition study suggested the application potential of the AWFR system for the pretreatment of decentralized domestic wastewater.

Keywords: decentralized domestic wastewater; seasonal temperature; anaerobic wool-felt filter reactor; high-throughput MiSeq sequencing

1. Introduction

Water, land, and energy are important resources for the rapid growth and development of the global economy. However, in recent years, many developing countries face challenges and pressures of water and land pollution, and energy shortages. Particularly in rural areas of China, water pollution problems are increasingly aggravated due to the direct discharge of a large amount of untreated domestic wastewater [1,2]. Although centralized biological processes have been well developed to be used in urban municipal wastewater treatment plants, they are not suitable for rural areas owing to the dispersed population, poor wastewater collection, and weaker economy in these areas. In addition, a concept of wastewater-to-resource is receiving increased attention by researchers and engineers working on wastewater treatment technology development [3]. Therefore, it is highly desirable to select a sustainable, robust and cost-effective process for the treatment of decentralized domestic wastewater in developing countries [4].

Simultaneous energy recovery and sustainable wastewater treatment make the application of anaerobic biotechnology in decentralized domestic wastewater treatment interesting [5–7]. Upflow anaerobic sludge blanket reactors (UASB) [8,9], anaerobic baffled reactor (ABR) [10], and anaerobic membrane bioreactor (AnMBR) [7,11,12] are commonly used to treat domestic wastewater. Despite the effective chemical oxygen demand (COD) removal by these anaerobic biotechnologies, some challenges still exist, such as complex three-phase separator and long solids' retention time of UASB, membrane fouling and high energy costs of AnMBR, and high land footprint of ABR. Compared to these high-rate anaerobic reactors, the anaerobic filters (AFs) have drawn attention because of the following advantages [13–15]: (1) simple design configuration with low capital and operating costs, (2) excellent capability of high biomass retention with carriers, (3) stable operation, (4) greater tolerance to hydraulic loading rate and organic loading rate, and (5) low footprint. The potential of using AFs for treating wastewater have been well developed not only for industrial wastewater, but also for domestic wastewater [5,16–18]. However, most of these studies focused on the effects of carriers, hydraulic retention time (HRT), and organic loading rate on the performance of AFs, and there were a few studies that focused on the effects of seasonal temperature [5,18]. Previous studies have demonstrated that anaerobic biotechnology strongly depends on operation temperature, and AFs are commonly operated under mesophilic conditions [5,19]. Nevertheless, considering the energy consumption and capital expenditure, it is economically unviable to heat anaerobic systems in decentralized domestic wastewater treatment of rural areas. Therefore, the application of AFs at seasonal temperature is more useful due to less energy demand. In context to China, this study is inevitable as the seasonal temperature varied throughout the whole year. Additionally, carriers are an important component of AFs, which determine the biomass retention capacity and the performance of the system. Therefore, the choices of appropriate carriers play an important role in AF system performance, particularly in the rural area of developing countries. A variety of natural materials, including zeolite [20], ceramic [21], rock [5,22] and coconut shells [23] have been adopted as biofilm carriers. Even though these natural materials are low-cost, they are susceptible to clogging and require significant operation attention due to biofilm growth [22]. Wool felt is a class of natural porous materials that is a common waste product of paper making factories in China. Compared to solid materials, wool felt is a class of soft filler with a large specific surface area of about 950 m^2/g that avoids clogging. Wool felt may be a viable option for AFs due to low cost and a large specific surface. Hence, wool felt has been applied before in a membrane reactor for the biosorption of heavy metals [24]. In response to those factors, a pilot-scale three-stage anaerobic wool-felt filter (AWFR) was designed to treat decentralized domestic wastewater.

It is well known that anaerobic biotechnology is a biological degradation process comprising mutual metabolic interaction among the bacterial and archaeal community. Microorganisms play important roles in the efficiency and stability of the anaerobic treatment [25]. However, most of the available studies have been operated as a "black box" without focusing on the role of the microbial communities [26]. To get comprehensive insights into the microbial community of the anaerobic process, molecular biology tools are used to determine the structures of the microorganisms in the system. To date, the molecular biology tools have been extensively developed. Nevertheless, most of the studies still applied the techniques of terminal restriction fragment length polymorphism (T-RFLP) [27], fluorescence in situ hybridization (FISH) [19,28], denaturing gradient gel electrophoresis (DGGE) [29–31] and quantitative polymerase chain reaction (q-PCR) [32]. Compared to the conventional ones, the high-throughput sequencing, especially Illumina MiSeq sequencing, is becoming one of the most popular molecular tools for microbial community analysis due to low-cost, fast turnaround time producing several gigabases of sequence and greater coverage [33]. Nevertheless, there is no study that focused on the microbial communities in the AWFR system.

The aim of this work was to evaluate the performance of a pilot-scale three-stage anaerobic wool-felt filter reactor (AWFR) in treating decentralized domestic wastewater seasonally. Additionally, this work also investigated the seasonal variations of microbial community in the AWFR system based on high-throughput MiSeq sequencing. The results provided a comprehensive understanding of the relationship between the microbial community and the performance of the AWFR system.

2. Materials and Methods

2.1. Experimental Setup and Operation

The experimental studies were conducted in a pilot-scale three-stage AWFR made of polyvinyl chlorine polymer. Each identical AWFR has a height of 2.5 m and an internal diameter of 170 mm as shown in Figure 1. In order to avoid clogging, each reactor with the effective volume of 50 L was packed with vertical wool felt carrier as biomass growth support media. The wool felt carriers had a high specific surface area (about 950 m^2/g) with a high porosity (>95%). The reactors were continuously fed with real decentralized domestic wastewater by using a peristaltic pump. Based on the seasonal temperature variations, the operation of the AWFR system was divided into five periods: start-up period (1–30 day, 15 ± 3.4 °C, $n = 30$), spring period (31–90 day, 21 ± 3.0 °C, $n = 60$), summer period (91–180 day, 31 ± 3.7 °C, $n = 90$), autumn period (181–270 day, 25 ± 5.2 °C, $n = 90$) and winter period (271–364 day, 10 ± 2.2 °C, $n = 94$). The main operating conditions are summarized in Table 1.

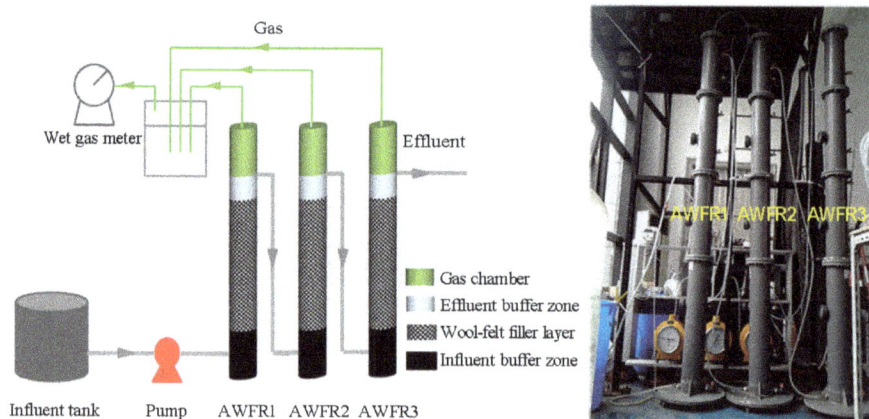

Figure 1. Schematic diagram and photo of the anaerobic wool-felt filter reactor (AWFR).

Table 1. The main operating conditions for the anaerobic wool-felt filter reactor (AWFR) system during the five operation stages.

Stage	Phase (Days)	Duration (Days)	Hydraulic Retention Time (HRT) (Days)	Hydraulic Loading Rate (HLR) (m^3/m^2/day)	Organic Loading Rate (OLR) (mgCOD/L/day) [1]	Temperature (°C) [1]
start-up	1–30	30	3	2.2	75 ± 13.6 ($n = 30$)	15 ± 3.4 ($n = 30$)
spring	31–90	60	2	4.4	143 ± 28.3 ($n = 60$)	21 ± 3.0 ($n = 60$)
summer	91–180	90	1	6.6	352 ± 61.8 ($n = 90$)	31 ± 3.7 ($n = 90$)
autumn	181–270	90	2	4.4	135 ± 32.9 ($n = 90$)	25 ± 5.2 ($n = 90$)
winter	271–364	94	3	2.2	82 ± 12.9 ($n = 94$)	10 ± 2.2 ($n = 94$)

[1] Values are given as mean ± standard deviation.

2.2. Decentralized Domestic Wastewater and Seed Sludge

The experimental setup was fed with real decentralized domestic wastewater, which was collected from the dormitories and restaurants of the Southeast University (Wuxi, China). The chemical characteristics of the decentralized domestic wastewater are presented in Table 2.

Table 2. The chemical characteristics of decentralized domestic wastewater (unit: mg/L, except for pH).

Parameter	Chemical Oxygen Demand (COD)	Total Phosphorus (TP)	Total Nitrogen (TN)	pH
Value [1]	284 ± 69.2	2.8 ± 0.8	32.2 ± 7.7	7.1 ± 0.2

[1] Values are given as mean ± standard deviation; number of measurements (*n*): *n* = 364 for COD, TP, TN and pH.

The seeding anaerobic sludge used in this study was obtained from the anaerobic digester of a municipal wastewater treatment plant located in Wuxi, China. Approximately 10 L anaerobic sludge was inoculated into the wool felt carrier of each reactor before startup of the AWFR system. The initial total suspended solids (TSS) and volatile suspended solids (VSS) of the seeding anaerobic sludge were 28.2 g/L and 14.3 g/L, respectively.

2.3. Analysis Methods

2.3.1. Chemical Analysis

Chemical oxygen demand (COD), total nitrogen (TN), total phosphorus (TP), total suspended solids (TSS) and volatile suspended solids (VSS) were measured according to the Standard Methods [34]. The concentrations of volatile fatty acids were analyzed using a gas chromatography GC 3900 (Tenghai, Shandong, China) equipped with an SE-30 capillary column (30 m × 0.32 mm × 0.25 μm) and a flame ionization detector. The operation temperatures of the injector port, column oven and detector were 200, 120 and 230 °C, respectively. Nitrogen gas was used as the carrier gas at a flow rate of 40 mL/min. The biogas production was measured by using an LML-1 wet gas meter (Changchun Automobile Filter Co., Ltd., Changchun, China). The biogas composition was determined using a gas chromatography GC 2001 (Tenghai, Shandong, China) equipped with a thermal conductivity detector and a 4 m × 3 mm inside diameter stainless-steel column packed with TDX-01 (80/100 mesh). The operation temperature of the injector port, column oven and detector were 150, 150 and 180 °C, respectively. Argon gas was used as the carrier gas at a flow rate of 25 mL/min. The pH and temperature were measured using a portable YSI-pH 100 meter (YSI Co., Yellow Springs, OH, USA).

2.3.2. COD Mass Balance Calculation

The COD mass balance of the AWFR system was conducted during spring, summer, autumn and winter. Seasonal periods were characterized by the following operation HRTs: 2 days (spring period covering Day 31–90), 1 day (summer period covering Day 91–180), 2 days (autumn period covering Day 181–270), and 3 days (winter period covering Day 271–364).

The COD mass balance was calculated using the Equation (1):

$$COD_{in} = COD_{VFAs} + COD_{CH4(g)} + COD_{CH4(s)} + COD_{others}, \tag{1}$$

where: COD_{in} (g/day) represents the average COD concentration of real decentralized domestic wastewater; COD_{VFAs} (g/day) represents the average COD concentration of acetate and propionate in the effluent; $COD_{CH4(g)}$ (g/day) represents the average COD concentration of methane produced in biogas; $COD_{CH4(s)}$ (g/day) represents the average COD concentration of methane dissolved in the effluent; COD_{others} (g/day) represents the organic matter that has been utilized for biomass formation, the complex organic matter that is not biodegradable, COD consumed by sulphate reducing bacteria (SRB), and COD removed and converted to CO_2.

Dissolved methane was calculated using Equations (2) and (3) suggested by a previous study [35]:

$$CH_4(s) = 4 \times K_H \times P_{gas} \times flow(Q), \tag{2}$$

$$K_H = 0.384 \times t + 36.44, \tag{3}$$

where K_H (mg/L/atm) is Henry's constant, P_{gas} (atm) is the partial pressure of the gas above the liquid, flow(Q) (L/day) is feed flow and t (°C) is the temperature.

2.3.3. Scanning Electron Microscopy (SEM)

Scanning electron microscopy (SEM) was used to investigate the development and structure of the anaerobic biofilm in the system. Representative samples (about 1 cm^2 size) were taken from the AWFR system on days 15, 90, 180, 270 and 340. The samples were firstly fixed with 3% (v/v) glutaraldehyde in 0.1 M phosphate buffer (pH 6.8) for 4 h at 4 °C. Then, these samples were washed with 0.1 M phosphate buffer for three times. Subsequently, these samples were dehydrated through graded ethanol (30%, 50%, 70%, 90%, and 100% v/v, 10 min for each concentration). After that, the samples were replaced by isoamyl acetate (twice, 10 min for each time), dried at a critical point and then coated with gold. Finally, these samples were examined via the scanning electron microscope (S-4800, Hitachi, Tokyo, Japan).

2.3.4. Microbial Community Analysis by Illumina MiSeq Sequencing

To assess the complete microbial community structures, anaerobic biofilm samples were collected from the AWFR system on days 15, 90, 180, 270 and 340. The total DNA from the anaerobic biofilm samples was extracted using the OMEGA Soil DNA Kit D5625-01 (Omega Bio-Tek, Norcross, GA, USA) based on the manufacturer's protocol. The quality and quantity of the extracted DNA were measured by a Nanodrop 1000 spectrophotometer (NanoDrop Technologies, Wilmington, DE, USA).

For gene libraries construction, the V3–V4 regions of 16S rDNA genes were PCR-amplified using BAC319F/806R (5′-ACTCCTACGGGAGGCAGCAG-3′/5′-GGACTACHVGGGTWTCTAAT-3′) and ARC349F/806R (5′-GYGCASCAGKCGMGAAW-3′/5′-GGACTACHVGGGTWTCTAAT-3′), respectively. PCR reactions were conducted in a total volume of 25 µL mixture containing 12.5 µL Premix Ex TaqTM Hot Start Version (TaKaRa, Dalian, China), 2.5 µL of each primer (1 µM), 25 ng template DNA and ddH$_2$O. The bacterial PCR amplification was performed under the following conditions: initial denaturing at 98 °C for 30 s, 30 cycles of denaturing (98 °C for 30 s), annealing (56 °C for 30 s), and elongation (72 °C for 45 s), and a final extension step at 72 °C for 10 min. Archaeal community PCR amplification condition was similar to the bacterial community except that the cycle's number was 35. The PCR products were confirmed by 2% agarose gel electrophoresis and purified using the AxyPrepDNA Gel Extraction Kit (Axygen, Union City, CA, USA) following the manufacturer's protocol. Finally, the purified amplicons were pooled in equimolar and paired-end sequenced using the Illumina MiSeq platform (Illumina Inc., San Diego, CA, USA).

After sequencing, the raw sequences were filtered for quality, trimmed and processed based on the Mothur (version 1.30.1, http://www.mothur.org) analytic pipeline. Briefly, the adapters, barcodes and primers were trimmed. The sequence reads containing ambiguous base calls, with homopolymers >6 bp and shorter than 200 bp were removed. Chimeras detected by UCHIME (version 4.2.40, http://drive5.com/usearch/manual/uchime_algo.html) were filtered out. The resulting high quality sequences were clustered into operation taxonomic units (OTU) at 97% similarity level by Mothur. Representative sequences selected for each OTU were assigned taxonomy using a Ribosomal Database Project (RDP) classifier with a confidence threshold of 80%. The raw sequencing data were deposited to the National Center for Biotechnology Information (NCBI) Short Read Archive (Accession number: SRP101990).

3. Results and Discussion

3.1. Bioreactor Performance

3.1.1. COD Removal

To evaluate the performance of the AWFR system, the seasonal COD removal efficiencies were investigated at different HRTs for 364 days. During the whole operation period, the influent COD concentration ranged from approximately 157 to 469 mg/L (Figure 2). At the beginning of the start-up period (days 1–30), the AWFR was operated in the temperatures ranged from 8 °C to 17 °C, with an HRT of three days. The COD removal efficiency fluctuated in the range of 24.6–51.8%. Based on the COD removal efficiency in the start-up period, we concluded that microbial community adapted to the operation conditions and the AWFR was successfully initiated. The COD removal efficiency increased gradually in spring (days 31–90), thus resulting in an average COD removal efficiency of 61 ± 7.9% (n = 60) at an HRT of two days, with average temperature variations of 21 ± 3.0 °C (n = 60). COD removal efficiency can improve with temperature. Thus, the HRT was adjusted to one day when the average temperature increased to 31 ± 3.7 °C (n = 90) in summer (days 91–180). Despite shortening the HRT, following temperature increase, the average effluent COD concentration decreased to 82.5 mg/L, and the average COD removal efficiency increased to 76 ± 7.2% (n = 90). The change in COD removal efficiency might be due to higher activities of microbial communities in summer. In order to keep COD removal efficiency above 50%, the HRT was adjusted from two days (autumn) to three days (winter). Even though the OLR decreased from 135 ± 32.9 mgCOD/L/day (n = 90) in autumn to 82 ± 12.9 mgCOD/L/day (n = 94) in winter, the COD removal efficiency still decreased from 57 ± 9.4% (autumn, average temperature 25 ± 5.2 °C, n = 90) to 52 ± 5.9% (winter, average temperature 10 ± 2.2 °C, n = 94). These findings indicated that the temperature played an important role in the AWFR system operation. Low temperature might increase the wastewater viscosity, slow down the rate of biological reaction, and decrease COD removal efficiency. Compared to the previous study with elastic fiber carriers in two-stage anaerobic filter for domestic wastewater treatment [12], the COD removal efficiency in this study was better, which may be because the wool felt carrier enhanced the agglomeration of the microbial community in the AWFR system. Overall, the performance of the AWFR system was relatively stable and the COD removal efficiency was higher in summer than in winter. To achieve better performance of the AWFR system at the temperature ranging from 8 °C to 35 °C, an HRT of three days is recommended.

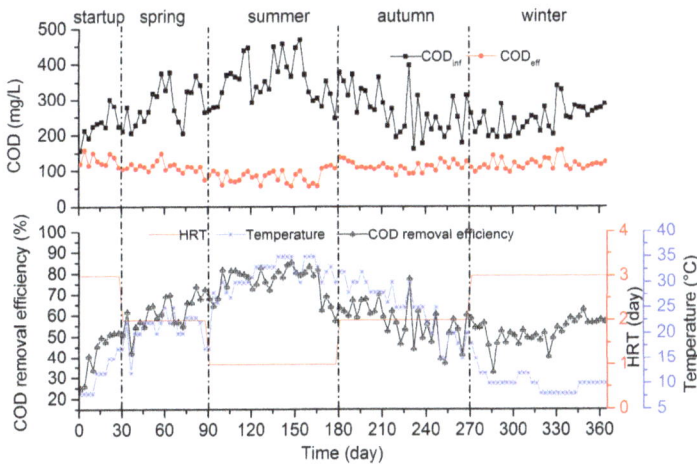

Figure 2. Influent and effluent concentrations of chemical oxygen demand (COD) and COD removal efficiency.

3.1.2. VFA Accumulation and Biogas Production

Since volatile fatty acids (VFA) is an important performance indicator of the anaerobic system, the variation of VFAs with respect to seasonal temperature variations was studied in the AWFR system. It was observed clearly that acetate and propionate were the main VFAs in the effluent as shown in Figure 3, indicating that organic matter was degraded to VFAs by both hydrolytic and acidogenic bacteria. However, no butyrate and valerate were detected, which might be due to their low concentrations below the detection threshold. During spring (days 31–90), the average effluent concentrations of acetate and propionate were 44.5 ± 8.2 mg/L ($n = 60$) and 3.8 ± 1.6 mg/L ($n = 60$), respectively. With the average temperature increasing to 31 ± 3.7 °C ($n = 90$) in summer (days 91–180), the average effluent concentrations of acetate and propionate drastically dropped to 30.1 ± 5.0 mg/L ($n = 90$) and 2.6 ± 0.7 mg/L ($n = 90$), respectively. Nevertheless, the concentrations of acetate and propionate increased from 45.6 ± 15.6 mg/L ($n = 90$) and 6.3 ± 2.6 mg/L ($n = 90$) to 62.1 ± 8.5 mg/L ($n = 94$) and 6.3 ± 3.2 mg/L ($n = 94$) in winter. Similar results were also observed with AFBR to treat domestic wastewater in winter [36]. The effluent concentrations of both acetate and propionate were much lower than those in winter even though the OLR in summer (352 ± 61.8 mgCOD/L/day, $n = 90$) was higher. This coincided with higher COD removal efficiency in summer than in winter, indicating that the conversion of VFAs to biogas was higher in summer than in winter because of high activities of the microbial community, especially for acetoclastic methanogens. Based on these findings, we further confirmed that the seasonal temperature variations affected the anaerobic metabolism of decentralized wastewater.

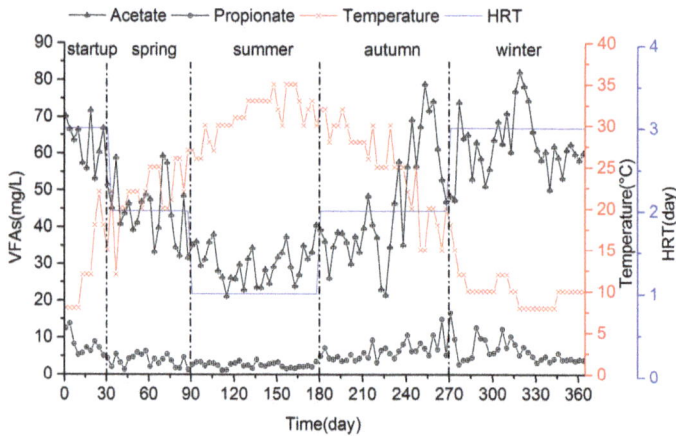

Figure 3. Volatile fatty acids (VFAs) production in the AWFR system during the whole operation.

The daily biogas production is shown in Figure 4. The daily biogas production in the start-up period was low, but after the formation of the stable anaerobic biofilm, it gradually increased with temperature increase, even though the HRT was reduced. The maximum daily biogas production of 10.7 L/day was achieved in summer, when the HRT was one day. This result corresponded well with a high reduction of COD concentration in effluent and a high COD removal efficiency during that period. However, the biogas production decreased with temperature decrease in winter, while acetate and propionate were accumulated. The lower biogas production in winter might be explained by poor degradation of organic matter and higher solubility of biogas at a lower temperature.

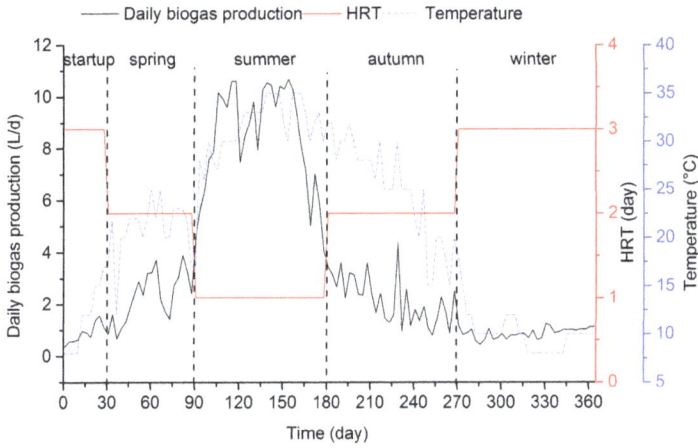

Figure 4. Biogas production in the system during the whole operation.

3.2. COD Mass Balance

Based on Equations (1)–(3), a COD mass balance was used to assess the bioconversion process of organic matter during spring, summer, autumn and winter (Figure 5). Approximately 43.5–52.5% of influent COD was converted to methane (including gaseous methane and dissolved methane). Similar COD conversions were reported for an anaerobic filter treating on-site domestic wastewater [37]. In addition, 10.2–31.0% of influent COD was converted to VFAs, while 25.5–37.3% of influent COD was transformed to biomass, CO_2 and non-biodegradable organic matters. During the spring, about 28.3% of influent COD was converted to gaseous methane and 18.7% of influent COD was accounted for VFAs. With an increase of temperature in summer, more organic matter was converted to gaseous methane (41.1% of influent COD) than VFAs (10.2% of influent COD). In contrast, more VFAs were accumulated in winter. Meanwhile, the results demonstrated that dissolved methane represented about 17.7%, 11.4%, 16.4%, and 23.7% of COD during spring, summer, autumn, and winter, respectively, indicating that more methane dissolved in effluent with a decrease of temperatures. As a potential energy resource, future studies should focus on the recovery of dissolved methane in the AWFR system to avoid the loss of dissolved methane.

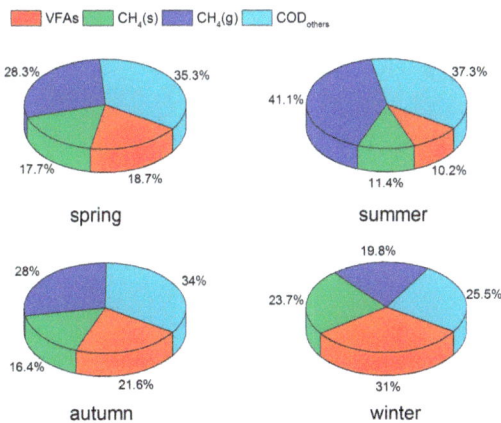

Figure 5. COD balance in different seasons.

3.3. Nutrient Removal

Nitrogen and phosphorous are two important nutrients for agricultural and landscaping reuse. The influent and effluent concentrations of TN and TP were monitored throughout the whole operation period. As shown in Figure 6a,b, the influent concentration of TN fluctuated from 19.6 to 54.2 mg/L throughout the operation period. In spring, with the temperature of 21 ± 3.0 °C ($n = 60$), the TN concentration in the effluent of the AWFR system was 30.2 ± 5.9 mg/L ($n = 60$), corresponding to a $15.7 \pm 2.5\%$ ($n = 60$) TN removal efficiency at an HRT of two days. When the temperature increased to 31 ± 3.7 °C ($n = 90$) in summer, the TN removal efficiency slightly increased to $19.0 \pm 2.7\%$ ($n = 90$) at an HRT of one day. Despite an HRT increase from one day in summer to three days in winter, the average TN removal efficiency decreased to $12.9 \pm 2.8\%$ ($n = 94$). Overall, the TN removal was low in this study because TN was probably removed by microbial assimilation only, and not via a nitrification–denitrification pathway. Meanwhile, a similar trend was observed for TP removal (Figure 6c,d). $25.0 \pm 3.5\%$ ($n = 90$) of TP was removed in summer at an HRT of one day, corresponding to an average effluent TP concentration of 2.2 ± 0.48 mg/L ($n = 90$). With the decrease of temperature from summer to winter, TP removal efficiency in winter was $16.1 \pm 4.4\%$ ($n = 94$) at an HRT of three days, which corresponded to an average TP concentration in the effluent of 2.0 ± 0.44 mg/L ($n = 94$). Most of the TN and TP remained in the effluent of the AWFR system. Therefore, future studies should focus on the ecological post-treatment of the AWFR system.

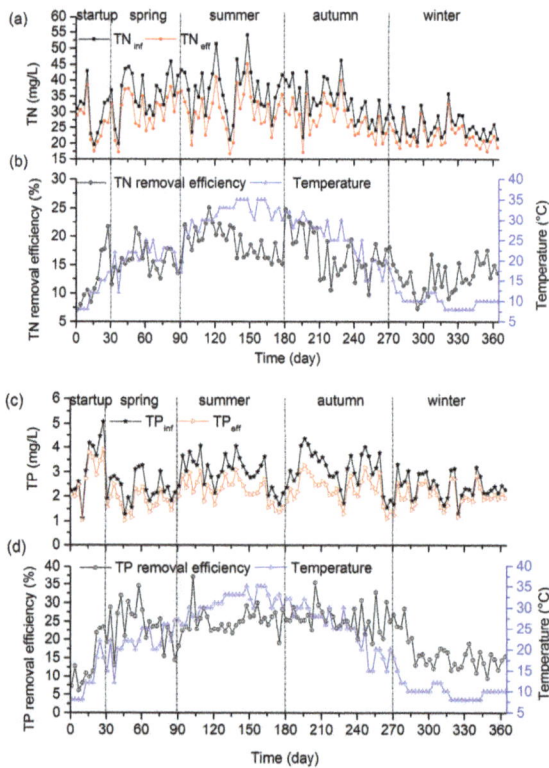

Figure 6. Total nitrogen (TN) (**a,b**) and total phosphorus (TP) (**c,d**) concentration in influent, effluent and the removal efficiency throughout the operation period.

3.4. Morphology and Structure of Anaerobic Biofilm Development

The biofilm in an anaerobic bioreactor comprises a complex microbial community. To observe the morphology and structures of the anaerobic biofilm, representative anaerobic biofilm samples (15, 90, 180, 270 and 340 days) were taken from the system for SEM analysis. Images of SEM revealed that the wool felt carrier had a highly porous and rough surface structure that played important roles in the anaerobic biofilm formation (Figure 7). Various types of microbes initially began to develop on the surface of the wool felt carrier primarily by Van der Waals and electrostatic forces (Figure 7a) [38,39]. From the anaerobic biofilm, cocci, bacillus, and filamentous bacteria were observed on the wool felt carrier. The anaerobic biofilm gradually became thicker (Figure 7c,e,g). A predominance of bacillus and filamentous bacteria deep into the biofilm matrix could be noticed. In addition, the filament-shaped *Methanosaeta-like* structures, long rod-shaped *Methanobacterium-like* structures, and *Methanolinea-like* structures were observed by SEM (Figure 7b,d,f,h,j), which were also observed in previous studies [31,40]. SEM observations were further confirmed by Illumina MiSeq sequencing analysis.

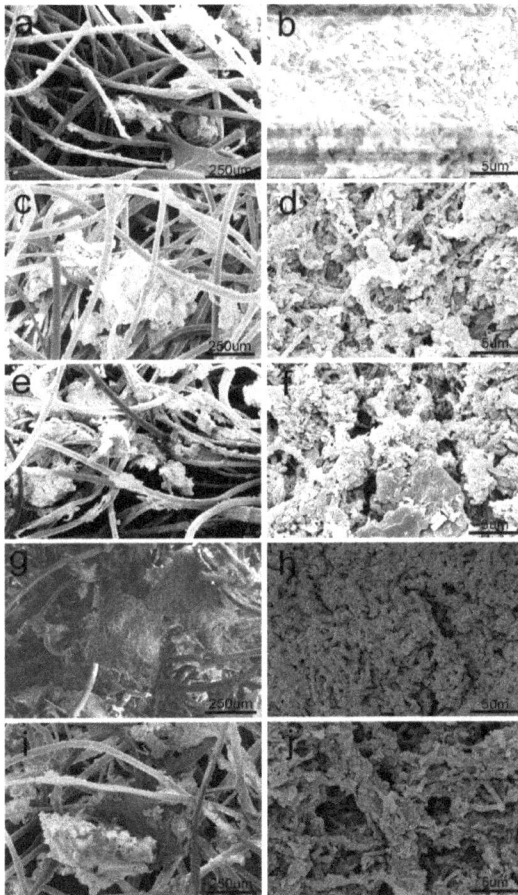

Figure 7. SEM of biofilm on the wool felt in the reactor: (**a,b**): start-up on day 15; (**c,d**): spring on day 90; (**e,f**): summer on day 180; (**g,h**): autumn on day 270; (**i,j**): winter on day 340.

3.5. Shifts in Microbial Community Structures with Seasonal Temperature

High COD removal and biogas production were achieved by the AWFR system in summer, but not in winter. To better elucidate the changes of the microbial community and explore the COD removal mechanism during the whole operation period, the composition and structures of the microbial community were analyzed. Biofilm samples on days 15, 90, 180, 270 and 340, which represented five different periods of the AWFR, were chosen for high-throughput sequencing by the Illumina MiSeq platform.

The archaeal community had a lower diversity and most of the archaeal sequences were assigned to the three main orders *Methanosarcinales, Methanobacteriales* and *Methanomicrobiales* throughout the operation period (Figure 8a). These orders occupied 92.8–96.5% of the total community and were closely related to acetotrophic and hydrogenotrophic methanogen. During the start-up period, the predominant orders *Methanosarcinales, Methanobacteriales* and *Methanomicrobiales* accounted for 49.3%, 21.4%, and 22.2%, respectively, of the total community. However, the relative abundance of the order *Methanosarcinales* (34.3%) decreased in spring, whereas the relative abundances of the orders *Methanobacteriales* (35.0%) and *Methanomicrobiales* (26.0%) increased. In contrast, the order *Methanosarcinales* showed an increase in abundance during summer, but a reduction in abundance in winter. The relative abundances of the orders *Methanobacteriales* and *Methanomicrobiales* significantly decreased in summer but largely increased in winter.

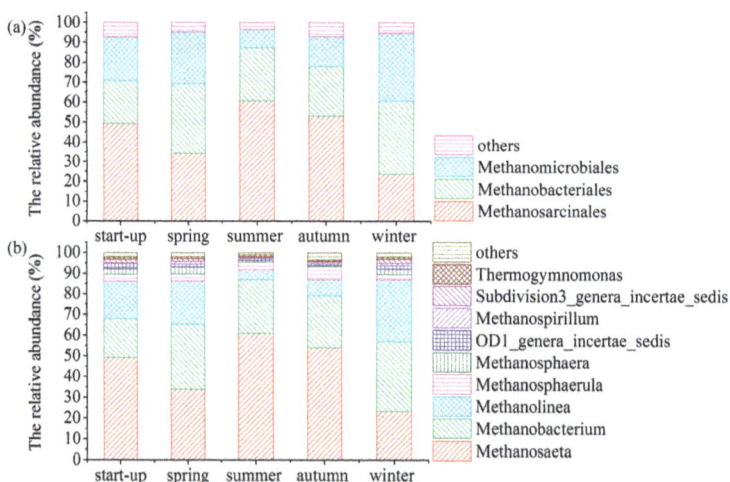

Figure 8. Archaeal community shifts at (**a**) order and (**b**) genus level.

To further validate the function of archaeal community, the microbial distribution at genus level was shown in Figure 8b. The results demonstrated that *Methanosaeta, Methanobacterium* and *Methanolinea* were the predominant genera throughout the operation period. Among these genera, *Methanosaeta* was affiliated with acetoclastic methanogen and the remaining genera *Methanobacterium* and *Methanolinea* were affiliated with hydrogenotrophic methanogens. *Methanosaeta* affiliated with the order *Methanosarcinales* had a high relative abundance (49.2%) in the start-up period, whereas the relative abundances of the genera *Methanobacterium* and *Methanolinea* were 18.7% and 18.3%, respectively. During summer, the genus *Methanosaeta* (61.0%) was the most abundant followed by the genera *Methanobacterium* (26.1%) and *Methanolinea* (4.8%). These findings indicated that acetoclastic methanogenesis was the main pathway, with some contribution from the hydrogenotrophic methanogens. These genera may have helped the AWFR system to maintain low VFA level, leading to relatively better COD removal efficiency and higher biogas production in summer. These results were in accordance with that in a previous study, achieving higher COD efficiency and biogas production in summer in a UASB treating domestic wastewater [8]. Nevertheless, during the winter period, the hydrogenotrophic genus *Methanobacterium* (33.8%) was the most dominant species followed by the hydrogenotrophic genus *Methanolinea* (29.9%) and the acetoclastic genus *Methanosaeta* (23.2%). Furthermore, other less abundant hydrogenotrophic genera were also detected in winter, including the genera *Methanospirllum*, *Methanosphaerula* and *Methanosphaera*, which are capable of using H_2/CO_2 to produce methane. Hydrogenotrophic genera (>63.7%) made up a large portion of the archaeal community in winter. These findings were in line with the previous observations of methanogen population shifts to favor hydrogen utilization under low temperatures [25,41–43]. It might be explained that lower temperature in winter affected the microbial membrane fluidity and inhibited the utilization of acetate than H_2 [44]. Although the relative abundance of the hydrogenotrophic genera was much higher than the acetoclastic genera in winter, COD removal and biogas production in winter were lower compared with in summer. The low temperature in winter may affect more the H_2/CO_2 methanogenesis than H_2/CO_2-dependent acetate production, and decrease the conversion rate of H_2/CO_2 to methane and increase the acetate accumulation [41,44].

In contrast to the archaeal community, the bacterial community in the AWFR system showed high diversity: 33 taxonomic categories at phyla level were identified (Table S1). This result was consistent with the previous studies [45–47]. This high diversity might suggest that a variety of bacterial communities participated in multiple metabolic pathways of organic matter degradation under seasonal temperature variation. The predominant bacteria in the AWFR system during the whole operation period were grouped into four phyla affiliated with *Firmicutes*, *Proteobacteria*, *Bacteroidetes*, and *Chloroflexi*, which occupied 89.3 ± 4.08% of the total phyla (Figure 9). These dominant phyla were also detected in the full-scale biogas digesters [47,48]. During the start-up period, the predominant microorganisms in the AWFR system belonged to the phylum *Proteobacteria*, *Firmicutes*, *Bacteroidetes* and *Chloroflexi*, with a relative abundance of 33.01%, 22.19%, 19.94% and 11.29%, respectively. With an increase of temperatures from spring to summer, the bacterial community structures remarkably changed. The abundance of the phylum *Firmicutes* increased from 37.26% to 64.97%, whereas the relative abundances of the phyla *Proteobacteria*, *Bacteroidetes*, and *Chloroflexi* decreased from 28.02% to 14.96%, 16.03% to 10.06%, and 6.00% to 3.87%, respectively. Interestingly, with a decrease of seasonal temperatures from summer to winter, the relative abundance of *Firmicutes* drastically decreased from 64.97% to 29.39%. In contrast, *Proteobacteria*, and *Bacteroidetes* obviously increased from 14.96% to 31.50% and 10.06% to 20.66%, respectively. Each phylum in our system likely possessed different tolerance and adaptation mechanisms to seasonal temperature variations. This was consistent with previous studies, stressing that temperature was the vital factor affecting the structures of the bacterial community in anaerobic digestion [48].

To gain further insight into the bacterial community structure in the AWFR system, some main bacteria at class level are shown in Figure 9b. During the start-up period, the dominant class was *Gammaproteobacteria* affiliated with the phylum *Proteobacteria*, with a relative abundance of 13.67%, followed by *Bacteroidia*, *Clostridia*, *Anaerolineae* and *Bacilli*, with a relative abundance of 12.87%, 11.88%, 10.96% and 9.89%, respectively. Among these predominant classes, *Bacteroidia* and *Clostridia* played significant roles in hydrolysis metabolism [47–49]. *Gammaproteobacteria* was the most representative class during the start-up period, which was also the main contributor to the phylum *Proteobacteria*. However, with an increased temperature during the summer period, *Bacilli* affiliated with the phylum *Firmicutes* was the most representative class, with a relative abundance of 57.17%, while the relative abundances of *Bacteroidia* and *Clostridia* were 6.61% and 7.43%, respectively. These results probably indicate that seasonal temperature in summer benefited the growth of *Bacilli* bacteria. *Bacteroidales* affiliated with the class *Bacteroidia* had been identified as hydrolyser and fermenter that participated in the conversation of cellulose and polysaccharide into VFAs [49]. *Clostridiales* affiliated with the class *Clostridia* was correlated with methane production, due to their diverse metabolism that utilized polysaccharide fermentation to produce VFAs for methanogenesis [47,48]. It was interesting to note that the class *Bacteroidales* was much more abundant in winter (12.45%) than in summer (6.61%), corresponding to an increase of VFAs in the winter. These results also corresponded with the decrease of COD removal efficiency and biogas production in the winter. Furthermore, the class *Deltaproteobacteria* was much less abundant in summer (1.94%) than in winter (7.81%). A similar tendency was shown by the class *Alphaproteobacteria* (4.92% in summer, and 9.57% in winter) and *Betaproteobacteria* (3.84% in summer, and 7.63% in winter).

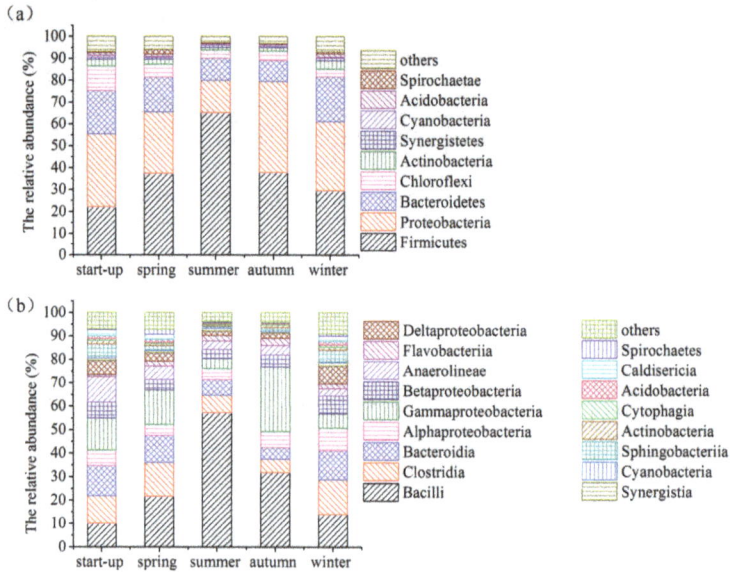

Figure 9. Bacterial community shifts at (**a**) phylum and (**b**) class level.

In order to better evaluate the structures of bacterial communities, the top 10 dominant genera in different periods of the system are presented in Figure 10. Among these genera, although the relative abundance of the bacteria at genus level differed depending on the operation conditions, it appeared that *Bacillus* was the most abundant genus throughout the whole operation period. *Bacillus* has been reported to possess the abilities of degrading various organic compounds [50–53]. The genus *Pseudomonas*, with a chemolithoautotrophic and heterotrophic functionality, has been reported to play an important role in organic compound degradation [54,55]. The presence of *Bacillus* and *Pseudomonas* in summer might be responsible for the degradation of organic matters by the AWFR system. This finding was consistent with the high COD removal efficiency and low effluent concentration of COD in summer. In addition, *VadinBC27_wastewater-sludge_group* was also observed throughout the operation period. It might be participating in the degradation of amino acids and some refractory organic matters [56–58]. The genus *Smithella*, which could degrade propionate to acetate [42,59], was dominant in start-up (3.13%) and winter (3.5%), and this was probably one of the reasons of acetate accumulation. The genus *Acinetobacter* was observed in autumn and winter, and was probably involved in the oxidation of organic matter or sulfides [13,60].

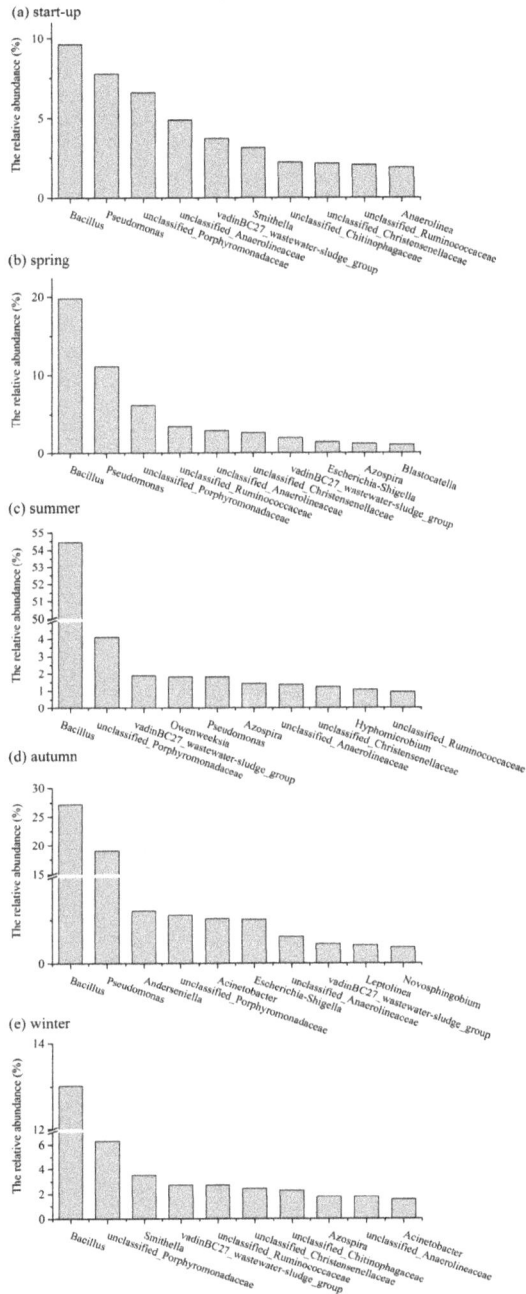

Figure 10. The top 10 bacterial genera during (**a**) start-up; (**b**) spring; (**c**) summer; (**d**) autumn; (**e**) winter period.

4. Conclusions

This study assesses the performance of the three-stage AWFR system for the treatment of decentralized domestic wastewater under seasonal variation of temperature. The COD removal efficiency changed with temperature: the average COD removal efficiency in summer and winter were 76 ± 7.2% (1-day HRT) and 52 ± 5.9% (3-day HRT), respectively. Although COD removal was lower in winter, approximately 43.5% of the influent COD was still converted to methane during that period. Miseq sequencing results suggested that seasonal temperature had a strong impact on the microbial community composition. The genera *Methanosaeta*, *Methanobacterium*, and *Methanolinea* were the predominant methanogens, whereas *Bacillus* was always the most abundant genus, which probably contributed to the fermentation processes throughout the whole operation period. Most of the nutrients, i.e., N and P, remained in the effluent, which could be treated by wetland or used for irrigation for agriculture. The AWFR system appears to be a sustainable option for the pretreatment of the decentralized domestic wastewater. However, longer HRT needs to be applied during winter.

Supplementary Materials: The following are available online at www.mdpi.com/2076-3417/7/6/605/s1, Table S1: The relative abundance of bacterial community at phylum level.

Acknowledgments: This research was supported by China National Water Pollution Control and Management Technology Major Projects (2012ZX07101-005). The authors would like to thank the editor and the anonymous reviewers for editing and review. We thank Liwei Sun, Ran Yu, Haq Nawaz Abbasi and John Leju Celestino Ladu for useful comments and manuscript polishing.

Author Contributions: Juanhong Li and Xiwu Lu conceived and designed the experiments, analyzed the data, and wrote the paper.

Conflicts of Interest: The authors declare no conflicts of interest.

References

1. Yu, R.; Wu, Q.; Lu, X. Constructed wetland in a compact rural domestic wastewater treatment system for nutrient removal. *Environ. Eng. Sci.* **2012**, *29*, 751–757. [CrossRef]

2. Dong, H.-Y.; Qiang, Z.-M.; Wang, W.-D.; Jin, H. Evaluation of rural wastewater treatment processes in a county of eastern China. *J. Environ. Monit.* **2012**, *14*, 1906–1913. [CrossRef] [PubMed]

3. McCarty, P.L.; Bae, J.; Kim, J. Domestic wastewater treatment as a net energy producer—Can this be achieved? *Environ. Sci. Technol.* **2011**, *45*, 7100–7106. [CrossRef] [PubMed]

4. Chernicharo, C.; Van Lier, J.; Noyola, A.; Ribeiro, T.B. Anaerobic sewage treatment: State of the art, constraints and challenges. *Rev. Environ. Sci. Bio/Technol.* **2015**, *14*, 649–679. [CrossRef]

5. López-López, A.; Albarrán-Rivas, M.G.; Hernández-Mena, L.; León-Becerril, E. An assessment of an anaerobic filter packed with a low-cost material for treating domestic wastewater. *Environ. Technol.* **2013**, *34*, 1151–1159. [CrossRef] [PubMed]

6. Smith, A.L.; Skerlos, S.J.; Raskin, L. Membrane biofilm development improves COD removal in anaerobic membrane bioreactor wastewater treatment. *Microb. Biotechnol.* **2015**, *8*, 883–894. [CrossRef] [PubMed]

7. Mei, X.; Wang, Z.; Miao, Y.; Wu, Z. Recover energy from domestic wastewater using anaerobic membrane bioreactor: Operating parameters optimization and energy balance analysis. *Energy* **2016**, *98*, 146–154. [CrossRef]

8. Bandara, W.M.; Kindaichi, T.; Satoh, H.; Sasakawa, M.; Nakahara, Y.; Takahashi, M.; Okabe, S. Anaerobic treatment of municipal wastewater at ambient temperature: Analysis of archaeal community structure and recovery of dissolved methane. *Water Res.* **2012**, *46*, 5756–5764. [CrossRef] [PubMed]

9. Chong, S.; Sen, T.K.; Kayaalp, A.; Ang, H.M. The performance enhancements of upflow anaerobic sludge blanket (UASB) reactors for domestic sludge treatment–a state-of-the-art review. *Water Res.* **2012**, *46*, 3434–3470. [CrossRef] [PubMed]

10. Feng, J.; Wang, Y.; Ji, X.; Yuan, D.; Li, H. Performance and bioparticle growth of anaerobic baffled reactor (ABR) fed with low-strength domestic sewage. *Front. Environ. Sci. Eng.* **2014**, *9*, 352–364. [CrossRef]

11. Smith, A.L.; Stadler, L.B.; Love, N.G.; Skerlos, S.J.; Raskin, L. Perspectives on anaerobic membrane bioreactor treatment of domestic wastewater: A critical review. *Bioresour. Technol.* **2012**, *122*, 149–159. [CrossRef] [PubMed]

12. Gouveia, J.; Plaza, F.; Garralon, G.; Fdzpolanco, F.; Pena, M. Long-term operation of a pilot scale anaerobic membrane bioreactor (AnMBR) for the treatment of municipal wastewater under psychrophilic conditions. *Bioresour. Technol.* **2015**, *185*, 225–233. [CrossRef] [PubMed]

13. Jo, Y.; Kim, J.; Hwang, S.; Lee, C. Anaerobic treatment of rice winery wastewater in an upflow filter packed with steel slag under different hydraulic loading conditions. *Bioresour. Technol.* **2015**, *193*, 53–61. [CrossRef] [PubMed]

14. Gannoun, H.; Khelifi, E.; Omri, I.; Jabari, L.; Fardeau, M.-L.; Bouallagui, H.; Godon, J.-J.; Hamdi, M. Microbial monitoring by molecular tools of an upflow anaerobic filter treating abattoir wastewaters. *Bioresour. Technol.* **2013**, *142*, 269–277. [CrossRef] [PubMed]

15. Gannoun, H.; Othman, N.B.; Bouallagui, H.; Moktar, H. Mesophilic and thermophilic anaerobic co-digestion of olive mill wastewaters and abattoir wastewaters in an upflow anaerobic filter. *Ind. Eng. Chem. Res.* **2007**, *46*, 6737–6743. [CrossRef]

16. Couto, E.D.A.D.; Calijuri, M.L.; Assemany, P.P.; Santiago, A.D.F.; Lopes, L.S. Greywater treatment in airports using anaerobic filter followed by uv disinfection: An efficient and low cost alternative. *J. Clean. Prod.* **2015**, *106*, 372–379. [CrossRef]

17. Wu, Y.; Zhu, W.; Lu, X. Identifying key parameters in a novel multistep bio-ecological wastewater treatment process for rural areas. *Ecol. Eng.* **2013**, *61*, 166–173. [CrossRef]

18. Tonon, D.; Tonetti, A.L.; Coraucci Filho, B.; Bueno, D.A.C. Wastewater treatment by anaerobic filter and sand filter: Hydraulic loading rates for removing organic matter, phosphorus, pathogens and nitrogen in tropical countries. *Ecol. Eng.* **2015**, *82*, 583–589. [CrossRef]

19. Lew, B.; Tarre, S.; Beliavski, M.; Green, M. Anaerobic degradation pathway and kinetics of domestic wastewater at low temperatures. *Bioresour. Technol.* **2009**, *100*, 6155–6162. [CrossRef] [PubMed]

20. Dutta, K.; Tsai, C.-Y.; Chen, W.-H.; Lin, J.-G. Effect of carriers on the performance of anaerobic sequencing batch biofilm reactor treating synthetic municipal wastewater. *Int. Biodeterior. Biodegrad.* **2014**, *95*, 84–88. [CrossRef]

21. Han, Z.; Chen, F.; Zhong, C.; Zhou, J.; Wu, X.; Yong, X.; Zhou, H.; Jiang, M.; Jia, H.; Wei, P. Effects of different carriers on biogas production and microbial community structure during anaerobic digestion of cassava ethanol wastewater. *Environ. Technol.* **2016**, 1–10. [CrossRef] [PubMed]

22. Bodkhe, S. Development of an improved anaerobic filter for municipal wastewater treatment. *Bioresour.Technol.* **2008**, *99*, 222–226. [CrossRef] [PubMed]

23. De Oliveira Cruz, L.M.; Stefanutti, R.; Coraucci Filho, B.; Tonetti, A.L. Coconut shells as filling material for anaerobic filters. *SpringerPlus* **2013**, *2*, 655. [CrossRef] [PubMed]

24. Esmaeili, A.; Beni, A.A. Novel membrane reactor design for heavy-metal removal by alginate nanoparticles. *J. Ind. Eng. Chem.* **2015**, *26*, 122–128. [CrossRef]

25. Seib, M.D.; Berg, K.J.; Zitomer, D.H. Influent wastewater microbiota and temperature influence anaerobic membrane bioreactor microbial community. *Bioresour. Technol.* **2016**, *216*, 446–452. [CrossRef] [PubMed]

26. Mckeown, R.M.; Hughes, D.; Collins, G.; Mahony, T.; Flaherty, V.O. Low-temperature anaerobic digestion for wastewater treatment. *Curr. Opin. Biotechnol.* **2012**, *23*, 444–451. [CrossRef] [PubMed]

27. Enright, A.-M.; Collins, G.; O'Flaherty, V. Temporal microbial diversity changes in solvent-degrading anaerobic granular sludge from low-temperature (15 °C) wastewater treatment bioreactors. *Syst. Appl. Microbiol.* **2007**, *30*, 471–482. [CrossRef] [PubMed]

28. Gomec, C.Y.; Letsiou, I.; Ozturk, I.; Eroglu, V.; Wilderer, P.A. Identification of archaeal population in the granular sludge of an uasb reactor treating sewage at low temperatures. *J. Environ. Sci. Heal. A.* **2008**, *43*, 1504–1510. [CrossRef] [PubMed]

29. Gao, W.; Leung, K.; Qin, W.; Liao, B. Effects of temperature and temperature shock on the performance and microbial community structure of a submerged anaerobic membrane bioreactor. *Bioresour. Technol.* **2011**, *102*, 8733–8740. [CrossRef] [PubMed]

30. Regueiro, L.; Carballa, M.; Lema, J.M. Outlining microbial community dynamics during temperature drop and subsequent recovery period in anaerobic co-digestion systems. *J. Biotechnol.* **2014**, *192*, 179–186. [CrossRef] [PubMed]

31. Abubakkar, S.; Kundu, K.; Sreekrishnan, T.R. Comparative study of the performance of an anaerobic rotating biological contactor and its potential to enrich hydrogenotrophic methanogens. *J. Chem. Technol. Biotechnol.* **2015**, *90*, 398–406. [CrossRef]

32. Takai, K.; Horikoshi, K. Rapid detection and quantification of members of the archaeal community by quantitative pcr using fluorogenic probes. *Appl. Environ. Microbiol.* **2000**, *66*, 5066–5072. [CrossRef] [PubMed]

33. McElhoe, J.A.; Holland, M.M.; Makova, K.D.; Su, M.S.-W.; Paul, I.M.; Baker, C.H.; Faith, S.A.; Young, B. Development and assessment of an optimized next-generation DNA sequencing approach for the mtgenome using the Illumina Miseq. *Forensic Sci. Int. Genet.* **2014**, *13*, 20–29. [CrossRef] [PubMed]

34. American Public Health Association (APHA). *Standard Methods for the Examination of Water and Wastewater*; American Public Health Association: Washington, DC, USA, 2005.

35. Krishna, G.V.T.G.; Kumar, P.; Kumar, P. Treatment of low strength complex wastewater using an anaerobic baffled reactor (ABR). *Bioresour. Technol.* **2008**, *99*, 8193–8200. [CrossRef] [PubMed]

36. Shin, C.; Mccarty, P.L.; Kim, J.; Bae, J. Pilot-scale temperate-climate treatment of domestic wastewater with a staged anaerobic fluidized membrane bioreactor (SAF-MBR). *Bioresour. Technol.* **2014**, *159*, 95–103. [CrossRef] [PubMed]

37. Sharma, M.K.; Khursheed, A.; Kazmi, A.A. Modified septic tank-anaerobic filter unit as a two-stage onsite domestic wastewater treatment system. *Environ. Technol.* **2014**, *35*, 2183–2193. [CrossRef] [PubMed]

38. Hermansson, M. The DLVO theory in microbial adhesion. *Colloids Surf. B: Biointerfaces* **1999**, *14*, 105–119. [CrossRef]

39. Fang, H.H.; Chan, K.-Y.; Xu, L.-C. Quantification of bacterial adhesion forces using atomic force microscopy (AFM). *J. Microbiol. Methods* **2000**, *40*, 89–97. [CrossRef]

40. Fernández, N.; Díaz, E.E.; Amils, R.; Sanz, J.L. Analysis of microbial community during biofilm development in an anaerobic wastewater treatment reactor. *Microb. Ecol.* **2008**, *56*, 121–132. [CrossRef] [PubMed]

41. Lettinga, G.; Rebac, S.; Zeeman, G. Challenge of psychrophilic anaerobic wastewater treatment. *Trends Biotechnol.* **2001**, *19*, 363–370. [CrossRef]

42. Mckeown, R.M.; Scully, C.; Enright, A.; Chinalia, F.A.; Lee, C.; Mahony, T.; Collins, G.; Oflaherty, V. Psychrophilic methanogenic community development during long-term cultivation of anaerobic granular biofilms. *ISME J.* **2009**, *3*, 1231–1242. [CrossRef] [PubMed]

43. Bialek, K.; Kumar, A.; Mahony, T.; Lens, P.N.L.; Flaherty, V.O. Microbial community structure and dynamics in anaerobic fluidized-bed and granular sludge-bed reactors: Influence of operational temperature and reactor configuration. *Microb. Biotechnol.* **2012**, *5*, 738–752. [CrossRef] [PubMed]

44. Fey, A.; Conrad, R. Effect of temperature on carbon and electron flow and on the archaeal community in methanogenic rice field soil. *Appl. Environ. Microbiol.* **2000**, *66*, 4790–4797. [CrossRef] [PubMed]

45. Smith, A.L.; Skerlos, S.J.; Raskin, L. Psychrophilic anaerobic membrane bioreactor treatment of domestic wastewater. *Water Res.* **2013**, *47*, 1655–1665. [CrossRef] [PubMed]

46. Mchugh, S.; Carton, M.W.; Collins, G.; Oflaherty, V. Reactor performance and microbial community dynamics during anaerobic biological treatment of wastewaters at 16–37 °C. *FEMS Microbiol. Ecol.* **2004**, *48*, 369–378. [CrossRef]

47. Sundberg, C.; Al-Soud, W.A.; Larsson, M.; Alm, E.; Yekta, S.S.; Svensson, B.H.; Sørensen, S.J.; Karlsson, A. 454 pyrosequencing analyses of bacterial and archaeal richness in 21 full-scale biogas digesters. *FEMS Microbiol. Ecol.* **2013**, *85*, 612–626. [CrossRef] [PubMed]

48. Lee, S.H.; Kang, H.J.; Lee, Y.H.; Lee, T.J.; Han, K.; Choi, Y.; Park, H.D. Monitoring bacterial community structure and variability in time scale in full-scale anaerobic digesters. *J. Environ. Monit.* **2012**, *14*, 1893–1905. [CrossRef] [PubMed]

49. Goux, X.; Calusinska, M.; Lemaigre, S.; Marynowska, M.; Klocke, M.; Udelhoven, T.; Benizri, E.; Delfosse, P. Microbial community dynamics in replicate anaerobic digesters exposed sequentially to increasing organic loading rate, acidosis, and process recovery. *Biotechnol. Biofuels* **2015**, *8*, 122. [CrossRef] [PubMed]

50. Chebbi, A.; Mnif, S.; Mhiri, N.; Jlaiel, L.; Sayadi, S.; Chamkha, M. A moderately thermophilic and mercaptan-degrading bacillus licheniformis strain can55 isolated from gas-washing wastewaters of the phosphate industry, Tunisia. *Int. Biodeterior. Biodegrad.* **2014**, *94*, 207–213. [CrossRef]

51. Nakkabi, A.; Sadiki, M.; Fahim, M.; Ittobane, N.; Ibnsoudakoraichi, S.; Barkai, H.; Abed, S.E. Biodegradation of Poly(ester urethane)s by Bacillus subtilis. *Int. J. Environ. Res.* **2015**, *9*, 157–162.

52. Patowary, K.; Saikia, R.R.; Kalita, M.C.; Deka, S. Degradation of polyaromatic hydrocarbons employing biosurfactant-producing Bacillus pumilus KS2. *Ann. Microbiol.* **2014**, *65*, 225–234. [CrossRef]

53. Xiao, Y.; Chen, S.; Gao, Y.; Hu, W.; Hu, M.; Zhong, G. Isolation of a novel beta-cypermethrin degrading strain Bacillus subtilis BSF01 and its biodegradation pathway. *Appl. Microbiol. Biotechnol.* **2014**, *99*, 2849–2859. [CrossRef] [PubMed]

54. Dhall, P.; Kumar, R.; Kumar, A. Biodegradation of sewage wastewater using autochthonous bacteria. *Sci. World J.* **2012**, *2012*, 861903. [CrossRef] [PubMed]

55. Antwi, P.; Li, J.; Boadi, P.O.; Meng, J.; Shi, E.; Xue, C.; Zhang, Y.; Ayivi, F. Functional bacterial and archaeal diversity revealed by 16S rRNA gene pyrosequencing during potato starch processing wastewater treatment in an UASB. *Bioresour. Technol.* **2017**, *235*, 348–357. [CrossRef] [PubMed]

56. Xie, Z.; Wang, Z.; Wang, Q.; Zhu, C.; Wu, Z. An anaerobic dynamic membrane bioreactor (AnMBR) for landfill leachate treatment: Performance and microbial community identification. *Bioresour. Technol.* **2014**, *161*, 29–39. [CrossRef] [PubMed]

57. Sun, W.; Yu, G.; Louie, T.S.; Liu, T.; Zhu, C.; Xue, G.; Gao, P. From mesophilic to thermophilic digestion: The transitions of anaerobic bacterial, archaeal, and fungal community structures in sludge and manure samples. *Appl. Microbiol. Biotechnol.* **2015**, *99*, 10271–10282. [CrossRef] [PubMed]

58. Tang, Y.; Shigematsu, T.; Morimura, S.; Kida, K. Microbial community analysis of mesophilic anaerobic protein degradation process using bovine serum albumin (BSA)-fed continuous cultivation. *J. Biosci. Bioeng.* **2005**, *99*, 150–164. [CrossRef] [PubMed]

59. Regueiro, L.; Lema, J.M.; Carballa, M. Key microbial communities steering the functioning of anaerobic digesters during hydraulic and organic overloading shocks. *Bioresour. Technol.* **2015**, *197*, 208–216. [CrossRef] [PubMed]

60. Resende, J.A.; Silva, V.L.D.; De Oliveira, T.L.R.; Fortunato, S.; Carneiro, J.D.C.; Otenio, M.H.; Diniz, C.G. Prevalence and persistence of potentially pathogenic and antibiotic resistant bacteria during anaerobic digestion treatment of cattle manure. *Bioresour. Technol.* **2014**, *153*, 284–291. [CrossRef] [PubMed]

applied
sciences

MDPI

Article

Removal of *Escherichia coli* by Intermittent Operation of Saturated Sand Columns Supplemented with Hydrochar Derived from Sewage Sludge

Jae Wook Chung [1,*], Oghosa Charles Edewi [1], Jan Willem Foppen [1], Gabriel Gerner [2], Rolf Krebs [2] and Piet Nicolaas Luc Lens [1]

[1]   UNESCO-IHE Institute for Water Education, P.O. BOX 3015, 2601 DA Delft, The Netherlands; charlesedewi@yahoo.com (O.C.E.); j.foppen@unesco-ihe.org (J.W.F.), p.lens@unesco-ihe.org (P.N.L.L.)
[2]   Institute of Natural Resource Sciences, Zurich University of Applied Sciences, Grüental, 8820 Wädenswil, Switzerland; gabriel_gerner@hotmail.com (G.G.); krbs@zhaw.ch (R.K.)
*   Correspondence: shoutjx@gmail.com; Tel.: +82-10-7206-6363

Received: 19 June 2017; Accepted: 10 August 2017; Published: 15 August 2017

Featured Application: Bacterial removal in water treatment using a sand column supplemented with adsorbents derived from hydrothermally treated sewage sludge.

Abstract: Hydrothermal carbonization (HTC) technology can convert various types of waste biomass into a carbon-rich product referred to as hydrochar. In order to verify the potential of hydrochar produced from stabilized sewage sludge to be an adsorbent for bacterial pathogen removal in water treatment, the *Escherichia coli*'s removal efficiency was determined by using 10 cm sand columns loaded with 1.5% (w/w) hydrochar. Furthermore, the removal of *E. coli* based on intermittent operation in larger columns of 50 cm was measured for 30 days. Since the removal of *E. coli* was not sufficient when the sand columns were supplemented with raw hydrochar, an additional cold-alkali activation of the hydrochar using potassium hydroxide was applied. This enabled more than 90% of *E. coli* removal in both the 10 cm and 50 cm column experiments. The enhancement of the *E. coli* removal efficiency could be attributed to the more hydrophobic surface of the KOH pre-treated hydrochar. The idle time during the intermittent flushing experiments in the sand-only columns without the hydrochar supplement had a significant effect on the *E. coli* removal ($p < 0.05$), resulting in a removal efficiency of 55.2%. This research suggested the possible utilization of hydrochar produced from sewage sludge as an adsorbent in water treatment for the removal of bacterial contaminants.

Keywords: *Escherichia coli*; bacterial removal; sewage sludge; chloride tracer; hydrothermal carbonization; hydrochar

---

1. Introduction

Hydrothermal carbonization (HTC) is considered an emerging technology for effective waste conversion and/or treatment. In the HTC, also known as "wet pyrolysis", process, feed stock (organic stock immersed in water) is heated in a pressure-resistant reactor. Autogenous pressure inside the reactor allows subcritical water temperatures (180–350 °C), and the feed stock is converted into a mixture of process water containing water-soluble organics and a carbonaceous solid called hydrochar [1]. Several distinctive features of HTC, such as minimal environmental impact, simplicity, cost effectiveness, low greenhouse gas emission, and energy efficiency, make the technology attractive [2].

In 1913, Bergius performed the first HTC experiment to facilitate the natural coalification process of organic feedstock under laboratory conditions [3]. Since then, extensive research has investigated

HTC for the conversion of a wide range of feedstock, from pure substances (e.g., cellulose and glucose) to more complex materials, such as fruit shells and paper [4–7]. The advantage of HTC over other pyrolysis techniques is that a feedstock with high moisture content can be directly converted into a solid carbonaceous material with high yield. The additional cost for feedstock drying in other conventional dry pyrolysis methods can be saved, and this enables various continuously generated waste materials, such as animal and human faeces, municipal sewage sludge, and activated sludge from wastewater treatment plants, to be used as potential feedstock [1].

The traditional disposal methods of sewage sludge have limitations, due to potential environmental risks that could result from various pollutants, such as pathogenic micro-organisms, organic pollutants, and heavy metals [8]. Recently, HTC has been suggested as a cost-effective and eco-friendly solution for this sewage sludge management challenge [9–11]. Also, the carbonaceous solid output (hydrochar) from the HTC process, with a competitive pollutant adsorption capacity, could replace commercial activated carbon in water treatment [12].

The research on the use of hydrochar derived from sewage as a low-cost adsorbent in water treatment is still in its infancy. It was reported that the hydrochars derived from industrial sludge and anaerobically digested sludge removed Pb(II) (q_m = 11 mg/g) in laboratory conditions [13]. Also, sewage sludge-derived hydrochar showed a comparable Pb(II) removal efficiency (q_m = 15 mg/g) [14]. For the removal of pathogenic contaminants, hydrochar derived from sewage sludge was able to achieve a 2 to >3 log removal of human pathogenic rotavirus and adenovirus [15]. However, the removal of faecal bacteria has not yet been reported. Therefore, the main objective of this research is to evaluate the hydrochar derived from sewage sludge as an additive adsorbent in sand filtration setups for *Escherichia coli* removal during water treatment.

2. Materials and Methods

2.1. Escherichia coli Suspension

As a surrogate of enteric bacterial pathogens, a wild-type *E. coli* strain UCFL-94 obtained from a grazing field was provided from previous research [16]. Previous investigations reported a relatively low sticking efficiency of the strain. UCFL-94 was inoculated in 50 mL nutrient broth (OXOID, Basingstoke, Great Britain), and the culture medium was stored at 37 °C for 24 h under agitation at 150 rpm using an orbital shaker. Fresh *E. coli* stocks were prepared every week during the experimental period. The influent for the *E. coli* removal experiments was prepared by dilution of the *E. coli* stock into artificial groundwater (AGW). AGW was prepared by dissolving 526 mg/L $CaCl_2 \cdot 2H_2O$, 184 mg/L $MgSO_4 \cdot 7H_2O$, 8.5 mg/L KH_2PO_4, 21.75 mg/L K_2HPO_4, and 17.7 mg/L Na_2HPO_4 in demineralized (DI) water [17]. The results from pH and electrical conductivity measurements on the AGW ranged between 6.6–6.8 and 1012–1030 µS/cm, respectively.

The influent was prepared at room temperature (23 ± 2 °C), and stabilized for >24 h before its use in the experiments. Since there was no significant change in the *E. coli* concentration of the influent during 4 days of observation (data not shown), the natural die-off of the *E. coli* was not considered. The *E. coli* concentration of the influent was controlled to be ~10^6 CFU/mL or ~10^3 CFU/mL for small and large column experiments, respectively. In this research, the *E. coli* concentration was measured by using the conventional heterotrophic plate counting method [18] employing Chromocult agar plates (ChromoCult® Coliform Agar, Merck, Darmstadt, Germany). After the incubation period of 24 h at 37 °C, the colonies on the agar plates were counted using a Colony Counter (IUL, Barcelona, Spain).

2.2. Hydrochar

The Zurich University of Applied Sciences (ZHAW, Wädenswil, Switzerland) provided the hydrochar stock used in this research. Briefly, the hydrothermal conversion of stabilized sewage sludge from a wastewater treatment plant was carried out with sulphuric and acetic acid supplements for 5 h at a median temperature of ~210 °C. The autogenous pressure inside the reactor ranged between

21 and 24 bar. The output of the hydrothermal reaction had the form of thick slurry. It was chilled to 20 °C and dehydrated using a membrane filter press. The resulting filter cake was oven-dried at 105 °C overnight, and manually powdered using a mortar and pestle. Finally, the hydrochar powders were rinsed by a few rounds of suspension in DI water and successive centrifugation at 2700 g for 3 min (236 HK, Hermle, Wehingen, Germany).

In order to increase the adsorptive capacity, the raw hydrochar was activated by KOH solution [17,19,20]. Washed hydrochar powders were suspended in a 1 M KOH solution at a concentration of 5 g hydrochar (dry weight)/L and stirred for 1 h at room temperature. The hydrochar-KOH suspension was washed as described earlier until a neutral pH was observed [17]. The KOH concentration for the activation was selected based on the results from preliminary experiments (data not shown) as described previously [17]. The concentration of each hydrochar suspension was determined by measuring the dry weight (at 105 °C). All hydrochar stocks were stored at 4 °C until needed.

2.3. Material Characterization

2.3.1. Zeta Potential

The zeta potential of hydrochar and *E. coli* in a certain pH range (4–10 for hydrochar and 5.5–8.5 for *E. coli*) was measured using a Zetasizer Nano ZS (Malvern, Worcestershire, UK) equipped with an auto-titration unit MPT-2. Hydrochar and *E. coli* samples were conditioned in AGW by several rounds of washing prior to the experiments. All of the test samples were diluted to a certain extent in order to have an adequate attenuator selection (6–8) of the instrument.

2.3.2. Elemental Composition

Both the raw and activated hydrochar samples were analyzed by X-ray fluorescence (XRF) spectroscopy in order to assess their elemental compositions using a SPECTRO-XEPOS XRF spectrometer (SPECTRO, Kleve, Germany). The instrument was equipped with a 10 mm^2 Si-Drift Detector with Peltier cooling and a spectral resolution (FWHM) at Mn Ka $\leq$155 eV for determination in the element range of Na–U (SPECTRO, 2014). The hydrochar samples were powdered (grain size <100 µm) using a milling instrument, MM 400 (Retsch, Haan, Germany), for 5 min at a frequency of 25 s^{-1}. Then, sample pellets with a 32 mm diameter were prepared by mixing of 4 g hydrochar powder with 0.9 g Licowax C micro powder PM (Clariant, Muttenz, Switzerland) and successive pressing under 15 tonnes pressure. The analysis was performed using the TurboQuant-screening method. Each side of the pellet was subjected to one exposure to X-ray radiation, and the mean results of both sides (two analyses) were recorded.

Also, the carbon (C), hydrogen (H), nitrogen (N), and oxygen (O) contents were measured based on the dry combustion method using a TruSpec analyzer (LECO, St. Joseph, MI, USA). The samples for the C, H, N, and O analysis were thoroughly dried at 105 °C and powdered using the mixer mill for 5 min at a frequency of 25 s^{-1}. Then, 100 mg of each sample was incinerated at 950 °C and recorded by a TruSpec CHN Macro Analyzer. Furthermore, the O content was assessed by combusting 3 mg of sample material at 1300 °C using the additional high-temperature pyrolysis furnace TruSpec Micro Oxygen Module. All of the samples were analyzed in duplicate, and the mean results were recorded.

2.3.3. Surface Functional Groups

The raw and activated hydrochar samples were analyzed by Photoacoustic Fourier transform Infrared Spectroscopy (FTIR-PAS) in order to investigate their surface functional groups. A Tensor 37 FTIR spectrometer (Bruker Optics, Fällanden, Switzerland) equipped with a photoacoustic optical cantilever microphone PA301 detector (Gasera, Turku, Finland) was used for recording infrared spectra in the range of 4000–400 cm^{-1}. In order to enhance the signal-to-noise ratio, an average determination of 32 single spectra was performed after the analyses. In addition, the interference resulting from CO_2

was minimized by subtracting the CO_2 spectrum from the sample spectrum. Prior to the measurements, samples were desiccated by drying at 105 °C for 1 h and successive storing in an exsiccator at room temperature. For each analysis, the PA301 sample cell was refreshed with helium gas of 99.999% purity.

2.3.4. Specific Surface Area and Pore Size Distribution

In order to investigate the surface area and pore size distribution of the hydrochars, gas sorption analyses were carried out using an Autosorb-iQ (Quantochrome, Boynton Beach, FL, USA). Prior to the analysis, the residual water and organic contents of the hydrochar samples were removed by placing them in dynamic vacuum conditions at 120 °C. Gas sorption experiments were performed at 77 K (−196 °C) employing N_2 gas as an adsorbing gas. The surface area was derived based on the multi-point Brunauer–Emmett–Teller (BET) method. Also, density functional theory (DFT) using the quench solid DFT (QSDFT) method was used for analyzing micro and mesoporous pore size distribution [21]. Macropores were not subjected to the analysis.

2.3.5. Surface Morphology

The surface morphology of the hydrochar variants was investigated by performing a Scanning Electron Microscopy (SEM) analysis using a Quanta 250 FEG (FEI, Hillsboro, OR, USA). The hydrochar sample was placed on an adhesive side of a carbon tape fixed on an aluminium stub. Successive flushing with nitrogen gas of 99.999% purity removed detached particles.

2.3.6. Hydrophobicity

The hydrophobic property of both hydrochar samples was investigated by measuring the static contact angle using a DSA100 (KRÜSS, Hamburg, Germany). In order to obtain a flat surface of a hydrochar sample, identical pelletizing procedures employed in XRF analysis were used (see Section 2.3.2), except for the addition of Licowax C [22]. The average results of triplicated measurements were recorded.

2.4. Column Experiments

2.4.1. Experimental Setup

A 99.1% pure-quartz sand (Kristall quartz-sand, Gebrüder Dorfner, Hirschau, Germany) was used as a packing material for the columns. A sieve analysis on the sand stock reported an effective grain size (D_{10}) of 0.45 mm, a uniformity coefficient of 2.0, and a maximum grain size of 2.0 mm [23]. The size of the hydrochar particles was smaller than 0.425 mm. In order to remove undesirable impurities, the sand stock was immersed in 5% HCl overnight and rinsed with DI water until the pH of the washing water became stabilized. Finally, the residual water was removed by draining and successive drying at 105 °C.

A certain amount of hydrochar suspension was added to the washed sand grains to achieve a 1.5% (dry w/w) hydrochar concentration, and the mixture was then thoroughly mixed. The hydrochar–sand mixture was then loaded into two types (small or large) of columns: Omnifit borosilicate glass columns (250 mm length × 25 mm inner diameter; Diba industries, Cambridge, UK) for supporting short (10 cm) column beds, and acrylic poly vinyl chloride (PVC) pipes with an inner diameter of 5.6 cm for supporting long (50 cm) column beds. During the packing process, the columns were manually tapped and agitated in order to prevent channelling in the column bed. Also, the packing material loaded in the column was compacted carefully using either a glass rod or a steel rod to avoid possible air entrapment. Each type of column was assembled using appropriate connectors and fitting materials. Throughout the experiments, the packing materials were tightly fixed in the columns by stoppers at both ends, without under-drain media and exposure to the atmosphere. Then, the columns were washed with DI water overnight using a Masterflex pump (model 77201-60, Vernon Hills, IL, USA). During the washing process, the DI water was fed into columns at an upward flow rate of 1 mL

(0.2 cm)/min for small columns, or 33.3 mL (1.35 cm)/min for large columns. Chloride tracer tests, performed on both the small and large columns (data not shown), reported a void ratio of ~40%.

The columns were vertically positioned using tripod stands or a steel frame rack. Prior to the main *E. coli* flushing experiment, chloride tracer tests were performed to estimate the pore volume (PV) and to verify a stable aqueous flow in the column bed. The columns were flushed with the tracer solution (0.02 M NaCl) and successively with the DI water. The tracer concentration in the effluent was measured using ion chromatography (ICS-1000, Dionex, Sunnyvale, CA, USA).

2.4.2. Small Column Experiments

The *E. coli* removal performance of the small columns (10 cm long filtration bed) packed with either sand, the sand–raw hydrochar mixture, or the sand–activated hydrochar mixture was investigated by column flushing experiments. For each type of column medium, a pair of columns loaded with fresh packing materials was used to generate two breakthrough curves (BTC). Each flushing consisted of a loading (feeding of 50 mL AGW seeded with *E. coli*) and successive deloading (feeding of 50 mL *E. coli*-free AGW) phase. The influent was fed into the column at an upward flow rate of 1 mL/min (0.2 cm/min). The *E. coli* concentration in the effluent was measured every 5 min. After the column flushing experiments, the vertical distribution of the hydrochar concentration was measured as described previously [17].

2.4.3. Large Column Experiments

Three pairs of large columns (50 cm long filtration bed) were prepared for testing the sand as well as the sand–raw and sand–activated hydrochar mixtures. The columns were subjected to intermittent daily flushings with *E. coli*-seeded AGW for 30 days. For each week, five consecutive daily flushings were performed, and 2 days of pause were given. This flushing regime was designed to give four sets of 24 h idle time (in the first 5 days of each week) and one set of 72 h idle time per week. For each flushing experiment, 1 L influent (~2 PV, calculated from the chloride tracer test, data not shown) was flushed into each column, and the effluent was sampled for every 100 mL.

During the flushing, the first PV of the influent expelled the residual pore water stored in the column media from the previous flushing experiment, and the second PV was stored in the column media until the next flushing. It is considered that the first five effluent samples reflect the effect of idle time in *E. coli* attenuation. The following five effluent samples represented the direct *E. coli* removal that occurred when passing through the column bed. One-way ANOVA and Tukey's test ($p < 0.05$) were used to verify the statistical differences of the results from the three types of setup. The statistical analyses were carried out using IBM SPSS Statistics for Windows Version 22.0 released 2013 (Armonk, NY, USA).

3. Results

3.1. Material Characterization

3.1.1. Zeta Potential

The zeta potential assessment performed on the test *E. coli* strain UCFL-94, raw, and activated hydrochar used in the column experiments showed negative values in the pH range of 6.6–6.8 (Figure 1). The zeta potential of raw hydrochar was slightly increased by KOH activation from -15 mV to -13 mV. Also, a negative zeta potential of -11 mV was measured for the *E. coli* UCFL-94 suspension.

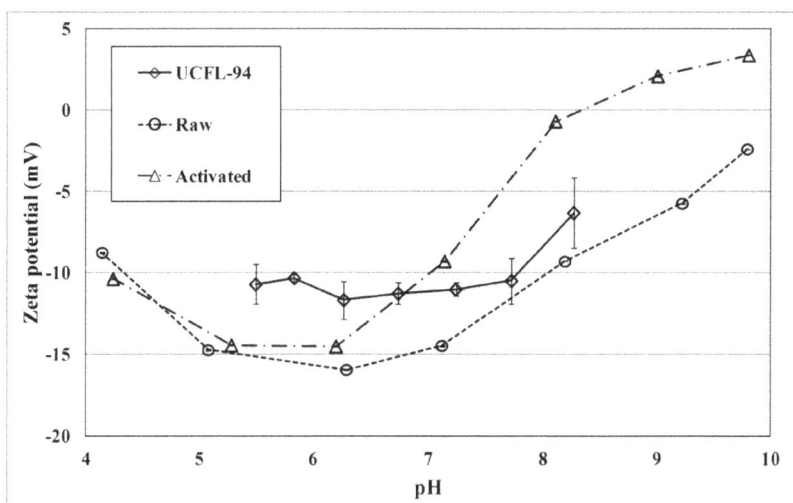

Figure 1. Zeta potential of the hydrochar and *Escherichia coli* strain over a function of pH. The lines represent the mean zeta potential value and error bars indicate the standard deviation.

3.1.2. Elemental Composition

Table 1 shows the elemental composition of both hydrochar samples obtained from the analyses using XRF and a TruSpec analyzer. The KOH activation increased the potassium (K) content by 0.4%, which supports the association of K on the surface of raw hydrochar. It was speculated that the increase in iron (Fe), Magnesium (Mg), and calcium (Ca) content was derived from impurities of the KOH reagent used for the activation, and/or due to an increase in relative concentration due to the loss of other elements. The decreases in aluminium (Al) and phosphorous (P) contents resulted from the washing out during the activation process. The alterations in the constitution of the carbon (C), hydrogen (H), Nitrogen (N), and oxygen (O) contents were considered insignificant.

Table 1. Elemental composition (%) of hydrochar variants used in this study.

	C	H	N	O	Ca	Mg	Al	K	Fe	P
Raw	28.6	3.6	2.0	22.3	5.1	0.9	2.6	0.5	5.0	4.5
Activated	29.4	3.7	1.9	21.1	5.8	1.0	2.0	0.9	5.5	2.6

Obtained from TruSpec analyzer (C, H, N, and O) or XRF (Ca, Mg, Al, K, Fe, and P).

3.1.3. Surface Functional Groups

The qualitative composition of surface functional groups was not significantly altered by the KOH activation. The raw and activated hydrochar samples showed similar peaks in the FTIR-PAS analyses (Figure 2). The spectra in the region between 1700 and 1300 cm^{-1} and the bands at 3800 cm^{-1} were assigned to the residual water content in the hydrochar samples. The bands in the range of 2930–2850 cm^{-1} corresponded to aliphatic C-H stretching vibrations [24–26]. The bands at 1600 and 1446 cm^{-1} referred to aromatic C=C stretching vibrations [26,27] and aliphatic CH$_2$ scissoring vibrations [28,29], respectively. A modest vibration shift was observed from the bands in the range of 1446–1400 cm^{-1} of the activated hydrochar samples. This referred to the deprotonation of the OH at the surface of hydrochar. Te bands in the 1110–1010 cm^{-1} region were derived from O stretching vibrations [25,27,28]. The bands at 780 cm^{-1} was derived from out-of-plane bending vibrations of aromatic C-H bonds [24,26].

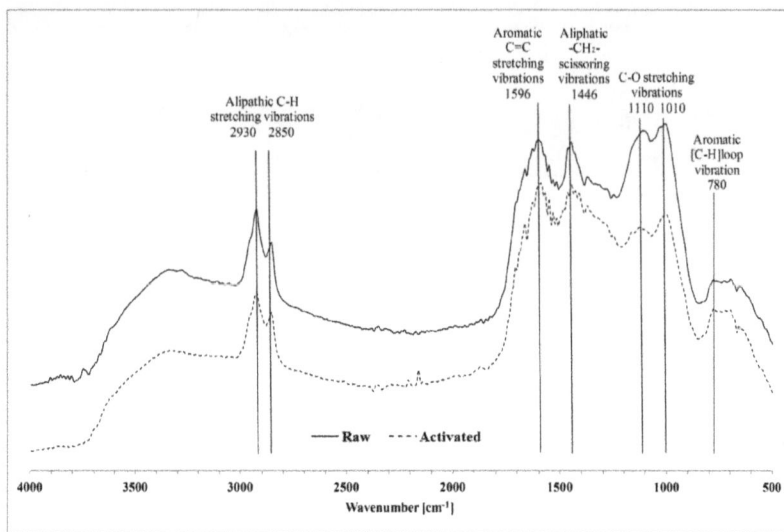

Figure 2. FTIR-PAS spectra of raw and activated hydrochar.

3.1.4. Specific Surface Area and Pore Size Distribution

The BET analysis performed on the raw and activated hydrochar showed a specific surface area of, respectively, 25.3 and 18.5 m^2/g. Figure 3 shows the results of the QSDFT pore size (differential pore volume) distribution, which demonstrates the pore volume composition attributed to the specific pore widths. The pore volume derived from the micro and mesopore range was negligible. A clear decrease in the surface area derived from the KOH activation was observable in the pore fraction, with a size >15 nm.

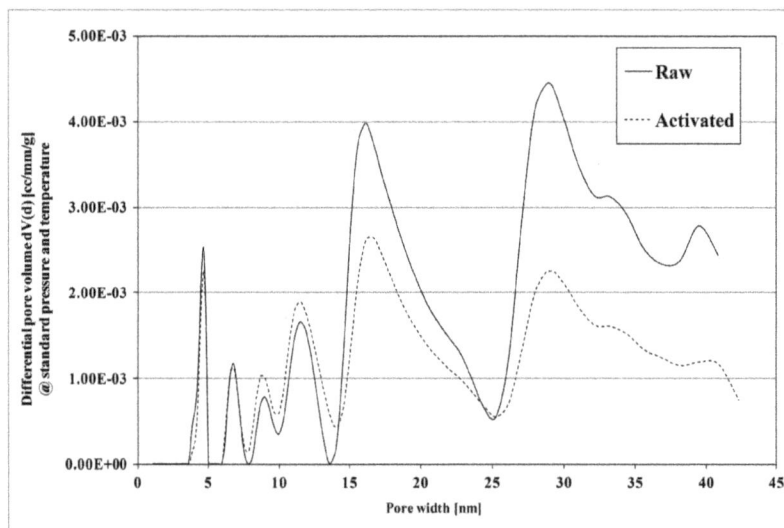

Figure 3. Pore size distribution curves obtained from quench solid density functional theory (QSDFT) analysis.

3.1.5. Surface Morphology

The SEM analyses for the raw and activated hydrochar samples were performed in order to investigate the morphological change derived from the KOH activation. The images from the SEM analyses indicated insignificant physical alterations (Figure 4). The surfaces of both types of hydrochar had a relatively rough structure, consisting of macropores (>50 nm) which could provide attachment sites for *E. coli* cells with a size of ~2 μm (measured by Zeta-sizer nano, Malvern, data not shown).

Figure 4. SEM images of raw (**left**) and activated (**right**) hydrochar.

3.1.6. Hydrophobicity

The results from the contact angle measurements on the raw and activated hydrochar samples indicated that the surfaces of both materials had hydrophobic characteristics. The hydrophobicity of the raw hydrochar was increased by KOH activation. The contact angle values of the raw and activated hydrochar were 126.5° (±2.9) (average of triplicate ± standard deviation) and 135.4° (±4.7), respectively.

3.2. E. coli Flushing Test

3.2.1. Breakthrough Analysis in Small Column Experiments

Figure 5 presents the BTCs obtained from the small column flushing experiments. All BTCs showed a clear pattern, consisting of a rising limb, a plateau phase, and a declining limb. The supplementation of raw hydrochar in the sand media resulted in early *E. coli* breakthrough. In contrast to the rising limb of the sand-only column observed after 15 min, the one from the raw hydrochar-amended column was observed already after 10 min. This was similar in the declining limbs: the one in the sand-only column started 5 min later than the column supplemented with the raw hydrochar. It was apparent that the decrease in the pore space of the sand media induced by the filling of pores by raw hydrochar amendment facilitated the *E. coli* transportation. While the effect of raw hydrochar addition in the sand media for the *E. coli* removal was insignificant, the amendments with the activated hydrochar showed an important increase in the removal efficiency. The C/C_0 ratios in the plateau phase of both the sand-only and raw hydrochar-amended columns were similar at ~0.9. However, the C/C_0 ratio of the sand column with the activated hydrochar supplement was only around 0.1. The average *E. coli* removal efficiencies of the sand and the raw and activated hydrochar-supplemented columns were 9.2%, 9.6%, and 90.1%, respectively. The measurements on the vertical hydrochar distribution in the column showed no significant migration of hydrochar particles during the flushing.

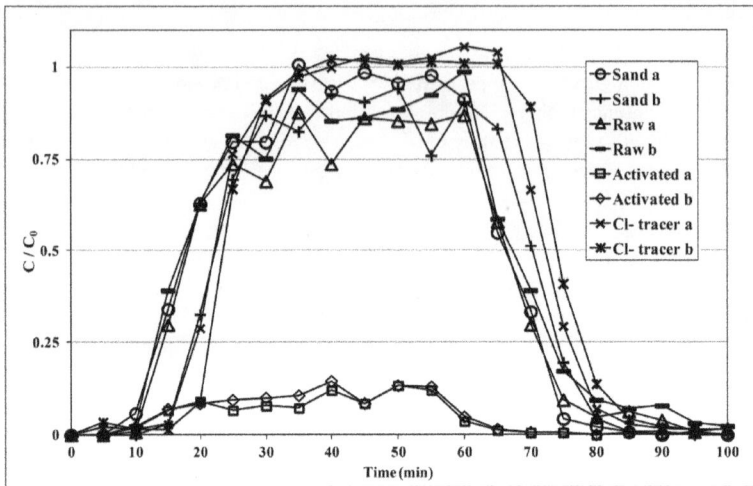

Figure 5. Breakthrough curves of *E. coli* from small column experiments carried out at a flow rate of 1 mL (0.20 cm)/min.

3.2.2. *E. coli* Removal Efficiency in Large Column Experiments

Figure 6 illustrates the *E. coli* removal efficiencies of the sand and raw and activated hydrochar-amended columns during the 30 days of experiments with intermittent operation. Similar to the results from the small column experiments, only the activated hydrochar amendment was effective at *E. coli* removal. The removal efficiency of the activated hydrochar-amended column was the highest (99.7%) on the first day, and declined gently until the last day of the experiment (78.9%). The sand-only and raw hydrochar-amended columns showed a comparable *E. coli* removal performance, in the range of 11.4–57.2%. Table 2 summarizes the overall *E. coli* removal efficiencies and the effect of the idle time during the large column experiments. Under all experimental conditions applied, the activated hydrochar-amended columns showed greater *E. coli* removal performances than the other columns. During 30 days of operation, the sand columns with activated hydrochar supplements showed an average total *E. coli* removal efficiency of 91.2%. In contrast, the sand-only and raw hydrochar-supplemented columns showed an average total *E. coli* removal efficiency of only 24.4% and 36.5%, respectively. The effect of the idle time on the *E. coli* removal was only observed in the sand-only columns; throughout the experimental period, the removal in the first PV was significantly greater than that of the second PV. The *E. coli* removal efficiency in the second PV, which represents the direct removal of *E. coli* when passing through the column media, was 17.2%. In contrast, the removal efficiencies in the first PV (stored in the sand media for 24 or 72 h and thus representing the effect of the idle time) were 52.1% and 66.9%, respectively. Throughout the experiments, the idle time had no clear effect on the *E. coli* removal performance of both types of hydrochar amendments, except for the 72 h idle time applied to the raw hydrochar-amended columns. Such an extended idle time increased the *E. coli* removal efficiency by 14.8%. The direct removal of *E. coli* (without idle time) was similar, at ~19% in the second PV in the sand-only and raw hydrochar-amended columns. This indicates a close correspondence with the results from the small column experiments.

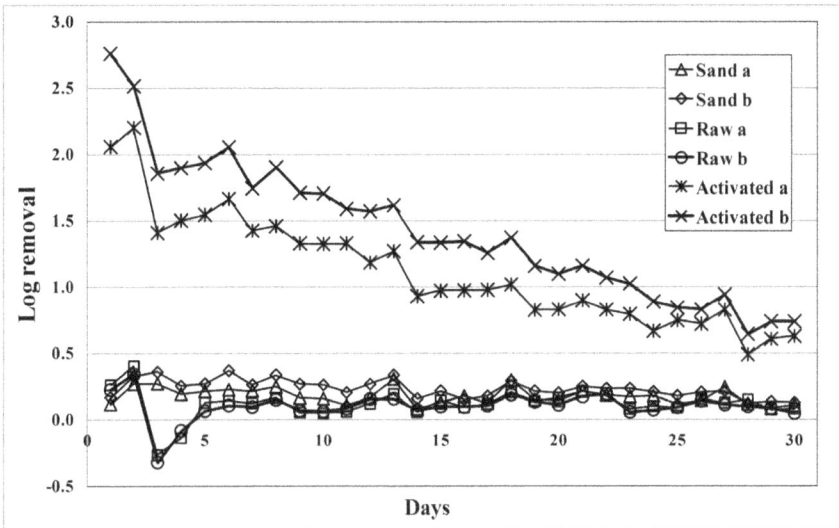

Figure 6. *E. coli* removal efficiencies from large columns during 30 days of flushing with a flow rate of 33.3 mL (1.35 cm)/min.

Table 2. Removal efficiency (average% ± standard deviation) of *E. coli* in large column experiments during 30 days of intermittent flushing in duplicate columns of each treatment.

| Content | Pore Volume | | | Total |
| | First | | Second | |
	24 h idle	72 h idle		
Sand	52.1 ± 13.3 ($n = 46$)	66.9 ± 14.7 ($n = 12$)	17.2 ± 8.64 [§] ($n = 60$)	36.5 ± 10.1 ($n = 60$)
Raw	23.0 ± 17.3 [†] ($n = 44$)	35.1 ± 12.9 ($n = 10$)	20.0 ± 12.5 [§†] ($n = 56$)	24.4 ± 10.5 [†] ($n = 56$)
Activated	92.8 ± 5.0 [‡] ($n = 46$)	92.6 ± 8.0 [‡] ($n = 12$)	90.0 ± 9.1 [‡] ($n = 60$)	± 7.5 [‡] ($n = 60$)

Within each column, values followed by the symbol [§] are not significantly different using Tukey's test at $p < 0.05$. Within each row, values followed by the same symbol [†] or [‡] are not significantly different using Tukey's test at $p < 0.05$.

4. Discussion

4.1. Effect of KOH Activation of Hydrochar on E. coli Removal

The performance of an adsorbent mainly depends on its surface, which provides internal pore structure and adsorptive sites [30]. Several attractive or repulsive forces between *E. coli* cells and the adsorbent surface regulate their mutual interaction [31]. Considering the characteristics of *E. coli*, the adsorptive removal will be facilitated when the adsorbent possesses a highly porous surface with a positive (less negative) surface charge and hydrophobic properties. It has been reported that the KOH activation of hydrochars derived from plant materials increased the surface roughness, resulting in enhanced heavy metal or *E. coli* removal from artificially contaminated influents [17,19]. However, the BET and SEM analyses carried out in this research could not explain the improvement in *E. coli* removal efficiency induced by the KOH activation. The modification in the surface morphology of the hydrochar by the KOH treatment was insignificant (Figure 4). Moreover, the specific surface area of the activated hydrochar was 27% less than that of the raw hydrochar (See Section 3.1.4).

A possible explanation for the decrease in the specific surface area derived from the KOH activation and subsequent washing processes can be the collapse of micro and mesopores, resulting in the development of a macroporous structure on the hydrochar surface. However, a large surface area of an adsorbent, mainly consisting of micro (<2 nm) and mesopores (2–50 nm), may not be a direct indicator for efficient *E. coli* removal, considering the size *E. coli* cells (~2 μm). Further investigations on the surface area and surface properties of the macropores are thus recommended in order to obtain a better understanding of the *E. coli* removal mechanisms by (KOH treated) biochar.

Because the investigations carried out to compare the characteristics of raw and activated hydrochar showed only minor differences, the improved *E. coli* removal performance obtained from the KOH activation could result from the increase in hydrophobicity of the hydrochar surface, which enforces the hydrophobic attraction. Hydrochar consists of a hydrophobic core and a hydrophilic outer surface [26,32]. The KOH activation carried out in this research would have brought more hydrophobic surfaces into contact with *E. coli* cells by removing the hydrophilic surface coatings formed by recondensation and repolymerization of water-soluble substances during the HTC process [27]. Another possible explanation for the advantageous effect of KOH activation can be an increase in the surface charge of the hydrochar, which would have weakened the electrostatic repulsion between *E. coli* cells and hydrochar surfaces.

4.2. Effect of Idling Time on E. coli Removal Efficiency

Previous research on slow-sand filtration units under intermittent operation reported a significant contribution of the idle time to the removal efficiency of bacteria [33]. The observations in this research on the effect of the idle time in the large sand column correspond well with these studies (Table 2). During the pause between daily flushing, bacterial surrogates residing in the sand bed were attenuated.

A more recent research [34] reported an important observation that the activities of the microbial community in the sand bed were more responsible for the viral attenuation than the physico-chemical processes. The attenuation of bacteriophages during the idle time increased along with the maturation of the filter bed, while the filters that operated under the suppression of microbial activities did not show any significant attenuation [34]. This may, however, not explain our results, because the bactericidal effect of the idle time observed in this research already existed from the first batch of the intermittent operation, and it was maintained at a comparable level throughout the 30 days of experiments (Table 2). This discrepancy might be due to differences in experimental conditions such as the type of influent/test micro-organisms, the size of sand grains, or the presence of standing water for oxygen infiltration into the filter bed.

Based on our observation, it could be speculated that the extended contact time between *E. coli* cells and sand surfaces provided more chances for bacterial attachment due to the motility of *E. coli* cells [35] or Brownian diffusion [36]. In contrast, the advantageous effect of the idle time on *E. coli* removal was negligible when either raw or activated hydrochar was supplemented in the sand bed. The provision of carbonaceous surfaces in the sand medium could have induced more desirable conditions for *E. coli* survival, by providing extra nutrients and more protection against external stress factors [37–40]. In order to clarify the *E. coli* inactivation mechanism during the idle time, more investigations on *E. coli*–surface interactions under static conditions are recommended.

4.3. Potential of HTC-Sand Filters for Pathogen Removal

Potable water can be provided either by centralized water treatment and supply systems or by decentralized (point-of-use) technologies that are installed on a house. Decentralized technologies have been recognized as appropriate options for poor rural communities. Biosand filters (BSF) are an example of a point-of-use technology that have been applied in developing countries [41]. Since the formation of a biofilm layer (Schmutzdecke) in a BSF plays a key role in removing microbial agents, insufficient pathogen removal during the startup (ripening) period is one of the main limitations of BSF. Previous research on BSF reported only 60–70% of *E. coli* removal during the ripening period, i.e., the first 3 weeks of operation [33,42]. Considering the superior *E. coli* removal efficiency (96.5%) in the same period, amendments of activated hydrochar to the sand bed would be an attractive option to overcome the inferior performance of BSF during the startup period. Since the experimental conditions employed in this research differed from the general BSF design and operational parameters, extrapolation of the results of this research to supplementation of BSF setups with hydrochar needs to be done carefully. Accordingly, it is recommended to perform more experiments employing (i) smaller sand grain sizes (<0.7mm), (ii) standing heads for ensuring sufficient oxygen supplementation, and (iii) influents with a higher microbial heterogeneity than that of the AGW used in this study, e.g., surface water from lakes, rivers, or canals [43,44].

Though it is reported that hydrothermal treatment decreased the environmental risk of abiotic contaminants, such as heavy metals and pharmaceuticals embedded in sewage sludge [45,46], the release of these contaminants from hydrochar has not yet been investigated. In case the hydrochar contains considerable amounts of these undesirable compounds, long-term monitoring on the effluent quality is recommended prior to the practical implementation of BSF supplemented with the hydrochar adsorbent. The properties of hydrochar are largely determined by parameters such as reaction temperature, time, and pressure, as well as catalyst and feedstock composition [1]. Further research on the optimization of these parameters can improve the stability of heavy metals in the hydrochar [45], and completely degrade the pharmaceutical residues during the hydrothermal treatment [46].

5. Conclusions

This research evaluated the use of hydrochar derived from stabilized sewage sludge as a low-cost adsorbent for pathogen removal in water treatment. The activation of hydrochar carried out at ambient temperatures using a strong alkaline solution increased the adsorptive performance of hydrochar by removing hydrophilic substances from the hydrochar surface, resulting in an increase in hydrophobicity of the biochar particles. Supplementation of activated hydrochar in a sand filtration unit (1.5%, w/w) with a 50 cm bed height yielded an *E. coli* removal efficiency exceeding 90% during 30 days of intermittent operation. Pathogen removal based on the use of hydrochar is a new concept which has not been extensively studied. Its practical implementation in developing countries still requires follow up investigations on the optimization of the process parameters and the durability of the filtration unit.

Acknowledgments: This research was funded by the Korean Church of Brussels, Mangu Jeja Church (Seoul, Korea), and the Netherlands Ministry of Development Cooperation (DGIS) through the UNESCO-IHE Partnership Research Fund. It was carried out in the framework of the research project 'Addressing the Sanitation Crisis in Unsewered Slum Areas of African Mega-cities' (SCUSA).

Author Contributions: Jae Wook Chung, Jan Willem Foppen, and Piet Nicolaas Luc Lens conceived and designed the experiments; Oghosa Charles Edewi performed the experiments; Jae Wook Chung and Oghosa Charles Edewi analyzed the data; Gabriel Gerner and Rolf Krebs contributed materials and analysis tools; Jae Wook Chung put together the initial drafts and finalized the paper.

Conflicts of Interest: The authors declare no conflict of interest.

References

1. Libra, J.A.; Ro, K.S.; Kammann, C.; Funke, A.; Berge, N.D.; Neubauer, Y.; Titirici, M.M.; Fühner, C.; Bens, O.; Kern, J.; et al. Hydrothermal carbonization of biomass residuals: A comparative review of the chemistry, processes and applications of wet and dry pyrolysis. *Biofuels* **2011**, *2*, 71–106. [CrossRef]
2. Titirici, M.-M.; White, R.J.; Falco, C.; Sevilla, M. Black perspectives for a green future: Hydrothermal carbons for environment protection and energy storage. *Energy Environ. Sci.* **2012**, *5*, 6796–6822. [CrossRef]
3. Funke, A.; Ziegler, F. Hydrothermal carbonization of biomass: A summary and discussion of chemical mechanisms for process engineering. *Biofuels Bioprod. Biorefin.* **2010**, *4*, 160–177. [CrossRef]
4. Lu, X.; Jordan, B.; Berge, N.D. Thermal conversion of municipal solid waste via hydrothermal carbonization: Comparison of carbonization products to products from current waste management techniques. *Waste Manag.* **2012**, *32*, 1353–1365. [CrossRef] [PubMed]
5. Nizamuddin, S.; Jayakumar, N.S.; Sahu, J.N.; Ganesan, P.; Bhutto, A.W.; Mubarak, N.M. Hydrothermal carbonization of oil palm shell. *Korean J. Chem. Eng.* **2015**, *32*, 1789–1797. [CrossRef]

6.   Erdogan, E.; Atila, B.; Mumme, J.; Reza, M.T.; Toptas, A.; Elibol, M.; Yanik, J. Characterization of products from hydrothermal carbonization of orange pomace including anaerobic digestibility of process liquor. *Bioresour. Technol.* **2015**, *196*, 35–42. [CrossRef] [PubMed]

7.   Islam, M.A.; Tan, I.A.W.; Benhouria, A.; Asif, M.; Hameed, B.H. Mesoporous and adsorptive properties of palm date seed activated carbon prepared via sequential hydrothermal carbonization and sodium hydroxide activation. *Chem. Eng. J.* **2015**, *270*, 187–195. [CrossRef]

8.   Fytili, D.; Zabaniotou, A. Utilization of sewage sludge in eu application of old and new methods—A review. *Renew. Sustain. Energy Rev.* **2008**, *12*, 116–140. [CrossRef]

9.   Escala, M.; Zumbuhl, T.; Koller, C.; Junge, R.; Krebs, R. Hydrothermal carbonization as an energy-efficient alternative to established drying technologies for sewage sludge: A feasibility study on a laboratory scale. *Energy Fuels* **2013**, *27*, 454–460. [CrossRef]

10.   Danso-Boateng, E.; Shama, G.; Wheatley, A.D.; Martin, S.J.; Holdich, R.G. Hydrothermal carbonisation of sewage sludge: Effect of process conditions on product characteristics and methane production. *Bioresour. Technol.* **2015**, *177*, 318–327. [CrossRef] [PubMed]

11.   Wirth, B.; Reza, T.; Mumme, J. Influence of digestion temperature and organic loading rate on the continuous anaerobic treatment of process liquor from hydrothermal carbonization of sewage sludge. *Bioresour. Technol.* **2015**, *198*, 215–222. [CrossRef] [PubMed]

12.   Smith, K.M.; Fowler, G.D.; Pullket, S.; Graham, N.J.D. Sewage sludge-based adsorbents: A review of their production, properties and use in water treatment applications. *Water Res.* **2009**, *43*, 2569–2594. [CrossRef] [PubMed]

13.   Alatalo, S.M.; Repo, E.; Makila, E.; Salonen, J.; Vakkilainen, E.; Sillanpaa, M. Adsorption behavior of hydrothermally treated municipal sludge & pulp and paper industry sludge. *Bioresour. Technol.* **2013**, *147*, 71–76. [PubMed]

14.   Spataru, A. The Use of Hydrochar as a Low Cost Adsorbent for Heavy Metal and Phosphate Removal from Wastewater. Master's Thesis, UNESCO-IHE, Delft, The Netherlands, 2014.

15.   Chung, J.W.; Foppen, J.W.; Gerner, G.; Krebs, R.; Lens, P.N.L. Removal of rotavirus and adenovirus from artificial ground water using hydrochar derived from sewage sludge. *J. Appl. Microbiol.* **2015**, *119*, 876–884. [CrossRef] [PubMed]

16.   Lutterodt, G.; Basnet, M.; Foppen, J.W.A.; Uhlenbrook, S. Determining minimum sticking efficiencies of six environmental *Escherichia coli* isolates. *J. Contam. Hydrol.* **2009**, *110*, 110–117. [CrossRef] [PubMed]

17.   Chung, J.W.; Foppen, J.W.; Izquierdo, M.; Lens, P.N.L. Removal of *Escherichia coli* from saturated sand columns supplemented with hydrochar produced from maize. *J. Environ. Qual.* **2014**, *43*, 2096–2103. [CrossRef] [PubMed]

18.   APHA. *Standard Methods for the Examination of Water and Wastewater*, 20th ed.; American Public Health Association: Washington, DC, USA, 1998.

19.   Regmi, P.; Moscoso, J.L.G.; Kumar, S.; Cao, X.Y.; Mao, J.D.; Schafran, G. Removal of copper and cadmium from aqueous solution using switchgrass biochar produced via hydrothermal carbonization process. *J. Environ. Manag.* **2012**, *109*, 61–69. [CrossRef] [PubMed]

20.   Sun, K.J.; Tang, J.C.; Gong, Y.Y.; Zhang, H.R. Characterization of potassium hydroxide (koh) modified hydrochars from different feedstocks for enhanced removal of heavy metals from water. *Environ. Sci. Pollut. Res.* **2015**, *22*, 16640–16651. [CrossRef] [PubMed]

21.   Neimark, A.V.; Lin, Y.; Ravikovitch, P.I.; Thommes, M. Quenched solid density functional theory and pore size analysis of micro-mesoporous carbons. *Carbon* **2009**, *47*, 1617–1628. [CrossRef]

22.   Jeong, S.-B.; Yang, Y.-C.; Chae, Y.-B.; Kim, B.-G. Characteristics of the treated ground calcium carbonate powder with stearic acid using the dry process coating system. *Mater. Trans.* **2009**, *50*, 409–414. [CrossRef]

23.   Matthess, G.; Bedbur, E.; Gundermann, K.O.; Loof, M.; Peters, D. Investigation on filtration mechanisms of bacteria and organic particles in porous-media. I. Background and methods. *Zentralblatt Fuer Hygiene Und Umweltmed.* **1991**, *191*, 53–97.

24.   Baccile, N.; Weber, J.; Falco, C.; Titirici, M.-M. Characterization of hydrothermal carbonization materials. In *Sustainable Carbon Materials from Hydrothermal Processes*; John Wiley & Sons, Ltd.: Hoboken, NJ, USA, 2013; pp. 151–211.

25.   Parshetti, G.K.; Liu, Z.G.; Jain, A.; Srinivasan, M.P.; Balasubramanian, R. Hydrothermal carbonization of sewage sludge for energy production with coal. *Fuel* **2013**, *111*, 201–210. [CrossRef]

26. Sevilla, M.; Fuertes, A.B. Chemical and structural properties of carbonaceous products obtained by hydrothermal carbonization of saccharides. *Chem. Eur. J.* **2009**, *15*, 4195–4203. [CrossRef] [PubMed]

27. Kumar, S.; Loganathan, V.A.; Gupta, R.B.; Barnett, M.O. An assessment of u(vi) removal from groundwater using biochar produced from hydrothermal carbonization. *J. Environ. Manag.* **2011**, *92*, 2504–2512. [CrossRef] [PubMed]

28. Bansal, R.C.; Goyal, M. *Activated Carbon Adsorption*; CRC Press, Taylor & Francis Group: Boca Raton, FL, USA, 2005; Volume 33487-2742, p. 33.

29. Silverstein, R.M.; Webster, F.X.; Kiemle, D.J. *Spectrometric Identification of Organic Compounds*, 7th ed.; John Wiley & Sons: Hoboken, NJ, USA, 2005; Volume 07030-577, pp. 82–88.

30. Unur, E. Functional nanoporous carbons from hydrothermally treated biomass for environmental purification. *Microporous Mesoporous Mater.* **2013**, *168*, 92–101. [CrossRef]

31. Foppen, J.W.A.; Schijven, J.F. Evaluation of data from the literature on the transport and survival of *Escherichia coli* and thermotolerant coliforms in aquifers under saturated conditions. *Water Res.* **2006**, *40*, 401–426. [CrossRef] [PubMed]

32. Jarrah, N.; van Ommen, J.G.; Lefferts, L. Development of monolith with a carbon-nanofiber-washcoat as a structured catalyst support in liquid phase. *Catal. Today* **2003**, *79*, 29–33. [CrossRef]

33. Elliott, M.; Stauber, C.; Koksal, F.; DiGiano, F.; Sobsey, M. Reductions of *E. Coli*, echovirus type 12 and bacteriophages in an intermittently operated household-scale slow sand filter. *Water Res.* **2008**, *42*, 2662–2670. [CrossRef] [PubMed]

34. Elliott, M.A.; DiGiano, F.A.; Sobsey, M.D. Virus attenuation by microbial mechanisms during the idle time of a household slow sand filter. *Water Res.* **2011**, *45*, 4092–4102. [CrossRef] [PubMed]

35. Maeda, K.; Imae, Y.; Shioi, J.I.; Oosawa, F. Effect of temperature on motility and chemotaxis of *Escherichia coli*. *J. Bacteriol.* **1976**, *127*, 1039–1046. [PubMed]

36. Li, G.; Tam, L.K.; Tang, J.X. Amplified effect of brownian motion in bacterial near-surface swimming. *Proc. Natl. Acad. Sci. USA* **2008**, *105*, 18355–18359. [CrossRef] [PubMed]

37. Gerba, C.P.; Schaiberger, G.E. Effect of particulates on virus survival in seawater. *J. Water Pollut. Control Federation* **1975**, *47*, 93–103.

38. Burton, G.A., Jr.; Gunnison, D.; Lanza, G.R. Survival of pathogenic bacteria in various freshwater sediments. *Appl. Environ. Microbiol.* **1987**, *53*, 633–638. [PubMed]

39. Sherer, B.M.; Miner, J.R.; Moore, J.A.; Buckhouse, J.C. Indicator bacterial survival in stream sediments. *J. Environ. Qual.* **1992**, *21*, 591–595. [CrossRef]

40. Howell, J.M.; Coyne, M.S.; Cornelius, P.L. Effect of sediment particle size and temperature on fecal bacteria mortality rates and the fecal coliform/fecal streptococci ratio. *J. Environ. Qual.* **1996**, *25*, 1216–1220. [CrossRef]

41. Sobsey, M.D.; Stauber, C.E.; Casanova, L.M.; Brown, J.M.; Elliott, M.A. Point of use household drinking water filtration: A practical, effective solution for providing sustained access to safe drinking water in the developing world. *Environ. Sci. Technol.* **2008**, *42*, 4261–4267. [CrossRef] [PubMed]

42. Stauber, C.E.; Elliott, M.A.; Koksal, F.; Ortiz, G.M.; DiGiano, F.A.; Sobsey, M.D. Characterisation of the biosand filter for *E. coli* reductions from household drinking water under controlled laboratory and field use conditions. *Water Sci. Technol.* **2006**, *54*, 1–7. [CrossRef] [PubMed]

43. Chan, C.C.V.; Neufeld, K.; Cusworth, D.; Gavrilovic, S.; Ngai, T. Investigation of the effect of grain size, flow rate and diffuser design on the cawst biosand filter performance. *Int. J. Serv. Learn. Eng. Humanit. Eng. Soc. Entrep.* **2015**, *10*, 1–23.

44. Buzunis, B.J. Intermittently operated slow sand filtration: A new water treatment process. In *Civil Engineering*; University of Calgary: Calgary, AB, Canada, 1995.

45. Huang, H.J.; Yuan, X.Z. The migration and transformation behaviors of heavy metals during the hydrothermal treatment of sewage sludge. *Bioresour. Technol.* **2016**, *200*, 991–998. [CrossRef] [PubMed]

46. vom Eyser, C.; Palmu, K.; Schmidt, T.C.; Tuerk, J. Pharmaceutical load in sewage sludge and biochar produced by hydrothermal carbonization. *Sci. Total Environ.* **2015**, *537*, 180–186. [CrossRef] [PubMed]

**applied
sciences**

MDPI

Article

Degradation of Trace Organic Contaminants by a Membrane Distillation—Enzymatic Bioreactor

Muhammad B. Asif [1], Faisal I. Hai [1,*], Jinguo Kang [1,2], Jason P. van de Merwe [3],
Frederic D. L. Leusch [3], Kazuo Yamamoto [4], William E. Price [2] and Long D. Nghiem [1]

[1]   Strategic Water Infrastructure Lab, School of Civil, Mining and Environmental Engineering,
      University of Wollongong, Wollongong NSW 2522, Australia; mba409@uowmail.edu.au (M.B.A.);
      jkang@uow.edu.au (J.K.); longn@uow.edu.au (L.D.N.)
[2]   Strategic Water Infrastructure Lab, School of Chemistry, University of Wollongong, Wollongong NSW 2522,
      Australia; wprice@uow.edu.au
[3]   Australian Rivers Institute and Griffith School of Environment, Griffith University, Gold Coast QLD 4222,
      Australia; j.vandemerwe@griffith.edu.au (J.P.v.d.M.); f.leusch@griffith.edu.au (F.D.L.L.)
[4]   Environmental Science Centre, Department of Urban Engineering, University of Tokyo, Tokyo 113-0033,
      Japan; yamamoto@esc.u-tokyo.ac.jp
*   Correspondence: faisal@uow.edu.au; Tel.: +61-2-42213054

Received: 31 July 2017; Accepted: 25 August 2017; Published: 28 August 2017

Abstract: A high retention enzymatic bioreactor was developed by coupling membrane distillation with an enzymatic bioreactor (MD-EMBR) to investigate the degradation of 13 phenolic and 17 non-phenolic trace organic contaminants (TrOCs). TrOCs were effectively retained (90–99%) by the MD membrane. Furthermore, significant laccase-catalyzed degradation (80–99%) was achieved for 10 phenolic and 3 non-phenolic TrOCs that contain strong electron donating functional groups. For the remaining TrOCs, enzymatic degradation ranged from 40 to 65%. This is still higher than those reported for enzymatic bioreactors equipped with ultrafiltration membranes, which retained laccase but not the TrOCs. Addition of three redox-mediators, namely syringaldehyde (SA), violuric acid (VA) and 1-hydroxybenzotriazole (HBT), in the MD-EMBR significantly broadened the spectrum of efficiently degraded TrOCs. Among the tested redox-mediators, VA (0.5 mM) was the most efficient and versatile mediator for enhanced TrOC degradation. The final effluent (i.e., membrane permeate) toxicity was below the detection limit, although there was a mediator-specific increase in toxicity of the bioreactor media.

Keywords: enzymatic membrane bioreactor (EMBR); laccase; membrane distillation; redox-mediators; trace organic contaminants (TrOCs)

1. Introduction

Laccase (EC 1.10.3.2), a copper-containing oxidoreductase enzyme, has been studied extensively for the degradation of recalcitrant compounds such as phenols and aromatic hydrocarbons [1–5]. In recent years, laccase-catalyzed degradation of trace organic contaminants (TrOCs) such as pharmaceuticals, pesticides, personal care products, industrial chemicals and steroid hormones has gained significant attention [6,7]. These TrOCs occur ubiquitously in municipal wastewater and have the potential to adversely affect aquatic ecosystems and human health [8–10].

TrOC degradation by laccase depends on a number of factors including pH, temperature, chemical structure of TrOCs and laccase properties [11–13]. In general, effective laccase-catalyzed degradation of TrOCs containing electron donating functional groups (EDGs) such as amine ($-NH_2$), alkoxy ($-OR$) or hydroxyl ($-OH$) was observed. On the other hand, degradation of TrOCs containing electron withdrawing functional groups (EWGs) such as halogen ($-X$), amide ($-CONR_2$) or nitro

(–NO$_2$) has been reported to be poor or unstable [11,14]. Degradation of TrOCs can be improved by adding different natural and synthetic redox-mediators that are low molecular weight compounds capable of exchanging electrons between laccase and TrOCs [15–17].

Initial studies have assessed the performance of laccase-catalyzed TrOC degradation in batch enzymatic bioreactors due to the concern of enzyme washout in a continuous flow system. In an attempt to prevent enzyme washout, an enzymatic membrane bioreactor (EMBR) was developed by coupling an ultrafiltration (UF) membrane to an enzymatic bioreactor [18,19]. Interestingly, during the operation of the EMBR, adsorption of some hydrophobic TrOCs (e.g., amitriptyline, oxybenzone and octocrylene) onto the enzyme gel layer over the membrane surface resulted in enhanced degradation of the adsorbed compounds [18]. In another study, removal of four non-phenolic TrOCs, namely atrazine, sulfamethoxazole, diclofenac and carbamazepine was improved by 15–25% following the addition of granular activated carbon (GAC) in EMBR. This was probably because simultaneous adsorption of laccase and TrOCs on GAC promoted the interaction of TrOCs with the active sites of laccase [20]. Results from previous studies indicate the complementarity of simultaneous laccase and TrOC retention within EMBR in contrast to only laccase retention by UF membranes utilized in the previously developed EMBRs. Hence, in this study, it is postulated that the integration of an enzymatic bioreactor with a high retention membrane could facilitate the degradation of resistant TrOCs by retaining both laccase and TrOCs.

Different configurations of conventional activated sludge-based high retention membrane bioreactors (HR-MBR), employing membrane distillation (MD), forward osmosis (FO) or nanofiltration (NF) membranes, have been investigated for advanced wastewater treatment [21–24]. Complete TrOC retention in HR-MBR improved the membrane permeate quality, but the poor removal of certain groups of TrOCs such as those containing EWGs led to their accumulation in the bioreactor. This indicates the necessity of formulating means to enhance biodegradation of TrOCs. In this context, it is noteworthy that recent reports confirm enhanced laccase-catalyzed degradation of selected TrOCs that are not amenable to degradation by conventional activated sludge [25,26]. However, the performance of a high retention—enzymatic membrane bioreactor for the removal of a wide range of TrOCs remains to be elucidated.

Among the high retention membrane systems, in MD, a vapor-liquid interface is developed around a hydrophobic micro-porous membrane that allows the water to pass through the membrane via diffusion due to vapor pressure gradient. Compared to conventional distillation processes such as fractional distillation, the MD process requires low temperature and could be operated by using low grade heat or solar energy [27,28]. Since the mass transfer in the MD process occurs in gaseous phase, it can theoretically achieve 100% retention of all non-volatile compounds [29]. The standalone MD process has been investigated for seawater desalination [30], industrial wastewater treatment [31], municipal wastewater treatment [32] and TrOC removal [29,33]. Thus, the MD process was selected for coupling to an enzymatic bioreactor in this study.

The aim of this study was to assess the performance of a laccase based membrane distillation—enzymatic membrane bioreactor (MD-EMBR) for the removal of TrOCs having diverse physicochemical properties (e.g., EDGs/EWGs, hydrophobicity and phenolic/non-phenolic moieties). A special focus was given to the improvement in TrOC degradation due to the addition of three redox-mediators, namely syringaldehyde (SA), violuric acid (VA) and 1-hydroxybenzotriazole (HBT) at different concentrations. In addition, performance of laccase-mediator systems was systematically compared based on TrOC degradation, enzyme stability and effluent toxicity.

2. Materials and Methods

2.1. Trace Organic Contaminants (TrOCs), Laccase and Mediators

A set of 30 TrOCs comprising 10 pharmaceuticals, four personal care products, six pesticides, four industrial chemicals, five steroid hormones and one phytoestrogen was selected based on their widespread occurrence in surface water bodies (see Supplementary Data Table S1). Key physicochemical properties of the TrOCs including molecular weight, water solubility, hydrophobicity (log D) and volatility (pK_H) are presented in Table 1. All TrOCs were purchased from Sigma Aldrich (Sydney, NSW, Australia), and were of analytical grade. A stock solution (25 mg/L) containing the mixture of 30 TrOCs was prepared in pure methanol, and kept in dark at −18 °C.

Laccase from genetically modified *Aspergillus oryzae* was supplied by Novozymes Australia Pty Ltd (Sydney, NSW, Australia). According to the supplier, the enzyme has a molecular weight, purity, activity and density of 56 KDa, 10% (w/w), 150,000 $\mu M_{(DMP)}$/min (measured using 2,6-dimethoxy phenol, DMP, as substrate) and 1.12 g/mL, respectively.

Two N–OH type redox-mediators, namely 1-hydroxybenzotriazole (HBT) and violuric acid (VA), and one phenolic redox-mediator, namely syringaldehyde (SA), were used in this study (see Supplementary Data Table S2). The selected mediators all follow hydrogen atom transfer (HAT) pathway for TrOC degradation [34], but the oxidation of phenolic and N–OH type redox-mediators by laccase produces highly reactive phenoxyl and aminoxyl radicals, respectively. The mediators were also purchased from Sigma Aldrich (Sydney, NSW, Australia). A separate stock solution (50 mM) of each redox-mediator was prepared, and stored at 4 °C in dark.

Table 1. Physicochemical properties of the selected TrOCs.

TrOCs	Chemical Formula	Molecular Weight	Log D at pH = 7	Water Solubility at 25 °C	Vapor Pressure	pK_H at pH 7
		g/mole		mg/L	(mmHg)	
Non-Phenolic Compounds						
Primidone	$C_{12}H_{14}N_2O$	218.25	0.83	1500	6.08×10^{-11}	13.93
Ketoprofen	$C_{16}H_{14}O_3$	254.28	0.19	554,000	3.32×10^{-8}	13.70
Naproxen	$C_{14}H_{14}O_3$	230.26	0.73	435,000	3.01×10^{-7}	12.68
Gemfibrozil	$C_{15}H_{22}O_3$	250.33	2.07	263,000	6.13×10^{-7}	12.11
Metronidazole	$C_6H_9N_3O_3$	171.15	−0.14	29,000	2.67×10^{-7}	11.68
Diclofenac	$C_{14}H_{11}Cl_2NO_2$	296.15	1.77	20,000	1.59×10^{-7}	11.51
Fenoprop	$C_9H_7Cl_3O_3$	269.51	−0.13	230,000	2.13×10^{-6}	11.48
Ibuprofen	$C_{13}H_{18}O_2$	206.28	0.94	928,000	1.39×10^{-4}	10.39
Ametryn	$C_9H_{17}N_5S$	27.33	2.97	140	1.72×10^{-6}	9.35
Clofibric acid	$C_{10}H_{11}ClO_3$	214.65	−1.06	100,000	1.03×10^{-4}	9.54
Carbamazepine	$C_{15}H_{12}N_2O$	236.27	1.89	220	5.78×10^{-7}	9.09
Octocrylene	$C_{24}H_{27}N$	361.48	6.89	0.36	2.56×10^{-9}	8.47
Amitriptyline	$C_{20}H_{23}N$	277.40	2.28	83	1.50×10^{-6}	8.18
Atrazine	$C_8H_{14}ClN_5$	215.68	2.64	69	1.27×10^{-5}	7.28
Propoxur	$C_{11}H_{15}NO_3$	209.24	1.54	800	1.53×10^{-6}	6.28
Benzophenone	$C_{13}H_{10}O$	182.22	3.21	150	8.23×10^{-4}	5.88
DEET	$C_{12}H_{17}NO$	191.3	2.42	1000	5.6×10^{-3}	5.85
Phenolic Compounds						
Enterolactone	$C_{18}H_{18}O_4$	288.38	2.53	200	3.29×10^{-13}	15.20
Estriol	$C_{18}H_{24}O_3$	298.33	1.89	32	1.34×10^{-9}	10.78
17α-Ethinylestradiol	$C_{20}H_{24}O_2$	269.40	4.11	3.9	3.74×10^{-9}	9.47
Oxybenzone	$C_{14}H_{12}O_3$	228.24	3.89	2700	5.26×10^{-6}	9.23
Estrone	$C_{18}H_{22}O_2$	270.37	3.62	5.9	1.54×10^{-8}	9.03
17β-Estradiol	$C_{18}H_{24}O_2$	272.38	4.15	3	9.82×10^{-9}	8.93
17β-Estradiol-17-acetate	$C_{20}H_{26}O_3$	314.42	5.11	1.9	9.88×10^{-9}	8.67
Bisphenol A	$C_{15}H_{16}O_2$	228.29	3.64	73	5.34×10^{-7}	8.66
Salicylic acid	$C_7H_6O_3$	138.12	−1.13	2240	8.2×10^{-5}	8.18
Pentachlorophenol	C_6HCl_5O	266.34	2.85	4800	3.49×10^{-4}	7.59
Triclosan	$C_{12}H_7Cl_3O_2$	289.54	5.28	19	3.26×10^{-5}	6.18
4-tert-Butylphenol	$C_{10}H_{14}O$	150.22	3.40	1000	0.0361	5.15
4-tert-Octylphenol	$C_{14}H_{22}O$	206.32	5.18	62	1.98×10^{-3}	5.06

2.2. Experimental Setup

The laboratory scale MD-EMBR setup comprised a glass enzymatic bioreactor (1.5 L) and an external direct contact membrane distillation system (Figure 1). The glass enzymatic bioreactor covered with aluminum foil was placed in a water bath, and the temperature of the water bath was maintained at $30 \pm 0.2\,^\circ$C using an immersion heating unit (Julabo, Seelbach, Germany). The enzymatic bioreactor was equipped with an air pump (ACO-002, Zhejiang Sensen Industry Co. Ltd., Zhoushan, China) to maintain a dissolved oxygen concentration of above 3 mg/L.

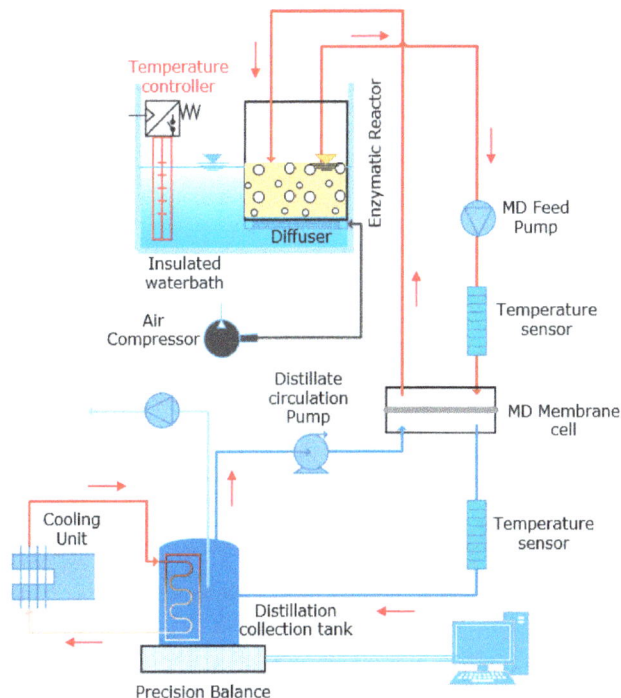

Figure 1. Schematic representation of the membrane distillation—enzymatic membrane bioreactor (MD-EMBR).

The external direct contact membrane distillation system contained an acrylic glass membrane cell, two circulation pumps (Micropump Inc., Vancouver, WA, USA) and a glass permeate tank (Figure 1). Feed and permeate flow channels were engraved on each block of the membrane cell. Length, width and height of each flow channel were 145, 95 and 3 mm, respectively.

A hydrophobic microporous polytetrafloroethylene (PTFE) membrane (GE, Minnetonka, MN, USA) was used during each experiment. The PTFE membrane has a nominal pore size of 0.22 μm, thickness of 175 μm, porosity of 70% and an active layer thickness of 5 μm [35].

2.3. Experimental Protocol

A series of experiments was carried out to evaluate the performance of MD-EMBR for TrOC degradation. At the start of the experiment, a mixture of the selected TrOCs (each at 20 μg/L) in Milli-Q water was added to the bioreactor. Laccase was added to the bioreactor for achieving an initial enzymatic activity of 95–100 $\mu M_{(DMP)}$/min. The media from the glass enzymatic bioreactor and water from the permeate tank were recirculated in their respective flow channels separated by the membrane. A chiller (SC100-A10, Thermo Scientific, Waltham, MA, USA) was used to regulate the temperature of the permeate tank at $10 \pm 0.1\,^\circ$C. The permeate tank was also placed on a precision balance (Mettler Toledo Inc., Columbus, OH, USA) to monitor permeate flux. The recirculation flow rate of both feed and the distillate was controlled at 1 L/min (corresponding to the cross flow velocity of 9 cm/s) using two rotameters.

Duplicate samples from the enzymatic bioreactor (100 mL each) and permeate tank (500 mL each) were taken after operating the MD-EMBR for 12 h. After evaluating the laccase-catalyzed degradation of TrOCs in MD-EMBR, the possible improvement in TrOC degradation was assessed with the addition of three redox-mediators (HBT, VA and SA) at two different concentrations (0.25 and 0.5 mM) via separate runs. Again duplicate samples from the enzymatic bioreactor and permeate tank were collected for the quantification of TrOCs.

Samples collected from the enzymatic bioreactor were diluted to 500 mL with Milli-Q water and were filtered through 0.45 μm glass fiber filter paper (Filtech, Wollongong, NSW, Australia). The pH of samples was adjusted to 2–2.5 using 4 M H_2SO_4 before solid phase extraction (SPE) and GC/MS analysis. For toxicity analysis, undiluted samples from the enzymatic bioreactor and permeate tank were collected in 2 mL amber vials at the end of each experiment, and stored at 4 °C until analysis.

2.4. Analytical Methods

2.4.1. TrOC Analysis

The concentration of TrOCs was measured using an analytical method involving SPE derivatization and quantitative determination by a Shimadzu GC/MS (QP5000) system as described by Hai et al. [36]. Limit of detection (LOD) for this method was compound specific and ranged from 1 to 20 ng/L (see Supplementary Data Table S1). Removal efficiencies by the enzymatic bioreactor (R_1) and the MD-EMBR (R_2) were calculated using Equations (1) and (2), respectively:

$$R_1 = 100 \times (1 - C_f/C_o) \tag{1}$$

$$R_2 = 100 \times (1 - C_p/C_o) \tag{2}$$

where, C_o and C_f are the concentration (ng/L) of specific TrOC in the enzymatic bioreactor at the beginning ($t = 0$ h) and end ($t = 12$ h) of experiment, respectively, while C_p is the concentration of specific TrOC in permeate at $t = 12$ h. The enzymatic transformation/degradation of TrOCs in the MD-EMBR was calculated using Equation (3):

$$C_o \times V_o = (C_f \times V_f) + (C_p \times V_p) + \text{biodegradation} \tag{3}$$

where, V_o, V_f and V_p represents the volume of feed (at $t = 0$ h), supernatant ($t = 12$ h) and permeate ($t = 12$ h), respectively.

2.4.2. Enzymatic Activity, ORP and Toxicity Assay

Laccase activity and effluent toxicity were examined as described elsewhere [18]. Laccase activity was measured by recording the change in absorbance at 468 nm due to the oxidation of 2,6-dimethoxyl phenol (DMP) in the presence of 100 mM sodium citrate (pH 4.5). Laccase activity expressed as $\mu M_{(DMP)}/\text{min}$ was then calculated from the molar extinction coefficient of 49.6/mM cm. Oxidation reduction potential (ORP) was measured at the start and end of each experiment using an ORP meter (WP-80D dual pH-mV meter, Thermo Fisher Scientific, Scoresby, VIC, Australia). Samples for toxicity analysis were collected from the enzymatic bioreactor and permeate tank at end of each experiment. Toxicity, expressed as a relative toxicity unit (rTU), was analyzed by measuring the inhibition of luminescence in the naturally bioluminescent bacteria, *Photobacterium leiognathi*, as previously described [37,38].

3. Results and Discussion

3.1. Overall Removal of TrOCs

In theory, MD membranes can retain all but the volatile organic compounds. In this study, the concentration of non-volatile ($pK_H > 9$) TrOCs in the permeate of the MD-EMBR was below the limit of detection of GC/MS. This is consistent with the observation in a previous study, where an MD membrane was coupled with an activated sludge bioreactor [22]. On the other hand, the MD system achieved 90–99% removal of relatively volatile TrOCs having $pK_H < 9$ (Figure 2). This compares favorably to their previously reported moderate to high removal (54–99%) by a standalone MD system [29]. In particular, removal of octocrylene ($pK_H = 8.47$), benzophenone ($pK_H = 5.88$), 4-tert-butylphenol ($pK_H = 5.15$), 4-tert-octylphenol ($pK_H = 5.06$) by the MD-EMBR was above 99%, compared to their 55–70% removal by the MD only [29]. These results suggest that the coupling of enzymatic degradation process to the MD system was favorable for achieving high TrOC removal.

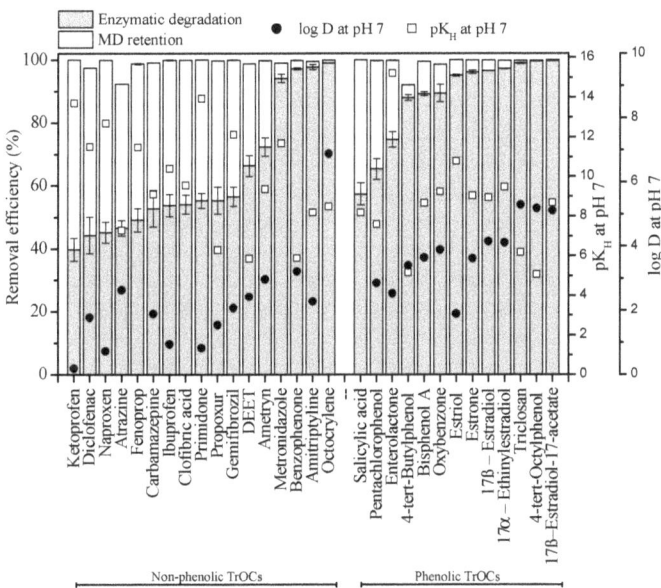

Figure 2. Overall removal and enzymatic degradation of 30 TrOCs in the MD-EMBR. Error bars indicate the standard deviation of duplicate samples. MD-EMBR operating conditions: the initial TrOC concentration and laccase activity was 20 μg/L and 95–100 $\mu M_{(DMP)}$/min, respectively; temperature of the enzymatic bioreactor and the permeate tank were kept at 30 and 10 °C, respectively; and cross-flow rate of media from the enzymatic bioreactor and distillate was 1 L/min (corresponding to a cross-flow velocity of 9 cm/s).

3.2. TrOC Degradation in Enzymatic Bioreactor

Degradation of a substrate by laccase involves the transfer of an electron from the substrate to laccase with concomitant reduction of atmospheric oxygen to water. The extent of degradation depends on, among others, the molecular properties (e.g., EWGs, EDGs or phenolic moiety) of the target substrate [11,39]. In this study, high degradation (87–99%) of 10 out 13 phenolic TrOCs was achieved by the MD-EMBR (Figure 2). These included five steroid hormones (estriol, estrone, 17β–estradiol, 17α–ethinylestradiol and 17β-estradiol-17-acetate (95–99%)), two industrial chemicals (4-tert-butylphenol, and 4-tert-octylphenol (87–99%)) and two personal care products (oxybenzone and triclosan (89–98%)). On the other hand, enzymatic degradation of some phenolic compounds, namely pentachlorophenol, enterolactone and salicylic acid, ranged from 55 to 75%. Their incomplete degradation, despite the presence of a strong EDG (i.e., hydroxyl group), can be attributed to the concomitant presence of an EWG (e.g., halogen) in their molecular structure (see Supplementary Data Table S1) [39].

Depending on the medium ORP, laccase can also degrade non-phenolic compounds. However, the reaction kinetics can be slow [17,39]. In this study, the enzymatic degradation of 17 non-phenolic TrOCs varied from 40–99% (Figure 2). Laccase catalyzed degradation of 13 compounds fell in the range of 40–65%, while the degradation of the remaining four TrOCs was in the range of 94–98%. The well degraded non-phenolic TrOCs include metronidazole, benzophenone, amitriptyline and octocrylene. High laccase-catalyzed degradation (80–99%) in continuous flow UF-EMBR has been previously reported [18,37] for benzophenone, amitriptyline and octocrylene, but not for metronidazole. Metronidazole contains both EWGs (i.e., $-NO_2$) and EDGs (i.e., methyl and hydroxyethyl) in its molecule (see Supplementary Data Table S1). High enzymatic degradation of metronidazole following its complete retention by the MD membrane in MD-EMBR can be attributed to the prolonged contact time that may have promoted the interaction of laccase with the EDGs of metronidazole.

An overall degradation of only 40–65% was achieved by the MD-EMBR for a number of non-phenolic TrOCs (Figure 2), however, these removal efficiencies in fact compare favorably with those reported in the literature [12,15,37]. For instance, laccase catalyzed degradation of carbamazepine, clofibric acid, fenoprop and atrazine has been reported to be less than 10% in both batch and continuous flow ultrafiltration based enzymatic bioreactors [15,18,40]. By contrast, 40–45% degradation of these TrOCs by the MD-EMBR was observed in this study. Since most of the selected non-phenolic TrOCs contain both EWGs and EDGs in their structure (see Supplementary Data Table S1), complete retention of these TrOCs in the enzymatic bioreactor may have facilitated the interaction of EDGs with nearby redox centers, thereby providing higher possibility of electron transfer to enzyme [17]. Previously, Nguyen et al. [20] reported that dosing of GAC into an UF-EMBR led to simultaneous adsorption of laccase and TrOCs on GAC, yielding significant improvement in the degradation of four non-phenolic TrOCs, namely atrazine, sulfamethoxazole, diclofenac and carbamazepine. Although our approach was different, it is conceivable that prolonged retention of TrOCs in the enzymatic bioreactor can improve their degradation.

It is noteworthy that phenolic TrOCs (e.g., triclosan, oxybenzone, bisphenol A and steroid hormones) can act as redox-mediators, and the fragments of phenoxyl radicals formed following their degradation by laccase can oxidize non-phenolic compounds [39]. Indeed Margot et al. [12] observed that degradation of diclofenac by laccase was significantly higher in the mixture of TrOCs containing diclofenac, bisphenol A and mefenamic acid than its degradation as a single compound. It is possible that complete retention of phenoxyl radicals formed due to the degradation of phenolic TrOCs aided better degradation of non-phenolic TrOCs by MD-EMBR as compared to previously developed UF-EMBR [18,37]. Further investigation would be required to substantiate this hypothesis but that is beyond the scope of this study.

This study confirms for the first time the improvement in TrOC degradation in an enzymatic bioreactor by coupling with it a high retention membrane (such as membrane distillation) as compared to a conventional ultrafiltration membrane. We used a direct contact membrane distillation module, but there may be case-specific scope of choice between different formats of membrane distillation. Future studies are recommended to assess the commercial viability of different configurations of MD such as vacuum MD and air gap MD, but that is beyond the scope of this study.

3.3. Impact of Mediator Addition on TrOC Degradation

As noted in Section 3.2, of the 30 TrOCs tested, MD-EMBR achieved high degradation (85–99%) for 14 compounds (10 phenolic and 4 non-phenolic compounds) but the degradation efficiency varied widely (40–70%) for the rest of the compounds. To improve the degradation of the latter group, three redox-mediators, namely SA, VA and HBT, were added at 0.25 and 0.5 mM concentrations each in separate runs. Depending on the redox-mediator type and concentration, degradation of phenolic compounds and non-phenolic compounds by the MD-EMBR was improved by 20–30% and 10–50%, respectively (Figure 3) as explained in the following sections.

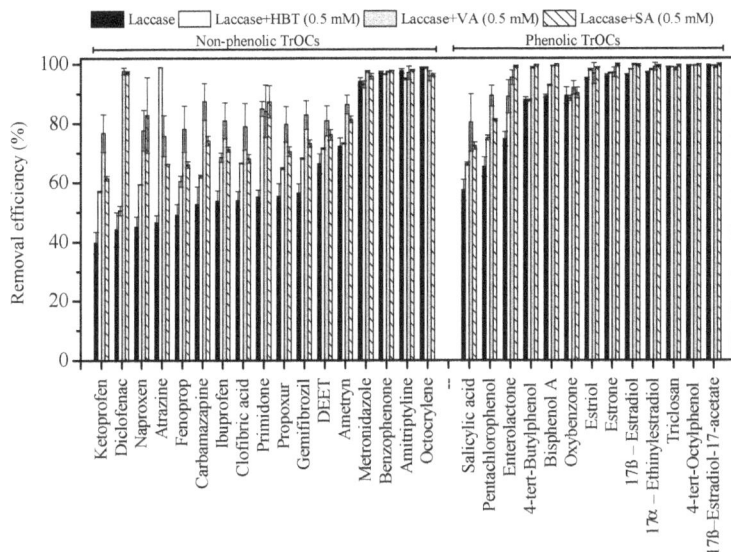

Figure 3. Enzymatic degradation of 30 TrOCs in the presence of three mediators, namely HBT, VA and SA (separately at 0.5 mM) in the MD-EMBR. Error bars indicate the standard deviation of duplicate samples. Operating conditions of the MD-EMBR are given in the caption of Figure 2.

3.3.1. Comparison of Redox-Mediators

To date, the impact of redox-mediator type on the improvement of TrOC degradation has been assessed mainly in small scale and batch tests [34,41,42]. For instance, Ashe et al. [34] investigated the performance of seven different redox-mediators including SA, HBT and VA for the degradation of four resistant TrOCs, namely atrazine, naproxen, oxybenzone and pentachlorophenol in 10 mL batch reactors. They achieved significant improvement (40–90%) at a concentration of 1 mM. Nguyen et al. [18] achieved enhanced (10–90%) removal of TrOCs in UF-EMBR using SA and HBT. However, this is the first study investigating the efficacy of SA, VA and HBT for enhanced degradation of a broad spectrum of TrOCs by an MD-EMBR.

All the tested redox-mediators enhanced the degradation of TrOCs. However, the best overall performance was shown by VA (Figure 3). In line with the findings of Nguyen et al. [37], degradation of the phenolic TrOCs that were already highly degraded by laccase (Figure 2) remained almost the same after the addition of redox-mediators. For the remaining phenolic TrOCs, VA (at 0.5 mM), compared to HBT and SA achieved better removal for two compounds, namely salicylic acid (80%) and pentachlorophenol (90%). Both VA and SA achieved above 95% degradation of enterolactone, which compares favorably with 45–70% degradation achieved in absence of mediators (Figure 3).

Of the 17 non-phenolic compounds, degradation of four compounds viz metronidazole, benzophenone, amitriptyline and octocrylene, was at least 90%, regardless of the mediator type (Figure 3). For the remaining compounds, VA (at 0.5 mM) achieved better degradation for 10 compounds compared to SA and HBT. SA (at 0.5 mM) performed the best for the degradation of two compounds, namely naproxen and primidone. It is well-known that the herbicide atrazine is resistant to laccase catalyzed degradation [18]. Compared to other redox-mediators, HBT was particularly efficient (>99%) for the degradation of atrazine. Although a superior ability of VA compared to other mediators for the degradation of non-phenolic TrOCs has been reported previously in a batch enzymatic bioreactor spiked with four TrOCs [34], the effectiveness of VA for the degradation of a broad spectrum of non-phenolic TrOCs is demonstrated for the first time in this study.

3.3.2. Impact of Mediator Concentration

Redox-mediator dose can affect TrOC degradation by changing the abundance, stability and reversibility of the generated radicals [43]. Therefore, the impact of two mediator concentrations (0.25 and 0.5 mM) on ORP, TrOC degradation, and enzyme stability was investigated.

Concentration-dependent improvement in the degradation of 18 TrOCs (5 phenolic and 13 non-phenolic compounds, Figure 4) was observed in MD-EMBR. The highest improvement in the degradation of TrOCs was achieved at 0.5 mM. Notably, increasing the concentration of SA, HBT and VA from 0.25 to 0.5 mM improved TOC degradation by up to 7%, 15% and 25%, respectively (Figure 4). This corresponds well with the respective increase of 2%, 5% and 15% of the reaction media ORP (Figure 5). On the other hand, degradation of 8 phenolic and 4 non-phenolic TrOCs in MD-EMBR was comparable at all the tested mediator concentrations (Supplementary data Figure S3). For instance, HBT achieved over 99% degradation of atrazine in MD-EMBR irrespective of the mediator concentration. This is consistent with HBT performance reported in case of UF-EMBR [18].

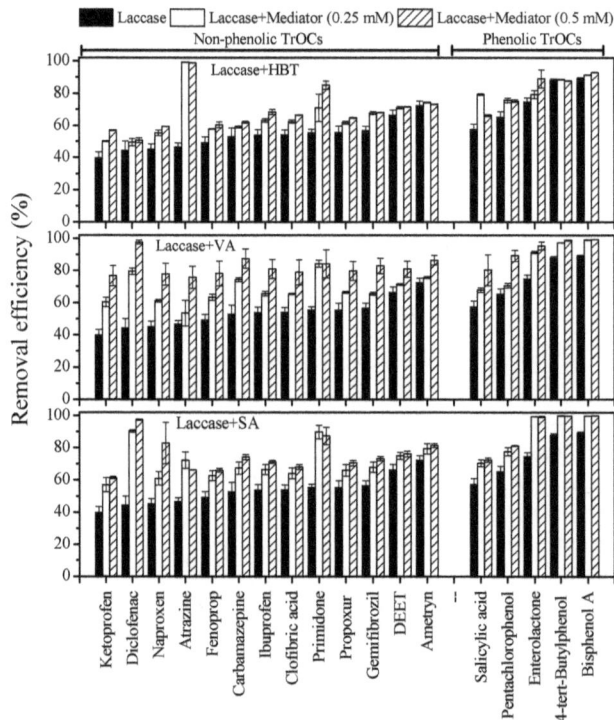

Figure 4. Impact of mediator concentration (0.25 and 0.5 mM) on the degradation of TrOCs in the MD-EMBR. Error bars indicate the standard deviation of duplicate samples. Operating conditions of the MD-EMBR are given in the caption of Figure 2. Only those TrOCs showing mediator concentration-dependent improvement in their degradation are shown here. For remaining TrOCs, results are given in Supplementary Data Figure S3.

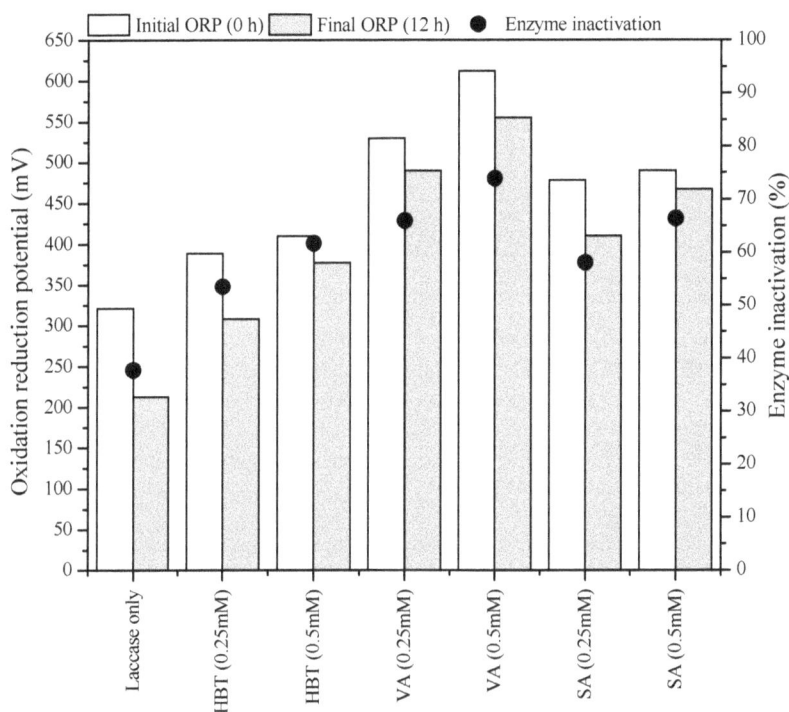

Figure 5. Effect of mediator type and concentration on oxidation reduction potential (ORP) and laccase inactivation in the MD-EMBR. Operating conditions of the MD-EMBR are given in the caption of Figure 2.

In general, the degradation of TrOCs that are easily amenable to laccase (Supplementary data Figure S3) does not improve significantly (less than 5% in this study), while the degradation of resistant TrOCs (Figure 4) may improve with the increase in mediator concentration, and may reach a plateau beyond a certain mediator concentration. However, the mediator concentration beyond which no improvement occurs may depend on the type of mediators as well as the target TrOCs [41,44].

3.3.3. Effect of Mediators on Enzyme Stability

In this study, a gradual inactivation of laccase was observed despite the absence of any known chemical inhibitors in the synthetic wastewater (Figure 5). In the absence of redox-mediators, a 37% laccase inactivation was observed over a period of 12 h. This was possibly due to the blockage of the active enzyme sites by the charged metabolites and/or hydraulic stress during membrane filtration [25,41]. Since the MD membrane can conceptually retain all nonvolatile organics including the transformation products/radicals, laccase inactivation with or without the presence of redox-mediators can be expected. The extent of laccase inactivation increased further when the mediators were added (61%, 66% and 73% for HBT, SA and VA, respectively, each at a concentration of 0.5 mM). The highly reactive radicals generated from mediators can enhance the degradation of TrOCs but at the same time may inactivate laccase [45]. Purich [17] suggested that the metabolites from the oxidation of substrate and/or mediators could react with enzyme to form non-productive complexes, thereby inactivating the enzyme.

The extent of laccase inactivation also depends on the concentration of redox-mediators. For instance, Khlifi-Slama et al. [45] observed a gradual increase in the inactivation of laccase from *Trametes trogii* following a stepwise increase in the concentration of HBT from 0.1 to 10 mM. In another study, increasing SA concentration from 0.1 to 1 mM resulted in aggravated inactivation of laccase from *Trametes versicolor* [42]. These results suggest that the degree of laccase inactivation is strongly influenced by redox-mediator concentration. Indeed, loss in laccase activity was increased by 7%, 9% and 11% in MD-EMBR due to the increase in the concentration of HBT, SA and VA, respectively, from 0.25 to 0.5 mM (Figure 5). Although laccase activity was greatly affected in the presence of redox-mediators, it was compensated by the improvement in TrOC degradation (Figure 3). For example, the highest drop in laccase activity was observed in the presence of VA (Figure 5), but it outperformed SA and HBT in terms of enhanced TrOC degradation (Figure 3).

3.4. Effluent Toxicity

The charged metabolites and highly reactive radicals produced following the oxidation of redox-mediators may improve TrOC degradation [18,46], but these can also cause an increase in effluent toxicity [18,47]. In this study, it was not possible to relate individual metabolites to specific parent compounds because we investigated a mixture of 30 TrOCs. Hence, the overall bacterial toxicity of the reaction mixture and permeate was evaluated at the end of each run. Of the three mediators tested, SA significantly increased the toxicity of the solution in the enzymatic bioreactor, whereas HBT and VA showed no effect on toxicity levels (Table 2). Compared to the background toxicity of the mixture of laccase and TrOCs in the enzymatic bioreactor of MD-EMBR (<1 to 1.8 rTU; $n = 2$), toxicity in the enzymatic bioreactor due to addition of HBT, VA and SA ranged from <1 to 1.7 rTU ($n = 2$), 3.3 to 3.9 rTU ($n = 2$) and 109 to 116 rTU ($n = 2$), respectively. Notably, the final effluent (i.e., membrane permeate) was not toxic to bacteria (<1 rTU) for any of the enzyme/mediator combinations, indicating that MD not only retained TrOCs and laccase but also the transformation byproducts and radicals responsible for inducing bacterial toxicity. This is an added advantage of coupling a high retention membrane to the enzymatic bioreactor.

Table 2. Toxicity of the reactor mixture and permeate following treatment of TrOCs with different mediators in MD-EMBR, expressed as relative toxic unit (rTU). Mediators were added separately at a concentration of 0.5 mM. The limit of detection of the toxicity assay was 10% inhibition of luminescence (i.e., 1 rTU). Toxicity in all permeate samples was below the limit of detection ($n = 2$).

Reaction Mixture	Toxicity of the Reactor Mixture (rTU)	Toxicity of the Permeate (rTU)
TrOCs + Laccase	<1–1.8	<1
TrOCs + Laccase + HBT (0.5 mM)	<1–1.7	<1
TrOCs + Laccase + VA (0.5 mM)	3.3–3.9	<1
TrOCs + Laccase + SA (0.5 mM)	109–116	<1

3.5. Permeate Flux

The driving force of permeate flux in MD is the difference between feed and distillate temperature. Ideally, feed and distillate temperature is maintained at over 50 and 20–25 °C, respectively to obtain a permeate flux of approximately 10 L/m² h [27,48]. In this study, however, to avoid thermal inhibition of laccase [46], temperature of the enzymatic reactor and permeate tank was kept at 30 and 10 °C, respectively. A stable permeate flux of around 4 L/m² h was observed during all experiments (Supplementary Data Figure S4), suggesting that membrane fouling did not occur during the operation period. This level of flux is consistent with the feed temperature employed. Notably, the average permeate flux (Figure 6) for laccase only, laccase-HBT, laccase-VA and laccase-SA variations was 3.69 ± 0.44 L/m² h ($n = 150$), 3.89 ± 0.63 L/m² h ($n = 283$), 3.92 ± 0.62 L/m² h ($n = 291$) and 3.86 ± 0.66 L/m² h ($n = 288$) LMH, respectively, confirming negligible impact of different type of mediator addition on membrane flux. In this study, the mass transfer coefficient (K_m) of the DCMD, which was calculated based on the method described by Nghiem et al. [49], ranged from 1.22 to 1.28 ($\times 10^{-3}$) L/m² h Pa. This value is in good agreement with that in previous studies [48,50].

Thus, this study shows both stable membrane hydraulic performance and improved enzymatic degradation of TrOCs following their complete retention by the MD membrane.

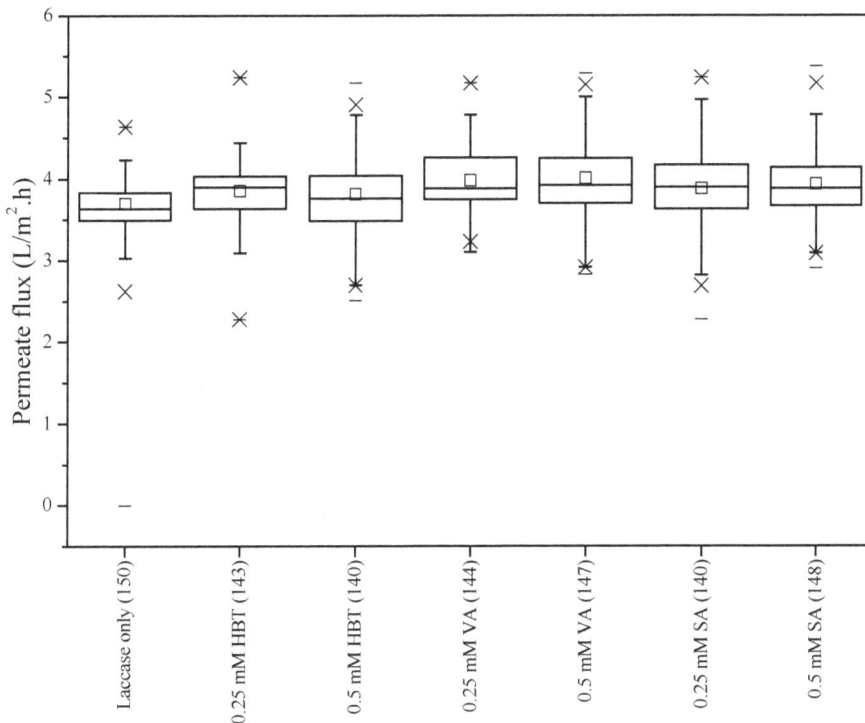

Figure 6. Permeate flux during the operation of MD-EMBR with and without the presence of redox-mediators. Box-and-whisker plot shows the interquartile range; median (horizontal line in the box); min and max (whiskers); average (small square in the box); and 1 and 99% percentiles (cross above and below the whiskers). Operating conditions for MD-EMBR: Temperature of the enzymatic bioreactor and the permeate tank were kept at 30 and 10 °C, respectively; cross-flow rate of water from enzymatic bioreactor and distillate was 1 L/min (corresponding to cross-flow velocity of 9 cm/s); the initial TrOC concentration and laccase activity was 20 µg/L and 95–100 $µM_{(DMP)}$/min, respectively; and each mediator was added at 0.25 or 0.5 mM concentration in separate runs.

4. Conclusions

Performance of an enzymatic bioreactor integrated with the MD system (MD-EMBR) was examined for the removal of 13 phenolic and 17 non-phenolic compounds. Based on permeate quality, MD-EMBR achieved 90–99% TrOC retention. Degradation of TrOCs varied (40–99%) depending on their molecular properties (electron withdrawing functional groups electron donating functional groups and phenolic moiety). High degradation (above 90%) of TrOCs containing EDGs in their chemical structure was observed in the MD-EMBR, while those containing EWGs in their molecular structure were moderately degraded (40–75%). Degradation of TrOCs was further improved by adding three redox-mediators, namely syringaldehyde (SA), violuric acid (VA) and 1-hydroxybenzotriazole (HBT). VA at 0.5 mM concentration was found to be the most effective mediator for improving the degradation of phenolic and non-phenolic TrOCs. Moreover, it was observed that the degradation of non-phenolic compounds in laccase-mediator system was strongly influenced by the tested concentration of the redox-mediators. Despite an increase in the toxicity of the reaction mixture caused by SA, the final effluent of the MD-EMBR was nontoxic.

Supplementary Materials: The following are available online at http://www.mdpi.com/2076-3417/7/9/879/s1, Table S1: Physicochemical properties of the selected trace organic contaminants (TrOCs), Table S2: Physicochemical properties of the selected redox-mediators, Figure S3: Impact of mediator concentration (0.25 and 0.5 mM) on the degradation of after an incubation time of 12 h in the MD-EMBR. Error bars indicate the standard deviation of duplicate samples. Degradation of these TrOCs did not improve by increasing mediator concentration, Figure S4: Permeate flux obtained during the operation of enzymatic membrane distillation (MD-EMBR) with and without the addition of redox mediators.

Acknowledgments: This research has been conducted with the support of the Australian Government Research Training Program Scholarship. Novozymes Pty. Ltd., Australia is thanked for the provision of enzyme solution. This study was partially funded by the GeoQuEST Research Centre, University of Wollongong, Australia.

Author Contributions: F.I.H. conceived and led the project. F.I.H. and M.B.A. planned the experiments in consultation with the coauthors. M.B.A. conducted the experiments. J.K., F.D.D.L. and J.P.M. analyzed TrOC and toxicity samples. F.I.H. and M.B.A. analyzed the data and prepared the manuscript with the contribution of K.Y., W.E.P., L.D.N., F.D.D.L. and J.P.M. to specific sections.

Conflicts of Interest: The authors declare no conflict of interest.

References

1. Hai, F.I.; Yamamoto, K.; Nakajima, F.; Fukushi, K. Application of a gac-coated hollow fiber module to couple enzymatic degradation of dye on membrane to whole cell biodegradation within a membrane bioreactor. *J. Membr. Sci.* **2012**, *389*, 67–75. [CrossRef]

2. Hai, F.I.; Yamamoto, K.; Fukushi, K. Hybrid treatment systems for dye wastewater. *Crit. Rev. Environ. Sci. Technol.* **2007**, *37*, 315–377. [CrossRef]

3. Liu, Y.; Yan, M.; Geng, Y.; Huang, J. Laccase immobilization on poly(p-phenylenediamine)/Fe$_3$O$_4$ nanocomposite for reactive blue 19 dye removal. *Appl. Sci.* **2016**, *6*, 232. [CrossRef]

4. Hai, F.I.; Yamamoto, K.; Fukushi, K. Development of a submerged membrane fungi reactor for textile wastewater treatment. *Desalination* **2006**, *192*, 315–322. [CrossRef]

5. Hai, F.I.; Yamamoto, K.; Nakajima, F.; Fukushi, K.; Nghiem, L.D.; Price, W.E.; Jin, B. Degradation of azo dye acid orange 7 in a membrane bioreactor by pellets and attached growth of *Coriolus versicolour*. *Bioresour. Technol.* **2013**, *141*, 29–34. [CrossRef] [PubMed]

6. Asif, M.B.; Hai, F.I.; Singh, L.; Price, W.E.; Nghiem, L.D. Degradation of pharmaceuticals and personal care products by white-rot fungi—A critical review. *Curr. Pollut. Rep.* **2017**, *3*, 88–103. [CrossRef]

7. Asif, M.B.; Hai, F.I.; Hou, J.; Price, W.E.; Nghiem, L.D. Impact of wastewater derived dissolved interfering compounds on growth, enzymatic activity and trace organic contaminant removal of white rot fungi—A critical review. *J. Environ. Manag.* **2017**, *201*, 89–109. [CrossRef] [PubMed]

8. Luo, Y.; Guo, W.; Ngo, H.H.; Nghiem, L.D.; Hai, F.I.; Zhang, J.; Liang, S.; Wang, X.C. A review on the occurrence of micropollutants in the aquatic environment and their fate and removal during wastewater treatment. *Sci. Total Environ.* **2014**, *473*, 619–641. [CrossRef] [PubMed]

9. Alexander, J.T.; Hai, F.I.; Al-aboud, T.M. Chemical coagulation-based processes for trace organic contaminant removal: Current state and future potential. *J. Environ. Manag.* **2012**, *111*, 195–207. [CrossRef] [PubMed]

10. Pal, A.; Gin, K.Y.-H.; Lin, A.Y.-C.; Reinhard, M. Impacts of emerging organic contaminants on freshwater resources: Review of recent occurrences, sources, fate and effects. *Sci. Total Environ.* **2010**, *408*, 6062–6069. [CrossRef] [PubMed]

11. Yang, S.; Hai, F.I.; Nghiem, L.D.; Price, W.E.; Roddick, F.; Moreira, M.T.; Magram, S.F. Understanding the factors controlling the removal of trace organic contaminants by white-rot fungi and their lignin modifying enzymes: A critical review. *Bioresour. Technol.* **2013**, *141*, 97–108. [CrossRef] [PubMed]

12. Margot, J.; Maillard, J.; Rossi, L.; Barry, D.A.; Holliger, C. Influence of treatment conditions on the oxidation of micropollutants by trametes versicolor laccase. *New Biotechnol.* **2013**, *30*, 803–813. [CrossRef] [PubMed]

13. Hai, F.I.; Nghiem, L.D.; Khan, S.J.; Price, W.E.; Yamamoto, K. Wastewater reuse: Removal of emerging trace organic contaminants. In *Membrane Biological Reactors: Theory, Modeling, Design, Management and Applications to Wastewater Reuse*; Hai, F.I., Yamamoto, K., Lee, C., Eds.; IWA Publishing: London, UK, 2014; pp. 165–205, ISBN 9781780400655.

14. Hai, F.I.; Tadkaew, N.; McDonald, J.A.; Khan, S.J.; Nghiem, L.D. Is halogen content the most important factor in the removal of halogenated trace organics by mbr treatment? *Bioresour. Technol.* **2011**, *102*, 6299–6303. [CrossRef] [PubMed]

15. Tran, N.H.; Urase, T.; Kusakabe, O. Biodegradation characteristics of pharmaceutical substances by whole fungal culture trametes versicolor and its laccase. *J. Water Environ. Technol.* **2010**, *8*, 125–140. [CrossRef]

16. Cañas, A.I.; Camarero, S. Laccases and their natural mediators: Biotechnological tools for sustainable eco-friendly processes. *Biotechnol. Adv.* **2010**, *28*, 694–705. [CrossRef] [PubMed]

17. Purich, D.L. *Enzyme Kinetics: Catalysis & Control: A Reference of Theory and Best-Practice Methods*; Elsevier: Amsterdam, The Netherlands, 2010; p. 759, ISBN 9780123809247.

18. Nguyen, L.N.; Hai, F.I.; Price, W.E.; Kang, J.; Leusch, F.D.; Roddick, F.; van de Merwe, J.P.; Magram, S.F.; Nghiem, L.D. Degradation of a broad spectrum of trace organic contaminants by an enzymatic membrane reactor: Complementary role of membrane retention and enzymatic degradation. *Int. Biodeterior. Biodegrad.* **2015**, *99*, 115–122. [CrossRef]

19. Lloret, L.; Eibes, G.; Feijoo, G.; Moreira, M.T.; Lema, J.M. Degradation of estrogens by laccase from myceliophthora thermophila in fed-batch and enzymatic membrane reactors. *J. Hazard. Mater.* **2012**, *213–214*, 175–183. [CrossRef] [PubMed]

20. Nguyen, L.N.; Hai, F.I.; Price, W.E.; Leusch, F.D.; Roddick, F.; Ngo, H.H.; Guo, W.; Magram, S.F.; Nghiem, L.D. The effects of mediator and granular activated carbon addition on degradation of trace organic contaminants by an enzymatic membrane reactor. *Bioresour. Technol.* **2014**, *167*, 169–177. [CrossRef] [PubMed]

21. Luo, W.; Hai, F.I.; Price, W.E.; Guo, W.; Ngo, H.H.; Yamamoto, K.; Nghiem, L.D. High retention membrane bioreactors: Challenges and opportunities. *Bioresour. Technol.* **2014**, *167*, 539–546. [CrossRef] [PubMed]

22. Wijekoon, K.C.; Hai, F.I.; Kang, J.; Price, W.E.; Guo, W.; Ngo, H.H.; Cath, T.Y.; Nghiem, L.D. A novel membrane distillation–thermophilic bioreactor system: Biological stability and trace organic compound removal. *Bioresour. Technol.* **2014**, *159*, 334–341. [CrossRef] [PubMed]

23. Luo, W.; Hai, F.I.; Kang, J.; Price, W.E.; Nghiem, L.D.; Elimelech, M. The role of forward osmosis and microfiltration in an integrated osmotic-microfiltration membrane bioreactor system. *Chemosphere* **2015**, *136*, 125–132. [CrossRef]

24. Choi, J.-H.; Dockko, S.; Fukushi, K.; Yamamoto, K. A novel application of a submerged nanofiltration membrane bioreactor (NF MBR) for wastewater treatment. *Desalination* **2002**, *146*, 413–420. [CrossRef]

25. Hai, F.I.; Nghiem, L.D.; Modin, O. Biocatalytic membrane reactors for the removal of recalcitrant and emerging pollutants from wastewater. In *Handbook of Membrane Reactors: Reactor Types and Industrial Applications*; Basile, A., Ed.; Woodhead Publishing: Cambridge, UK, 2013; Volume 2, pp. 763–807, ISBN 9780857094155.

26. Modin, O.; Hai, F.I.; Nghiem, L.D.; Basile, A.; Fukushi, K. Gas-diffusion, extractive, biocatalytic and electrochemical membrane biological reactors. In *Membrane Biological Reactors: Theory, Modeling, Design, Management and Applications to Wastewater Reuse*; Hai, F.I., Yamamoto, K., Lee, C., Eds.; IWA Publishing: London, UK, 2014; pp. 299–334, ISBN 9781780400655.

27. Alkhudhiri, A.; Darwish, N.; Hilal, N. Membrane distillation: A comprehensive review. *Desalination* **2012**, *287*, 2–18. [CrossRef]

28. Qtaishat, M.R.; Banat, F. Desalination by solar powered membrane distillation systems. *Desalination* **2013**, *308*, 186–197. [CrossRef]

29. Wijekoon, K.C.; Hai, F.I.; Kang, J.; Price, W.E.; Cath, T.Y.; Nghiem, L.D. Rejection and fate of trace organic compounds (TrOCs) during membrane distillation. *J. Membr. Sci.* **2014**, *453*, 636–642. [CrossRef]

30. Duong, H.C.; Duke, M.; Gray, S.; Cooper, P.; Nghiem, L.D. Membrane scaling and prevention techniques during seawater desalination by air gap membrane distillation. *Desalination* **2016**, *397*, 92–100. [CrossRef]

31. Li, J.; Wu, J.; Sun, H.; Cheng, F.; Liu, Y. Advanced treatment of biologically treated coking wastewater by membrane distillation coupled with pre-coagulation. *Desalination* **2016**, *380*, 43–51. [CrossRef]

32. Phattaranawik, J.; Fane, A.G.; Pasquier, A.C.; Bing, W. A novel membrane bioreactor based on membrane distillation. *Desalination* **2008**, *223*, 386–395. [CrossRef]

33. Asif, M.B.; Nguyen, L.N.; Hai, F.I.; Price, W.E.; Nghiem, L.D. Integration of an enzymatic bioreactor with membrane distillation for enhanced biodegradation of trace organic contaminants. *Int. Biodeterior. Biodegrad.* **2017**, in press. [CrossRef]

34. Ashe, B.; Nguyen, L.N.; Hai, F.I.; Lee, D.-J.; van de Merwe, J.P.; Leusch, F.D.; Price, W.E.; Nghiem, L.D. Impacts of redox-mediator type on trace organic contaminants degradation by laccase: Degradation efficiency, laccase stability and effluent toxicity. *Int. Biodeterior. Biodegrad.* **2016**, *113*, 169–176. [CrossRef]

35. Nghiem, L.D.; Cath, T. A scaling mitigation approach during direct contact membrane distillation. *Sep. Purif. Technol.* **2011**, *80*, 315–322. [CrossRef]

36. Hai, F.I.; Tessmer, K.; Nguyen, L.N.; Kang, J.; Price, W.E.; Nghiem, L.D. Removal of micropollutants by membrane bioreactor under temperature variation. *J. Membr. Sci.* **2011**, *383*, 144–151. [CrossRef]

37. Nguyen, L.N.; van de Merwe, J.P.; Hai, F.I.; Leusch, F.D.; Kang, J.; Price, W.E.; Roddick, F.; Magram, S.F.; Nghiem, L.D. Laccase-syringaldehyde-mediated degradation of trace organic contaminants in an enzymatic membrane reactor: Removal efficiency and effluent toxicity. *Bioresour. Technol.* **2016**, *200*, 477–484. [CrossRef] [PubMed]

38. Van de Merwe, J.P.; Leusch, F.D. A sensitive and high throughput bacterial luminescence assay for assessing aquatic toxicity—The blt-screen. *Environ. Sci. Process. Impacts* **2015**, *17*, 947–955. [CrossRef] [PubMed]

39. D'Acunzo, F.; Galli, C.; Gentili, P.; Sergi, F. Mechanistic and steric issues in the oxidation of phenolic and non-phenolic compounds by laccase or laccase-mediator systems. The case of bifunctional substrates. *New J. Chem.* **2006**, *30*, 583–591. [CrossRef]

40. Hata, T.; Shintate, H.; Kawai, S.; Okamura, H.; Nishida, T. Elimination of carbamazepine by repeated treatment with laccase in the presence of 1-hydroxybenzotriazole. *J. Hazard. Mater.* **2010**, *181*, 1175–1178. [CrossRef] [PubMed]

41. Lloret, L.; Eibes, G.; Moreira, M.; Feijoo, G.; Lema, J. On the use of a high-redox potential laccase as an alternative for the transformation of non-steroidal anti-inflammatory drugs (NSAIDs). *J. Mol. Catal. B Enzym.* **2013**, *97*, 233–242. [CrossRef]

42. Nguyen, L.N.; Hai, F.I.; Kang, J.; Leusch, F.D.; Roddick, F.; Magram, S.F.; Price, W.E.; Nghiem, L.D. Enhancement of trace organic contaminant degradation by crude enzyme extract from trametes versicolor culture: Effect of mediator type and concentration. *J. Taiwan Inst. Chem. Eng.* **2014**, *45*, 1855–1862. [CrossRef]

43. D'Acunzo, F.; Baiocco, P.; Galli, C. A study of the oxidation of ethers with the enzyme laccase under mediation by two N–OH–type compounds. *New J. Chem.* **2003**, *27*, 329–332. [CrossRef]

44. Mizuno, H.; Hirai, H.; Kawai, S.; Nishida, T. Removal of estrogenic activity of iso-butylparaben and *n*-butylparaben by laccase in the presence of 1-hydroxybenzotriazole. *Biodegradation* **2009**, *20*, 533–539. [CrossRef] [PubMed]

45. Khlifi-Slama, R.; Mechichi, T.; Sayadi, S.; Dhouib, A. Effect of natural mediators on the stability of trametes trogii laccase during the decolourization of textile wastewaters. *J. Microbiol.* **2012**, *50*, 226–234. [CrossRef] [PubMed]

46. Kim, Y.-J.; Nicell, J.A. Impact of reaction conditions on the laccase-catalyzed conversion of bisphenol A. *Bioresour. Technol.* **2006**, *97*, 1431–1442. [CrossRef] [PubMed]

47. Marco-Urrea, E.; Pérez-Trujillo, M.; Vicent, T.; Caminal, G. Ability of white-rot fungi to remove selected pharmaceuticals and identification of degradation products of ibuprofen by trametes versicolor. *Chemosphere* **2009**, *74*, 765–772. [CrossRef]

48. Duong, H.C.; Hai, F.I.; Al-Jubainawi, A.; Ma, Z.; He, T.; Nghiem, L.D. Liquid desiccant lithium chloride regeneration by membrane distillation for air conditioning. *Sep. Purif. Technol.* **2017**, *177*, 121–128. [CrossRef]

49. Nghiem, L.D.; Hildinger, F.; Hai, F.I.; Cath, T. Treatment of saline aqueous solutions using direct contact membrane distillation. *Desalination Water Treat.* **2011**, *32*, 234–241. [CrossRef]

50. Duong, H.C.; Cooper, P.; Nelemans, B.; Cath, T.Y.; Nghiem, L.D. Optimising thermal efficiency of direct contact membrane distillation by brine recycling for small-scale seawater desalination. *Desalination* **2015**, *374*, 1–9. [CrossRef]

applied
sciences

MDPI

Article

Formulation of Laccase Nanobiocatalysts Based on Ionic and Covalent Interactions for the Enhanced Oxidation of Phenolic Compounds

Maria Teresa Moreira *, Yolanda Moldes-Diz, Sara Feijoo, Gemma Eibes, Juan M. Lema and Gumersindo Feijoo

Department of Chemical Engineering, Institute of Technology, University of Santiago de Compostela, 15782 Santiago de Compostela, Spain; yolanda.moldes@usc.es (Y.M.-D.); s.feijoomoreira@student.tudelft.nl (S.F.); gemma.eibes@usc.es (G.E.); juan.lema@usc.es (J.M.L.); gumersindo.feijoo@usc.es (G.F.)
* Correspondence: maite.moreira@usc.es; Tel.: +34-881-816-792

Received: 20 July 2017; Accepted: 16 August 2017; Published: 18 August 2017

Featured Application: Development and potential use of nanobiocatalysts for the removal of phenolic compounds as well as other related xenobiotics present in industrial wastewaters.

Abstract: Oxidative biocatalysis by laccase arises as a promising alternative in the development of advanced oxidation processes for the removal of xenobiotics. The aim of this work is to develop various types of nanobiocatalysts based on laccase immobilized on different superparamagnetic and non-magnetic nanoparticles to improve the stability of the biocatalysts. Several techniques of enzyme immobilization were evaluated based on ionic exchange and covalent bonding. The highest yields of laccase immobilization were achieved for the covalent laccase nanoconjugates of silica-coated magnetic nanoparticles (2.66 U mg^{-1} NPs), formed by the covalent attachment of the enzyme between the aldehyde groups of the glutaraldehyde-functionalized nanoparticle and the amino groups of the enzyme. Moreover, its application in the biotransformation of phenol as a model recalcitrant compound was tested at different pH and successfully achieved at pH 6 for 24 h. A sequential batch operation was carried out, with complete recovery of the nanobiocatalyst and minimal deactivation of the enzyme after four cycles of phenol oxidation. The major drawback associated with the use of the nanoparticles relies on the energy consumption required for their production and the use of chemicals, that account for a major contribution in the normalized index of 5.28×10^{-3}. The reduction of cyclohexane (used in the synthesis of silica-coated magnetic nanoparticles) led to a significant lower index (3.62×10^{-3}); however, the immobilization was negatively affected, which discouraged this alternative.

Keywords: laccase; nanocatalyst; immobilization; phenol; sequential batch reactor

---

1. Introduction

Laccase is a high potential oxidative enzyme with broad substrate specificity towards aromatic compounds, which makes it a promising candidate for the degradation of xenobiotics containing hydroxyl and amine groups [1–3]. However, the relatively low stability of the free enzyme arises as a major technical hurdle that hampers its large-scale application [4]. Beyond the potentiality of protein engineering and directed evolution to change enzyme conformation [5], enzyme immobilization can be applied to enhance the protein stability by the prevention of autolysis or proteolysis, rigidification of the enzyme structure via multipoint covalent attachment, and generation of favorable microenvironments [6–8]. This method has been demonstrated to improve the activity and stability of the biocatalyst in both aqueous and organic phases, provided that the support permits the diffusion of the substrate to the active site of the enzyme [9]. Furthermore, it may facilitate the simple recovery of the enzymes by centrifugation, sedimentation, or other physical separation methods and reuse in continuous systems. However, immobilized enzymes can also encounter several drawbacks, such as mass transfer limitations or interaction between the enzyme and the support that may reduce its catalytic potential [10].

Conceptually, there are two basic methods for enzyme immobilization, as the enzyme-support link can take place by physical or chemical interactions. Physical coupling methods include the entrapment of the enzyme within a tridimensional matrix, its encapsulation in an organic or inorganic polymer, and its adsorption to the support surface by ionic exchange [11], whereas covalent bonding assures the irreversible binding of the enzyme to the support matrix.

Among a wide range of alternatives, the large specific surface area characteristic of nanomaterials makes this type of support an ideal candidate for enzyme immobilization [12]. The efficiency of ionic exchange depends on the pH and ionic strength of the medium as well as the hydrophobic nature of the nanoparticle surface [13–15]. Regarding covalent bonding, nanoparticles may provide a homogeneous core-shell structure, which can be functionalized to react with nucleophilic groups on the enzyme [16]. Most enzymes are covalently attached to the lysine amino groups, which are typically present on the protein surface [17]. Several factors, including pH, ionic strength, protein concentration, additives, and nanoparticles structure (porous or non-porous material) may affect the biocatalyst and the effectiveness of covalent bonding between the enzyme and the support [16,18].

The immobilization of laccase on different types of nanoparticles such as silver and gold nanoparticles [19], chitosan-coated magnetic nanoparticles [20], and carbon nanotubes [21] has been demonstrated in recent years, although few processes have been used for practical applications at full-scale [22]. The main aim of this work is to perform the efficient immobilization of laccase on different types of magnetic and non-magnetic nanoparticles. Two different immobilization procedures will be followed: ionic exchange between the enzyme and the nanoparticle, and covalent bonding of the enzyme protein to the surface of the nanoparticle using glutaraldehyde or carbodiimide as cross-linkers [23,24]. Glutaraldehyde, a bifunctional and versatile agent, may react with different enzyme moieties, principally involving primary amino groups of proteins, although it may eventually react with other groups such as thiols, phenols, and imidazoles [25]. On the other hand, carbodiimide is used to form amide linkage between carboxylates and amino terminal groups from the enzyme [26]. The catalytic activity of the different nanobiocatalysts will be evaluated in terms of the biotransformation potential of phenol as the model compound. Once the successful immobilization of laccase is proved, we will aim to examine how the application of life cycle principles may be helpful in the reformulation of the production scheme of the most suitable support.

2. Materials and Methods

2.1. Chemicals and Nanoparticles for Enzyme Immobilization

(3-Aminopropyl)triethoxysilane (APTES) (≥98%), 2,2′-azinobis(3-ethylbenzothiazoline-6-sulphonic acid) (ABTS) (≥98%), glutaraldehyde (25%), 3-(Ethyliminomethyleneamino)-*N*,*N*-dimethylpropan-1- amine (EDC) (≥98%), and fumed silica nanoparticles were purchased from Sigma-Aldrich (St. Louis, MO, USA). Non-coated magnetite nanoparticles, single-core silica-coated magnetic nanoparticles (FeO-2206W), multi-core silica-coated magnetic nanoparticles (S-57), polyacrylic acid nanoparticles (FeO-2204W and FeO-36), and polyethyleneimine-coated magnetic nanoparticles (VOZ-19) were supplied by Nanogap (Ames, Spain). Detailed characteristics of the nanoparticles evaluated are presented in Table 1.

Table 1. Characteristics of the different nanoparticles.

Type of Nanoparticles	Size (nm)	Concentration (mg NPs mL^{-1})
Fumed silica nanoparticles (fsNP)	7	59
Silica-coated magnetic nanoparticles		
FeO-2206W (single-core)	21.5 ± 2.1	5
S-57 (multi-core)	11.8 ± 2.4	10.9
Polyacrylic acid (PAA) magnetic nanoparticles		
FeO-2204W	10.1 ± 2.4	20.5
FeO-36	23.1 ± 4.9	16.2
Polyethylenimine (PEI) magnetic nanoparticles		
VOZ-19	10 ± 1.2	56
Non-coated magnetite nanoparticles	9.9 ± 1.4	17.4

2.2. Laccase Activity

Laccase activity from *Trametes versicolor* (activity ~10 U mg^{-1}, Sigma-Aldrich, St. Louis, MO, USA) was measured according to Zimmerman et al. [23]. Following this protocol, 50 μL of sample was added to 150 μL of 0.267 mM ABTS (in McIlvaine buffer; pH 3) in 96-well plates. The ABTS oxidation was monitored by measuring the absorbance at 420 nm for 7 min (with intervals of 6 s), with a molar extinction coefficient of the cation radical of 36,800 M^{-1} cm^{-1} [24]. One unit U of activity was defined as the amount of enzyme capable of producing 1 μmol of the cation radical per min.

2.3. Functionalization of Laccase onto Silica and Silica-Coated Magnetic Nanoparticles

The immobilization process for fumed silica nanoparticles (fsNP) and silica-coated magnetic nanoparticles (smNP) requires their previous functionalization, in which reactive groups are added based on the modification of their surface by the addition of (3-aminopropyl)triethoxysilane (APTES) [23]. The protocol starts with the incubation of the nanoparticles in phosphate buffer (100 mM, pH 7) and APTES (0.8 mmol APTES g^{-1} nanoparticles) under agitation (100 rpm) for 12 h at room temperature. The residual APTES concentration in the supernatants was monitored as follows: 50 μL of 5.3 mM glutaraldehyde solution was added to 150 μL supernatant. The yellow coloration due to the imine bond resulting from the chemical reaction of APTES with glutaraldehyde was measured spectrophotometrically at 390 nm. After four washing steps, no residual APTES was detected.

2.4. Immobilization of Laccase onto Silica and Polyethylenimine Nanoparticles

The amino-functionalized nanoparticles were then used to perform the immobilization of laccase according to the sorption-assisted immobilization (SAI) protocol [23], where the amino-functionalized nanoparticles and laccase (15 mg mL^{-1}) were incubated in phosphate buffer (pH 7, 100 mM) at 4 °C and 100 rpm for 2 h. Next, glutaraldehyde was added dropwise to the mixture of nanoparticles and laccase, and the solution was incubated for an additional 18 h. The unreacted glutaraldehyde and the excess and unstable bound enzymes were washed away.

The immobilization procedure for the polyethylenimine-coated magnetic nanoparticles (PEI-mNPs) was identical to the one previously described for silica nanoparticles except for the step of functionalization with APTES (not required here). The enzymatic activity of both NP-laccase conjugates and supernatants was measured in these immobilization processes as well as in the following ones to determine the activity yield, washing loss, and enzyme load. Variable concentrations of glutaraldehyde and laccase activity were used in the immobilization process: 4–8 mmol g^{-1} NPs and 0.9–1.88 U mg^{-1} NPs, respectively.

2.5. Immobilization of Laccase onto Polyacrylic Acid-Coated Magnetic Nanoparticles

The functionalization of the nanoparticles was conducted according to the method described by Nobs et al. [24]. The nanoparticles (5 mg mL^{-1}) were suspended in 2-(*N*-morpholino)ethanesulfonic acid (MES) buffer 0.1 M (pH 4.7), and EDC (12 mg mL^{-1}) and N-hydroxysuccinimide (NHS, 33 mg mL^{-1}) were added with gentle agitation (100 rpm) at room temperature (25 °C) to complete the reaction after 24 h. The unreacted NHS and EDC were removed by repeated washing and centrifugation (4000 rpm, 6 min), and were resuspended in MES buffer (0.1 M, pH 4.7). The amino-functionalized nanoparticles were then used to perform the immobilization of laccase by the aforementioned SAI method [23] with 8 mmol glutaraldehyde g^{-1} NPs and 1.88 U laccase mg^{-1} NPs.

2.6. Immobilization of Laccase by Ionic Exchange on Magnetite Nanoparticles

Laccase immobilization in magnetite nanoparticles (lacking any external coating) was carried out by ionic exchange of the enzyme with magnetite nanoparticles. Laccase was added (0.55 U mg^{-1} NPs) to previously washed nanoparticles and incubated at 4 °C, 100 rpm, and pH 5 for 4 h. After incubation, the nanobiocatalyst was washed five times in sodium phosphate buffer before storage.

2.7. Biotransformation of Phenol by Laccase Immobilized onto fsNPs and Single-Core Silica-Coated Magnetic Nanoparticles in Batch Operation

The oxidation of phenol by the enzymatic system was investigated in a reaction medium containing phenol (10 mg L^{-1}) dissolved in phosphate buffer (100 mM, pH 7) and immobilized laccase (1000 U L^{-1}) onto fsNPs or smNPs (FeO-2206W) in 10-mL flasks. In parallel, experiments with free laccase as well as controls lacking laccase with functionalized fsNPs and mNPs were also carried out. Samples were withdrawn at specific time intervals for 24 h to monitor phenol removal.

2.8. Consecutive Cycles of Batch Biotransformation of Phenol by Laccase Immobilized onto Single-Core Silica-Coated Magnetic Nanoparticles

Variable pH values (5–7) were investigated to perform the biotransformation of phenol (10 mg L^{-1}) by laccase immobilized on silica-coated magnetic nanoparticles (1000 U L^{-1}). Acetate buffer (100 mM) was applied for pH 5, while in the case of pH 6 and 7, phosphate buffer (100 mM) was used. Thereafter, the operation of the enzymatic system was conducted in a tank reactor (100 mL) under stirring at room temperature for several consecutive cycles. The reaction medium consisted of phenol (10 mg L^{-1}), phosphate buffer (100 mM, pH 6), and a single initial pulse of laccase (1000 U L^{-1}) immobilized onto FeO-2206W smNP. The effluent of the reactor was withdrawn at the end of the cycle and the nanobiocatalyst was recovered by an external magnetic field before a new cycle started.

2.9. Phenol Analysis

Phenol concentration was determined by high-performance liquid chromatography (HPLC) at a detection wavelength of 270 nm on a Jasco XLC HPLC (Jasco Analítica, Madrid, Spain). This equipment was coupled with a diode detector 3110 MD, a 4.6 × 150 nm Gemini reversed-phase column (3 μm C18 110 Å) from Phenomenex (supplied by Jasco Analítica, Madrid, Spain), and an HP ChromNav data processor. A 25-μL sample volume was injected into the column. The mobile phase contained 50% acetonitrile and 50% water. The flow rate was fixed at 0.4 mL min^{-1} under isocratic conditions.

2.10. Life Cycle Assessment Methodology

Life Cycle Assessment (LCA) is a methodology that aims to analyze products, processes, and/or services from an environmental point of view, and should be part of the decision-making process toward sustainability [27]. The guidelines established by International Organization for Standardization (ISO) standards [28] have been considered to perform the LCA study. The environmental profiles of the silica-coated mNPs production were determined according to the production route described in a previous paper [29]. In a typical synthesis of silica-coated mNPs, polyoxyethylene(5)nonylphenyl ether (Igepal CO-520) and cyclohexane are mechanically stirred before the addition of oleic-acid magnetite nanoparticles (2.5% wt in cyclohexane). Finally, ammonium hydroxide solution and tetraethyl orthosilicate (TEOS) are added consecutively to form a transparent red solution of reverse micro-emulsion. The core-shell nanoparticles are precipitated with isopropanol (IPA) to disrupt the reverse microemulsion and are then washed extensively with IPA and deionized water. Finally, the core-shell nanoparticles are re-dispersed in deionized water. In the case of the production of magnetic nanoparticles with a thin silica-coating, the procedure is similar except for the concentration of cyclohexane (0.5%, five times lower). Inventory data for the foreground systems were obtained from a semi-pilot unit and data from the background system (production of electricity, chemicals, and wastewater treatment) were taken from Ecoinvent database® version 3 [30–32] and, when possible, updated for Spain [33]. The environmental assessment was conducted using characterization factors from ReCiPe Midpoint methodology [34] and the following impact categories were considered in the analysis: climate change, ozone depletion, terrestrial acidification, freshwater eutrophication, marine eutrophication, human toxicity, photochemical oxidant formation, terrestrial ecotoxicity, freshwater ecotoxicity, marine ecotoxicity, and fossil depletion. SimaPro version 7.3.3 (PRé Consultants, Amersfoort, The Netherlands) was the software used for the computational implementation of the life cycle inventory data and the computation of the environmental profiles [35].

3. Results and Discussion

3.1. Immobilization of Laccase onto Different Types of Nanoparticles

Several strategies of laccase immobilization were evaluated on magnetic and non-magnetic nanoparticles to obtain various types of nanobiocatalysts. The enzymatic activities of both NP-laccase conjugates and supernatants were measured to determine the activity yield, washing loss, and enzyme load (Table 2). The tradeoff analysis of the different outcomes will be critical to identify the most suitable option for its further use.

Table 2. Activity yield, washing loss, and enzyme loading for the optimal doses in the immobilization processes.

Different Types of Nanoparticles	Washing Loss (%)	Activity Yield (%)	Enzyme Loading (U mg^{-1} NPs)
Covalent immobilization			
Fumed silica nanoparticles (fsNP)	5.6 ± 1.3	100 ± 6.1	1.78 ± 0.07
Silica-coated magnetic nanoparticles			
FeO-2206W (single-core)	16.4 ± 2.81	99.7 ± 0.35	2.66 ± 0.65
S-57 (multi-core)	66.63 ± 1.67	31.3 ± 0.76	0.42 ± 0.05
Polyacrylic acid (PAA) magnetic nanoparticles			
FeO-2204W	96.83 ± 1.4	2.55 ± 6.5	0.11 ± 0.34
FeO-36	99.7 ± 2.34	0.12 ± 2.3	0.01 ± 0.24
Polyethylenimine (PEI) magnetic nanoparticles			
VOZ-19	27.18 ± 0.08	80.5 ± 0.21	1.54 ± 0.03
Ionic exchange immobilization			
Non-coated magnetite nanoparticles	45.3 ± 0.9	58.5 ± 1.5	0.69 ± 0.05

Covalent bonding produces stronger bonds between the enzyme and the support, allowing its reuse more easily than with other available immobilization methods [36,37] and preventing the leaching of enzymes from the support [38,39]. In this study, different coatings as well as single- and multi-core nanoparticles were evaluated for laccase immobilization (Table 2). The covalent bonding between the supports with carboxylic groups (polyacrylic acid) did not result in satisfactory immobilization (yields lower than 5%). This may be due to excessive crosslinking of the protein molecule (due to the presence of both NH_2 and COOH groups in the enzyme and also to the instability of carbodiimide, which led to very low activity yields, as was also observed in other studies with activity yields below 14%) [40,41].

However, laccase was successfully immobilized in other nanoparticles such as PEI-mNPs, which showed an activity yield higher than 80% (Table 2). The stability of this biocatalyst was inferior to laccase immobilized onto fsNPs and FeO-2206-W smNPs (Table 2). In the specific case of PEI-mNPs, the enzyme activity after three months was 50%, which was significantly lower than those of fsNPs and FeO-2206-W (93% and 99%, respectively). Similar results were observed for a previous report with silica nanoparticles, using remarkably higher dosages of APTES and glutaraldehyde than the values considered in this research [23]. The rationale behind the high activity yields is attributed to the fact that immobilized laccases on this type of support would have high affinity for standard substrates such as ABTS. For instance, Arca-Ramos et al. [42] reported the hyperactivation of laccase from *T. versicolor* after the formation of covalent bonds with silica nanoparticles; whereas Matijosyte et al. [11] described a similar behavior for laccase from *T. villosa* (activity recovery up to 148%) after the formation of cross-linking aggregates (CLEAS®, CLEATechnologies, Delft, The Netherlands).

The process of immobilization by ionic exchange is based on the interaction of the charged groups of the enzyme with the groups of opposite charges in the support. It provides a weak bond between the enzyme and the support so that the native structure of the enzyme is unaltered. Moreover, the bonding is reversible and it is sensitive to changes in the pH and ionic strength, which can lead to the recovery of the support [9]. When this approach was considered for the immobilization of laccase, not only limited yield was evidenced, but also the change of basic pH led to enzyme desorption. When performing the immobilization of laccase at different pH values, the best results were observed when the immobilization process was performed at pH 5 with an activity yield higher than 50% (Table 2), possibly because the point of zero charge (PZC) of the magnetite is between pH 6.5–7.9 [43], while the isoelectric point of laccase is at pH 3 [44]. The main drawback is that laccase stability decreases with lower pH, which was evidenced by the reduction of the immobilized enzyme. Considering the best results of activity yield and enzyme load, fsNPs and FeO-2206W smNPs were selected for the following experiments of phenol biotransformation.

3.2. Biotransformation of Phenol by Laccase Immobilized onto fsNPs and Silica-Coated Magnetic Nanoparticles in Batch Operation

The capacity of free and immobilized enzymes onto fsNPs and FeO-2206W smNPs to transform phenol was assessed in batch operation. The results showed that the higher phenol transformation was achieved by free laccase (>75%), whereas phenol conversion was around 23% and 48% for fsNPs and FeO-2206W smNPs, respectively (Figure 1). Lower activity of immobilized laccase towards phenolic substrates has also been previously reported. For instance, Arca-Ramos et al. [42] found that bisphenol A degradation rate was much slower for immobilized laccases (from 6- to 26-fold lower than that of the free enzyme). This lower reaction rate was related to the potential aggregation of the nanoparticles which could reduce substrate accessibility. Wang et al. [45] studied phenol degradation by immobilized laccase on magnetic silica nanoparticles, and similar results were observed at pH 7. The rate of phenol conversion for laccase immobilized onto FeO-2206W smNPs ($2.01 \mu M h^{-1}$) is almost two times higher than that for fsNPs. Controls with phenol lacking laccase but with functionalized nanoparticles were performed, with no decrease in phenol concentration in all cases after 24 h.

Kurniawati and Nicell [46] reported that laccase can be inactivated due to the presence of free radicals in the reaction medium generated from phenol transformation (not by the substrate). However, this effect was only evident at phenol concentrations of 2000 μM (188 mg/L), almost 20-fold higher than that used in the present work. Hence, no enzymatic activity changes occurred in any of the experiments with a noticeable enzyme deactivation when performing the experiment with free laccase.

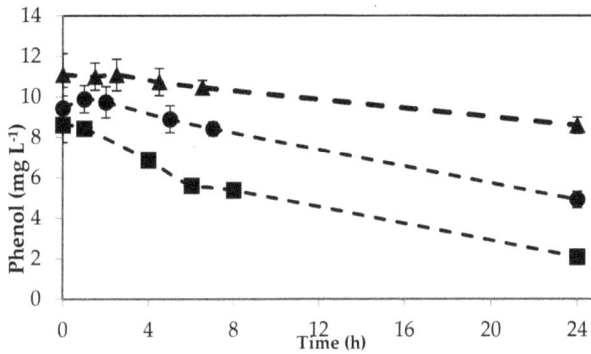

Figure 1. Phenol transformation with free enzyme (■), Tv immobilized onto fsNPs (▲), and onto FeO-2206W (•) over 24 h at pH 7.

According to the results, the nanobiocatalyst with FeO-2206W as a support seems to be the most adequate, as it achieved higher phenol biotransformation yields. Moreover, the separation of the nanobiocatalyst should be much simpler under a magnetic field, while intense centrifugation should be required when considering fumed silica nanoparticles. Accordingly, the single-core silica-coated nanobiocatalyst was used to prove its potential of reuse for phenol biotransformation in sequential batches.

3.3. Sequential Batch Biotransformation of Phenol by Laccase Immobilized onto Silica-Coated Magnetic Nanoparticles

Due to the remarkable effect of pH on phenol conversion reported in previous works [45,47], the biotransformation of phenol was assessed at different pH levels (5, 6, and 7). The conversion efficiencies and rates are shown in Table 3. An improvement of phenol conversion was observed when the pH was decreased to 6 or 5, which lead to an increase of 16%. A similar pH range was found suitable for phenol by immobilized enzyme onto silica-coated magnetic nanoparticles [45]. Furthermore, the enzymatic activity was maintained constant in all the experiments. The immobilized laccase retained 95%, 97%, and 100% of its initial activity at pH 5, 6, and 7, respectively, after incubation at room temperature for 24 h.

Table 3. Phenol biotransformation at variable pH levels by laccase immobilized onto silica-coated magnetic nanoparticles (FeO-2206W) for 24 h.

pH	Phenol Biotransformation (%)	Biotransformation Rate (mg $L^{-1} h^{-1}$)
5	67.9	0.383
6	63.9	0.326
7	48.1	0.189

The reusability of the nanobiocatalyst was assessed in consecutive cycles of 24 h. It was observed that phenol transformation was higher than 60% and was maintained constant after four cycles (Figure 2). In other reports, phenol was almost entirely biotransformed in consecutive cycles with laccase immobilized on magnetic mesoporous silica nanoparticles with a dose of enzyme 18 times higher [45]. The immobilized laccase retained 97% of its initial activity after the consecutive batch treatments of phenol with magnetic separation.

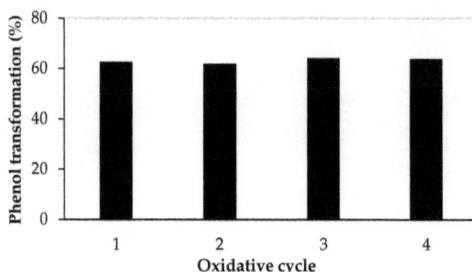

Figure 2. Phenol transformation in subsequent cycles of enzymatic treatment with laccase immobilized onto FeO-2206W (pH 6).

3.4. Environmental Indicators of the Single-Core Magnetic Nanoparticle

The development of new processes must comply with sustainability criteria. The methodology applied for the holistic assessment of the most suitable support was based on a life cycle perspective [48], which would include all aspects of activities during the life of a product, such as the extraction of raw materials and resources, production processes, use of products, recovery, recycling of some fractions, and the final disposal at the end-of-life stage.

In this study, inventory data for the foreground systems (direct inputs and outputs for each scenario) such as electricity requirements (estimated with power and operational data from the different units: reactors, dryers, heaters) as well as the use of chemicals and water were average data from semi-pilot scale experiments, obtained by on-site measurements of production processes developed for a time period of three months. The environmental assessment was conducted using characterization factors from ReCiPe Midpoint methodology [34], and the impact categories for the different mNP production routes are displayed considering one gram of FeO-2206W mNPs as a functional unit (Table 4).

Table 4. Normalized environmental impacts associated with the production of silica-coated mNPs (FeO-2206W) and silica thin shell per g of mNP.

Impact Category	Scenarios	
	Silica-Coated mNPs	Silica Thin Shell
Climate change	1.43×10^{-4}	1.03×10^{-4}
Ozone depletion	3.69×10^{-6}	2.42×10^{-6}
Terrestrial acidification	1.83×10^{-4}	1.30×10^{-4}
Freshwater eutrophication	9.11×10^{-4}	5.42×10^{-4}
Marine eutrophication	2.12×10^{-5}	1.45×10^{-5}
Human toxicity	5.93×10^{-4}	3.98×10^{-4}
Photochemical oxidant formation	1.33×10^{-4}	9.05×10^{-5}
Terrestrial ecotoxicity	6.18×10^{-6}	4.37×10^{-6}
Freshwater ecotoxicity	1.22×10^{-3}	8.36×10^{-4}
Marine ecotoxicity	1.45×10^{-3}	9.84×10^{-4}
Fossil depletion	6.20×10^{-4}	5.18×10^{-4}
Normalized index	5.28×10^{-3}	3.62×10^{-3}

The normalization results show that the impacts associated with the consumption of energy are dominant, but chemicals used in the formulations and for re-dispersion are also relevant. Regarding electricity, it is consumed for stirring, and the mechanical agitation required to obtain the transparent red solution of reverse micro-emulsion until complete reaction is remarkable (97% of the total electrical requirements). Regarding chemicals, the cyclohexane required in the formulation is the environmental critical chemical, since it is responsible for more than 85% of burdens derived from chemicals contributions regardless the impact category. Aiming to reduce the impacts, we also considered the use of a lower dose of cyclohexane to obtain a thin silica coating on the nanoparticle. Although the environmental impact was reduced, the immobilization of the enzyme was negatively affected by 40%, which was detrimental to the overall efficiency of the process. However, it should be highlighted that the production systems have been assessed at the pilot scale and optimizations should be required for large-scale application.

4. Conclusions

The study compared the immobilization of a commercially available *T. versicolor* laccase onto glutaraldehyde-activated, sulfo-NHS/EDC-activated magnetic and non-magnetic nanoparticles by covalent binding and onto magnetic nanoparticles by ion exchange, as well as its use as a nanobiocatalyst for phenol biotransformation. In summary, the most efficient biotransformation and the best activity yield was obtained by using laccase immobilized onto silica-coated magnetic nanoparticles. The magnetic nanobiocatalyst achieved a phenol biotransformation higher than 60%. One major outcome of this study is that the immobilized laccase is magnetically recoverable and can actually be reused in repeated cycles of phenol removal. The easy recovery of the nanobiocatalyst from the reaction media is a remarkable advantage from an operational perspective. Regarding the environmental impacts associated with the production of silica-coated magnetic nanoparticles, the use of energy and chemicals used in the formulations and for re-dispersion are the major contributors. Aiming to reduce the impacts, the use of a lower dose of cyclohexane implied lower environmental impact but negatively affected the immobilization yield of the enzyme, which discouraged this modification in the production process.

Acknowledgments: This work was financially supported by the Spanish Ministry of Economy and Competitiveness (CTQ2013-44762-R and CTQ2016-79461-R, program co-funded by FEDER). The authors belong to the Galician Competitive Research Group GRC 2013-032, program co-funded by FEDER. Yolanda Moldes-Diz thanks the Spanish Ministry of Economy and Competitiveness for her predoctoral fellowship.

Author Contributions: Maria Teresa Moreira, Juan M. Lema and Gumersindo Feijoo conceived and designed the experiments; Yolanda Moldes-Diz performed the experiments; Sara Feijoo performed the LCA study, Maria Teresa Moreira and Gemma Eibes analyzed the data and revised the different versions of the manuscript; Maria Teresa Moreira. and Yolanda Moldes-Diz wrote the paper.

Conflicts of Interest: The authors declare no conflict of interest. The founding sponsors had no role in the design of the study; in the collection, analyses, or interpretation of data; in the writing of the manuscript, and in the decision to publish the results.

References

1. Kunamneni, A.; Ballesteros, A.; Plou, F.J.; Alcalde, M. Fungal laccase—A versatile enzyme for biotechnological applications. In *Communicating Current Research and Educational Topics and Trends in Applied Microbiology*; Mendez-Vilas, A., Ed.; Formex: Badajoz, Spain, 2007; Volume 1, pp. 233–245. ISBN 978-84-611-9422-3.

2. Thurston, C.F. The structure and function of fungal laccases. *Microbiology* **1994**, *140*, 19–26. [CrossRef]

3. Wesenberg, D.; Kyriakides, I.; Agathos, S.N. White-rot fungi and their enzymes for the treatment of industrial dye effluents. *Biotechnol. Adv.* **2003**, *22*, 161–187. [CrossRef] [PubMed]

4. Misson, M.; Zhang, H.; Jin, B. Nanobiocatalyst advancements and bioprocessing applications. *J. R. Soc. Interface* **2015**, *12*, 20140891. [CrossRef] [PubMed]

5. Illanes, R. *Enzyme Biocalysis : Principles and Applications*; Springer: Berlin, Germany, 2008.

6. Hernandez, R.; Fernandez-Lafuente, J.M. Control of protein immobilization: Coupling immobilization and site-directed mutagenesis to improve biocatalyst or biosensor performance. *Enzyme Microb. Technol.* **2011**, *48*, 107–122. [CrossRef] [PubMed]

7. Rodrigues, R.C.; Ortiz, C.; Berenguer-Murcia, A.; Torres, R.; Fernández-Lafuente, R. Modifying enzyme activity and selectivity by immobilization. *Chem. Soc. Rev.* **2013**, *42*, 6290–6307. [CrossRef] [PubMed]

8. Sheldon, R.A. Enzyme Immobilization: The Quest for Optimum Performance. *Adv. Synth. Catal.* **2007**, *349*, 1289–1307. [CrossRef]

9. Duran, N.; Rosa, M.A.; D´Annibale, A.; Gianfreda, L. Applications of laccases and tyrosinases(phenoloxidases) immobilized on different supports: A review. *Enzyme Microb. Technol.* **2002**, *31*, 907–931. [CrossRef]

10. Cui, J.; Jia, S.; Liang, L.; Zhao, Y.; Feng, Y. Mesoporous CLEAs-silica composite microparticles with high activity and enhanced stability. *Sci. Rep.* **2015**, *5*, 14203. [CrossRef] [PubMed]

11. Matijosyte, I.; Arends, I.W.C.E.; Vries, S.; Sheldon, R.A. Preparation and use of cross-linked enzyme aggregates (CLEAs) of laccases. *J. Mol. Catal. B Enzyme* **2010**, *62*, 142–148. [CrossRef]

12. Qu, X.; Alvarez, P.J.J.; Li, Q. Applications of nanotechnology in water and wastewater treatment. *Water Res.* **2013**, *47*, 3931–3946. [CrossRef] [PubMed]

13. Ahn, M.-Y.; Zimmerman, A.R.; Martinez, C.E.; Archibald, D.D.; Bollag, J.-M.; Dec, J. Characteristics of Trametes villosa laccase adsorbed on aluminium hydroxide. *Enzyme Microb. Technol.* **2007**, *41*, 141–148. [CrossRef]

14. Qiu, H.; Xu, C.; Huang, X.; Ding, Y.; Qu, Y.; Gao, P. Adsorption of laccase on the surface of nanoporous gold and the electron transfer between them. *J. Phys. Chem. C* **2008**, *112*, 14781–14785. [CrossRef]

15. Fernandez-Fernandez, M.; Sanroman, M.A.; Moldes, D. Recent developments and applications of immobilized laccase. *Biotechnol. Adv.* **2013**, *31*, 1808–1825. [CrossRef] [PubMed]

16. Arroyo, M. Immobilized enzymes: Theory, methods of study and applications. *ARS Pharm.* **1998**, *39*, 23–39.

17. Brady, D.; Jordaan, J. Advances in enzyme immobilisation. *Biotechnol. Lett.* **2009**, *31*, 1639–1650. [CrossRef] [PubMed]

18. García-Galán, C.; Berenguer-Murcia, A.; Fernandez-Lafuente, R.; Rodrigues, R.C. Potential of Different Enzyme Immobilization Strategies to Improve Enzyme Performance. *Adv. Synth. Catal.* **2011**, *353*, 2885–2904. [CrossRef]

19. Mazur, M.; Krysiński, P.; Michota-Kamińska, A.; Bukowska, J.; Rogalski, J.; Blanchard, G.J. Immobilization of laccase on gold, silver and indium tin oxide by zirconium–phosphonate–carboxylate (ZPC) coordination chemistry. *Bioelectrochemistry* **2007**, *71*, 15–22. [CrossRef] [PubMed]

20. Kalkan, N.; Aksoy, S.; Aksoy, E.; Hasirci, N. Preparation of chitosan-coated magnetite nanoparticles and application for immobilization of laccase. *J. Appl. Polym. Sci.* **2011**, *123*, 707–716. [CrossRef]

21. Feng, W.; Ji, P. Enzymes immobilized on carbon nanotubes. *Biotechnol. Adv.* **2011**, *29*, 889–895. [CrossRef] [PubMed]

22. DiCosimo, R.; McAuliffe, J.; Polouse, A.J.; Bohlmann, G. Industrial use of immobilized enzymes. *Chem. Soc. Rev.* **2013**, *42*, 6437–6474. [CrossRef] [PubMed]

23. Zimmermann, Y.S.; Shahgaldian, P.; Corvini, P.F.X.; Hommes, G. Sorption-assisted surface conjugation: A way to stabilize laccase enzyme. *Appl. Microbiol. Biotechnol.* **2011**, *92*, 169–178. [CrossRef] [PubMed]

24. Nobs, L.; Buchegger, F.; Gurny, R.; Allemann, E. Surface modification of poly(lactic acid) nanoparticles by covalent attachment of thiol groups by means of three methods. *Int. J. Pharm.* **2003**, *29*, 327–337. [CrossRef]

25. Barbosa, O.; Ortiz, C.; Berenguer-Murcia, A. Glutaraldehyde in bio-catalysts design: A useful crosslinker and a versatile tool in enzyme immobilization. *RSC Adv.* **2014**, *4*, 1583–1600. [CrossRef]

26. Zucca, P.; Sanjust, E. Inorganic materials as supports for covalent enzyme immobilization: Methods and mechanisms. *Molecules* **2014**, *19*, 14139–14194. [CrossRef] [PubMed]

27. Baumann, H.; Tillman, A.M. *The Hitch Hilker's Guide to LCA: An Orientation in Life Cycle Assessment Methodology and Application*; Studentlitteratur: Lund, Sweden, 2004; ISBN 9144023642.

28. *Environmental Management—Life Cycle Assessment—Principles and Framework*; ISO 14040:2006; International Organization for Standardization: Geneva, Switzerland, 2006.

29. Feijoo, S.; Gonzalez-García, S.; Moldes-Diz, Y.; Vazquez-Vazquez, C.; Feijoo, G.; Moreira, M.T. Comparative life cycle assessment of different synthesis routes of magnetic nanoparticles. *J. Clean. Prod.* **2017**, *43*, 528–538. [CrossRef]

30. Dones, R.; Bauer, C.; Bolliger, R.; Burger, B.; Faist, E.M.; Frischknecht, R.; Heck, T.; Jungbluth, N.; Röder, A.; Tuchschmid, M. *Life Cycle Inventories of Energy Systems: Results for Current Systems in Switzerland and Other UCTE Countries*; Ecoinvent Report No. 5; Swiss Centre for Life Cycle Inventories: Düberdorf, Switzerland, 2007; Available online: https://www.researchgate.net/profile/Deborah_Andrews2/publication/271710820_The_life_cycle_assessment_of_a_UK_data_centre/links/5620b9ec08aed8dd194054ea.pdf (accessed on 7 June 2016).

31. Althaus, H.J.; Chudacoff, M.; Hischier, R.; Jungbluth, N.; Osses, M.; Primas, A. *Life Cycle Inventories of Chemicals*; Ecoinvent Report No.8 v2.0 EMPA; Swiss Centre for Life Cycle Inventories: Düberdorf, Switzerland, 2007.

32. Doka, G. *Life Cycle Inventories of Waste Treatment Services*; Ecoinvent Report No. 13; Swiss Centre for Life Cycle Inventories: Dübendorf, Switzerland, 2003.

33. Red Eléctrica de España. Avance del Informe del Sistema Eléctrico Español 2014. Available online: http://www.ree.es/sites/default/files/downloadable/avance_informe_sistema_electrico_2014.pdf (accessed on 10 June 2016).

34. Goedkoop, M.J.; Heijungs, R.; Huijbregts, M.; Schryver, A.D.; Struijs, J.; Zelm, R. A Life Cycle Impact Assessment Method Which Comprises Harmonised Category Indicators at the Midpoint and the Endpoint Level. Available online: http://www.leidenuniv.nl/cml/ssp/publications/recipe_characterisation.pdf (accessed on 7 June 2016).

35. PRé Consultants. 2017. Available online: http://www.pre.nl (accessed on 1 July 2017).

36. Sheldon, R.A. Cross-linked enzyme aggregates (CLEAs): Stable and recyclable biocatalysts. *Biochem. Soc. Trans.* **2007**, *35*, 1583–1587. [CrossRef] [PubMed]

37. Ovsejevi, K.; Manta, C.; Batista-Viera, F. Reversible covalent immobilization of enzymes via disulfide bonds. *Methods Mol. Biol.* **2013**, *1051*, 89–116. [CrossRef] [PubMed]

38. Guisán, J.M. *Immobilization of Enzymes and Cells*, 2nd ed.; Humana Press Inc.: Clifton, NJ, USA, 2006; ISBN 978-1-58829-290-2.

39. Flickinger, M.; Drew, S. *Encyclopedia of Bioprocess Technology: Fermentation, Biocatalysis and Bioseparation Fermentation, Biocatalysis and Bioseparation*; Wiley: New York, NY, USA, 1999; ISBN 0-471-13822-3.

40. Majumder, A.B.; Mondal, K.; Singh, T.P.; Gupta, M.N. Designing cross-linked lipase aggregates for optimum performance as biocatalysts. *Biocatal. Biotransform.* **2008**, *26*, 235–242. [CrossRef]

41. Kumar, S.; Jana, A.K.; Maiti, M.; Dhamija, I. Carbodiimide-mediated immobilization of serratiopeptidase on amino-, carboxyl-functionalized magnetic nanoparticles and characterization for target delivery. *J. Nanopart. Res.* **2014**, *16*, 2233. [CrossRef]

42. Arca-Ramos, A.; Ammann, E.M.; Gasser, C.A.; Nastold, P.; Eibes, G.; Feijoo, G.; Lema, J.M.; Moreira, M.T.; Corvini, P.F.X. Assessing the use of nanoimmobilized laccases to remove micropollutants from wastewater. *Environ. Sci. Pollut. Res.* **2016**, *23*, 3217–3228. [CrossRef] [PubMed]

43. Corgié, S.; Kahawong, P.; Duan, X.; Bowser, D.; Edward, J.B.; Walker, L.P.; Giannelis, E.P. Self-assembled complexes of horseradish peroxidase with magnetic nanoparticles showing enhanced peroxidase activity. *Adv. Funct. Mater.* **2012**, *22*, 1940–1951. [CrossRef]

44. Claus, H.; Faber, G.; König, H. Redox-mediated decolorization of synthetic dyes by fungal laccases. *Appl. Microbiol. Biotechnol.* **2002**, *59*, 672–678. [CrossRef] [PubMed]

45. Wang, F.; Hu, Y.; Guo, C.; Huang, W.; Liu, C.-Z. Enhanced phenol degradation in cooking wastewater by immobilized laccase on magnetic mesoporous silica nanoparticles in a magnetically stabilized fluidized bed. *Bioresour. Technol.* **2012**, *110*, 120–124. [CrossRef] [PubMed]

46. Kurniawati, S.; Nicell, J.A. A Comprehensive Kinetic Model of Laccase-Catalyzed Oxidation of Aqueous Phenol. *Biotechnol. Prog.* **2009**, *25*, 763–773. [CrossRef] [PubMed]

47. Bayramoglu, G.; Arica, M.Y. Enzymatic removal of phenol and p-chlorophenol in enzyme reactor: Horseradish peroxidase immobilized on magnetic beads. *Hazard. Mater.* **2008**, *156*, 148–155. [CrossRef] [PubMed]

48. Grieger, K.D.; Laurent, A.; Miseljic, M.; Christensen, F.; Baun, A.; Olsen, S.I. Analysis of current research addressing complementary use of life-cycle assessment and risk assessment for engineered nanomaterials: Have lessons been learned from previous experience with chemicals? *J. Nanopart. Res.* **2012**, *14*, 958. [CrossRef]

applied
sciences

MDPI

Article

Removal of Crotamiton from Reverse Osmosis Concentrate by a TiO$_2$/Zeolite Composite Sheet

Qun Xiang [1], Shuji Fukahori [2], Naoyuki Yamashita [3], Hiroaki Tanaka [3] and Taku Fujiwara [4,*]

[1] The United Graduate School of Agricultural Sciences, Ehime University, 3-5-7 Tarumi, Matsuyama, Ehime 790-8566, Japan; s16dre09@s.kochi-u.ac.jp
[2] Paper Industry Innovation Center of Ehime University, 127 Mendoricho Otsu, Shikokuchuo, Ehime 799-0113, Japan; fukahori.shuji.mj@ehime-u.ac.jp
[3] Research Center for Environmental Quality Management, Kyoto University, 1-2 Yumihama, Otsu, Shiga 520-0811, Japan; yamashita@biwa.eqc.kyoto-u.ac.jp (N.Y.); htanaka@biwa.eqc.kyoto-u.ac.jp (H.T.)
[4] Research and Education Faculty, National Sciences Cluster, Agriculture Unit, Kochi University, 200 Monobe Otsu, Nankoku, Kochi 783-8502, Japan
*   Correspondence: fujiwarat@kochi-u.ac.jp; Tel.: +81-88-864-5163

Received: 29 June 2017; Accepted: 24 July 2017; Published: 31 July 2017

Abstract: Reverse osmosis (RO) concentrate from wastewater reuse facilities contains concentrated emerging pollutants, such as pharmaceuticals. In this research, a paper-like composite sheet consisting of titanium dioxide (TiO$_2$) and zeolite was synthesized, and removal of the antipruritic agent crotamiton from RO concentrate was studied using the TiO$_2$/zeolite composite sheet. The RO concentrate was obtained from a pilot-scale municipal secondary effluent reclamation plant. Effective immobilization of the two powders in the sheet made it easy to handle and to separate the photocatalyst and adsorbent from purified water. The TiO$_2$/zeolite composite sheet showed excellent performance for crotamiton adsorption without obvious inhibition by other components in the RO concentrate. With ultraviolet irradiation, crotamiton was simultaneously removed through adsorption and photocatalysis. The photocatalytic decomposition of crotamiton in the RO concentrate was significantly inhibited by the water matrix at high initial crotamiton concentrations, whereas rapid decomposition was achieved at low initial crotamiton concentrations. The major degradation intermediates were also adsorbed by the composite sheet. This result provides a promising method of mitigating secondary pollution caused by the harmful intermediates produced during advanced oxidation processes. The cyclic use of the HSZ-385/P25 composite sheet indicated the feasibility of continuously removing crotamiton from RO concentrate.

Keywords: paper-like composite sheet; zeolite; photocatalysis; reverse osmosis concentrate; pharmaceutical; inhibitory effect; intermediate

1. Introduction

Reverse osmosis (RO) is a well-established technology for water desalination, the production of potable water, and more recently, tertiary wastewater treatment [1,2]. With increasing global water demand, it is predicted that the global market value of RO system components will reach 8.1 billion USD by 2018 [3]. Along with the purification of wastewater, the RO process produces a concentrate containing high levels of rejected pollutants (about 15–20% of the influent volume) [4]. Some of the emerging pollutants, such as pharmaceuticals and personal care products, are very persistent in sewage effluent, resulting in raised awareness of the environmental risk of RO concentrates [1,5,6]. Genotoxicity evaluation using the *SOS/umu* test has provided direct evidence that RO concentrates have much higher toxicological risk than RO influents [7]. Therefore, suitable technology needs to be developed for treating RO concentrates before discharging them into receiving water or recycling for other purposes. This requirement is especially important for large-scale RO treatment systems [8].

In a number of recent studies, TiO_2 photocatalysis has been used to treat pharmaceuticals in wastewater [9]. The nonselective oxidation ability of hydroxyl radicals enables effective degradation of various organic pollutants. However, the photocatalysis of target compounds can be inhibited by coexisting materials, such as inorganic ions and organic matter, in the wastewater [10–12]. In addition, toxic intermediates may be produced during photocatalysis, and the effects of pharmaceutical degradation products in the environment are of concern. Furthermore, when TiO_2 or nano-TiO_2 powder in water is exposed to ultraviolet (UV) radiation, radicals that are harmful to aquatic organisms are produced [13]. Therefore, the effective recovery of catalyst powder after wastewater treatment should be taken into consideration.

Wastewater treatment frequently involves adsorption processes, and various types of adsorbents have been developed to remove different pollutants [14–19]. The high-silica Y-type zeolite HSZ-385, which is a hydrophobic zeolite, has been used to remove sulfonamide antibiotics from wastewater and selectively removes sulfonamides even in the presence of high concentrations of coexisting materials [20]. However, after adsorption, the contaminants are permanently transferred to the sorbent and not destroyed, which can lead to problems with saturation of the adsorbent.

Attempts have been made to synthesize TiO_2-adsorbent composites that perform both photocatalysis and adsorption to remove pharmaceuticals from wastewater [21–23]. This synergistic effect has been confirmed for TiO_2 and zeolite in a TiO_2/zeolite composite powder that was used to remove sulfonamide antibiotics [23]. Wu et al. condensed nano-TiO_2 on the surfaces of carbon spheres through hydrothermal treatment to generate core–shell structures, and found that visible light absorption was enhanced compared with pure TiO_2 because of the interface formed between the two materials [21]. The activated carbon fiber felt (ACFF) in the TiO_2/ACFF porous composites significantly enhances the photocatalytic property of toluene by hindering the recombination of electron-hole pairs, reducing the TiO_2 band gap energy, and accelerating toluene adsorption [24]. Using a papermaking technique, Fukahori et al., prepared a paper-like composite sheet from TiO_2 and zeolite powder [25]. Under UV irradiation, bisphenol A was effectively degraded through the synergistic effect of the TiO_2 photocatalyst and zeolite adsorbent in these sheets [25]. In addition, the degradation intermediates of bisphenol A, which may be harmful to the environment, were temporarily captured by zeolite in the composite sheet and eventually decomposed through photocatalysis [26]. However, these studies were conducted using ultrapure water as the solvent, and the inhibitory effects of other components of the wastewater matrix have not been investigated.

In this study, we synthesized a TiO_2/zeolite composite sheet to remove of crotamiton from RO concentrate, and to recover the catalyst and adsorbent after water treatment. Crotamiton is a scabicide and antipruritic agent that has frequently been detected in sewage effluent in Japan because of its stable nature and wide consumption [27–29]. The effect of coexisting matter from the wastewater matrix on inhibiting crotamiton degradation was evaluated. In addition, the behavior of crotamiton degradation intermediates during photocatalysis was investigated.

2. Materials and Methods

2.1. Materials

HSZ-385 (surface area 600 m^2/g, mean particle size 4 μm, SiO_2/Al_2O_3 ratio 100:1) was purchased from Tosoh Ltd. (Tokyo, Japan). TiO_2 powder (P-25, 50 m^2/g, anatase) was purchased from Degussa (Dusseldorf, Germany) and F-type zeolite powder (F9, SiO_2/Al_2O_3 ratio 2.1:1) was purchased from Wako Pure Chemical Industries, Ltd. (Tokyo, Japan). Crotamiton (purity > 97%) and isotope-labelled surrogate crotamiton-d7 (purity 94.5%) were purchased from Sigma-Aldrich (St Louis, MO, USA) and Hayashi Pure Chemical (Osaka, Japan), respectively. Crotamiton-d7 was dissolved in methanol (purity > 99.8%; Kanto Chemical Co., Inc., Tokyo, Japan) to prepare an internal standard solution, which was stored at −20 °C. All other chemicals used were of reagent grade.

Composite sheets consisting of TiO_2 and zeolite (HSZ-385 or F-9) were prepared using a papermaking technique. TiO_2, zeolite (3.125 g each), and polyethylene terephthalate fiber (6.25 g) were suspended in water (1 L); a cationic flocculant [poly-(amideamine) epichlorohydrin, 0.05% of total solid] and an anionic flocculant (anionic polyacrylamide, 0.084% of total solid) were sequentially added and the final suspension was stirred. Hand sheets with a grammage of 200 g/m^2 were prepared according to JIS P8222 [30]. The sheets were dried at 120 °C. The mass ratio of TiO_2 to zeolite in the composite sheet was 1:1. The TiO_2/zeolite composite sheet used in this study contained 4 mg/cm^2 of TiO_2 and zeolite. Characterization of the TiO_2/zeolite composite sheet was performed by scanning electron microscopy-energy dispersive X-ray spectroscopy (SEM-EDS: ProX; Phenom World) as shown in Figure 1. The SEM and EDS images revealed the uniform distribution of TiO_2 and zeolite powder in the composite sheet.

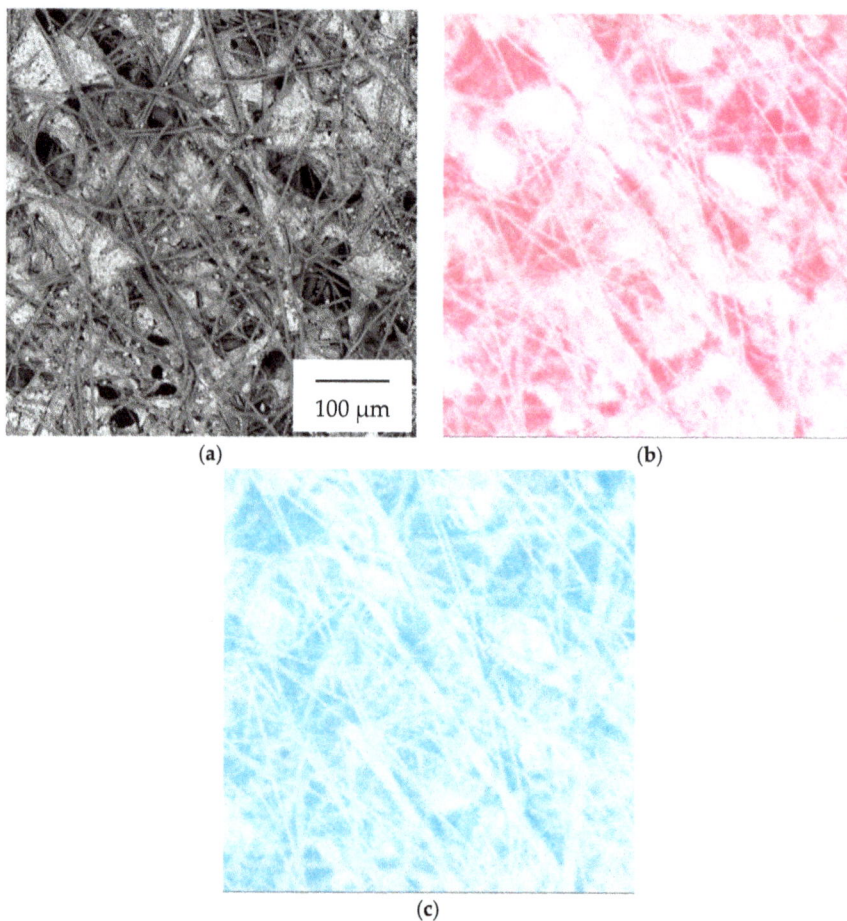

Figure 1. Scanning electron microscopy-energy dispersive X-ray spectroscopy (SEM-EDS) mapping images of TiO_2/zeolite composite sheet: SEM images (**a**); EDS mapping of Si (**b**) and Ti (**c**).

RO concentrate was collected from a nanofiltration/RO pilot-scale plant for municipal secondary effluent reclamation on 7 March 2017 in Japan, and stored at 4 °C. The RO concentrate was analyzed and the results are as shown in Table 1. Details of the quantitative analyses are given in the Supplementary materials. To clarify the mechanism for removing of crotamiton with the composite sheet, crotamiton solutions were prepared using either RO concentrate or ultrapure water (Millipore, Tokyo, Japan).

Table 1. Water quality analysis of the reverse osmosis concentrate.

Parameter	Value	Ion	mg/L
pH	7.8	Na^+	223
Conductivity (mS/m)	170	NH_4^+	25.7
		K^+	26.4
TOC (mgC/L) [a]	10.1	Mg^{2+}	22.2
		Ca^{2+}	45.5
COD_{cr} (mg/L) [b]	22	Cl^-	316
UV absorbance (λ = 365 nm) (1/cm) [c]	0.049	NO_2^-	15.6
		NO_3^-	46.8
Alkalinity (mgCaCO$_3$/L)	158	SO_4^{2-}	87.5

[a] TOC, total organic carbon; [b] COD_{cr}, chemical oxygen demand; [c] UV, ultraviolet.

2.2. Quantitative Analyses

To determine the concentration of crotamiton in the RO concentrate solution, solid phase extraction (SPE) was carried out. The cartridges (Oasis HLB, 60 mg, 3 mL, Waters, Milford, MA, USA) were conditioned with 2 mL of methanol, followed by 2 mL of ultrapure water. Aqueous samples spiked with the internal standard solution were then loaded onto the cartridges. Next, the cartridges were washed with 2 mL of ultrapure water and dried with a GL-SPE vacuum manifold system (GL Science, Tokyo, Japan) for 30 min. The analyte was eluted first with 1 mL of 10% methanol and then with 4 mL of methanol. The average recovery rate of crotamiton was 97 $\pm$ 1.7% (mean $\pm$ standard deviation, n = 3).

The concentrations of crotamiton were determined with the internal standard addition method using liquid chromatography tandem mass spectrometry (LC/MS/MS, Acquity UPLC-Xevo TQ; Waters) after SPE. The intermediates were identified from the mass spectral patterns obtained by LC/MS/MS.

2.3. Methods

Adsorption experiments were carried out by submerging the TiO$_2$/zeolite composite sheets (2 $\times$ 5.5 cm^2) at a depth of 4 cm in 50 mL of the crotamiton solution (10 mg/L or 120 µg/L) at pH 7.0 $\pm$ 0.1 without ultraviolet irradiation (Figure S1). The mixture was stirred at a moderate speed at 25 °C. After a set treatment time, the treated solutions were passed through a DISMIC-13HP 0.2-µm membrane filter (Toyo Roshi Kaisha, Tokyo, Japan) to determine the crotamiton concentrations in the aqueous phase (C_t).

For the adsorption and photocatalytic degradation experiments, UV irradiation was applied perpendicular to the sheet surface (Figure S1) with a FL287-BL365 UV lamp (Raytronics, Tokyo, Japan), which had a maximum output wavelength of 365 nm. The UV intensity at the center of the reactor was controlled at 1000 µW/cm^2 using a UV-340C light meter (Custom, Tokyo, Japan). The other experimental conditions were the same as for the adsorption experiment. After a set irradiation time, the treated solutions were passed through 0.2-µm membrane filters, and the crotamiton concentrations (C_t) were then determined.

To measure the mass of crotamiton in the sheet, the composite sheet was soaked in methanol (purity > 99.8%). After ultrasonication for 60 min (38 kHz, 120 W; US-3KS; SND Co., Ltd., Nagano, Japan), the treated solutions were filtered through 0.2-µm membrane filters and analyzed by

LC/MS/MS (C_t'). The recovery rate for desorption was $107 \pm 1\%$. The mass of crotamiton in the treated solution ($M_{in\ water}$) was calculated using Equation (1), in which V is the solution volume.

$$M_{in\ water} = C_t \times V \qquad (1)$$

Similarly, the mass of crotamiton in the composite sheet ($M_{in\ sheet}$) was calculated using Equation (2).

$$M_{in\ sheet} = C_t' \times V \qquad (2)$$

The total mass of crotamiton remaining in the system ($M_{in\ system}$) was calculated using Equation (3).

$$M_{in\ sheet} = C_t' \times V \qquad (3)$$

3. Results and Discussion

3.1. Adsorption of Crotamiton by the HSZ-385/P25 Composite Sheet

The HSZ-385/P25 composite sheet was applied to the adsorption of crotamiton in the RO concentrate. In preliminarily experiments, we confirmed that crotamiton was rapidly adsorbed by HSZ-385 zeolite powder (Figure S2). We also confirmed that crotamiton was not adsorbed by P25 [31]. The crotamiton concentrations were plotted against time (Figure 2). Similar performances of the sheet in RO concentrate and ultrapure water revealed that other components in the RO concentrate (Table 1) did not obviously affect the adsorption of crotamiton by the composite sheet within the 24-hr treatment period.

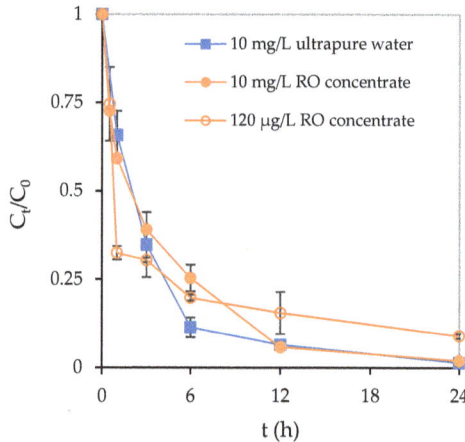

Figure 2. Adsorption of crotamiton using the HSZ-385/P25 composite sheet. Results are means ± standard deviations (*n* = 2).

It has been reported that inorganic ions and organic materials affect the adsorption of pharmaceuticals [32]. Nevertheless, the adsorption of sulfonamide antibiotics from livestock urine and bisphenol A from landfill leachate by HSZ-385 was not affected by coexisting ions [14,20]. Meanwhile, even if the organic carbon content of porcine urine was two orders of magnitude higher than those of the sulfonamides, the sulfonamides were also effectively removed [20]. For bisphenol A, the removal efficiency decreased slightly when more than 50 mg/L humic acid was added [14]. Umar et al., reported that humic-like and fulvic acid-like matter in the RO concentrate were the major contributors to the color of the concentrate [33]. The RO concentrate used in this research appeared to be light brown. In the present study, when the HSZ-385/P25 composite sheet was used to adsorb the raw RO concentrate without crotamiton spiking, only approximately 10% of the total organic carbon (TOC) was removed. Therefore, in the RO concentrate, crotamiton would be removed by adsorption on the composite sheet prior to the raw organic matter. In addition, the removal results for different initial crotamiton concentrations in the RO concentrate were similar, which implies that crotamiton can be adsorbed by the HSZ-385/P25 composite sheet over a wide range of initial concentrations.

We previously investigated the mechanism involved in the adsorption of sulfonamides to HSZ-385. We found that HSZ-385 adsorbed neutral sulfonamides more effectively than non-neutral sulfonamides, and that hydrophobic interactions played important roles in the adsorption process [34]. Crotamiton is hydrophobic (logK_{OW} = 2.73) and has no ionizable functional groups. Therefore, hydrophobic interactions may play an important role in the adsorption of crotamiton by the HSZ-385/P25 composite sheet.

3.2. Photocatalytic Degradation of Crotamiton by the F9/P25 Composite Sheet

To clarify the photocatalysis of crotamiton by the F9/P25 composite sheet, the composite sheet was synthesized to be similar to the HSZ-385 composite sheet. Both F9 zeolite and the F9/P25 composite sheet did not remove crotamiton by adsorption (Figure S2). The F9 zeolite is a hydrophilic zeolite, whereas the Y-type zeolite HSZ-385 is a hydrophobic zeolite. This is further evidence that crotamiton is removed by HSZ-385 mainly through hydrophobic interactions.

The photocatalytic degradation of crotamiton over time by the F9/P25 composite sheet is shown in Figure 3. Direct photolysis of crotamiton was not observed [31]. The removal efficiency of crotamiton from the RO concentrate was much lower than that from the ultrapure water. After 24 hr of UV irradiation, the majority of the crotamiton in the ultrapure water was degraded. In contrast, ca. 50% of the crotamiton was degraded in the RO concentrate. Linear relationships were found between ln (C_t/C_0) and UV irradiation time (t) (Figure 3). Therefore, the first-order kinetic model shown in Equation (4) was used to evaluate the photocatalysis of crotamiton. In that equation, k_1 is the pseudo-first-order rate constant. The k_1 value for crotamiton removal from the RO concentrate by the F9/P25 composite sheet was 0.048 hr^{-1}, and was only half of that in the ultrapure water (0.092 hr^{-1}). Obviously, the lower rate constant reflects the effect of other components in the RO concentrate on the photocatalytic degradation of crotamiton.

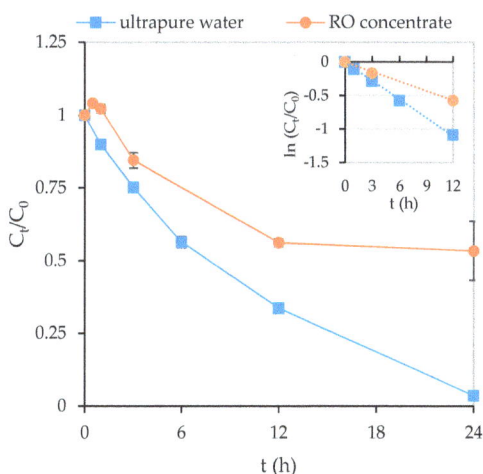

Figure 3. Photocatalytic degradation of crotamiton using the F9/P25 composite sheet with ultraviolet irradiation (C_0 = 10 mg/L). The inset shows the fitting results of the first-order kinetic model. Results are means ± standard deviations (n = 2).

$$\ln (C_t/C_0) = -k_1 t \tag{4}$$

UV absorbance is an important parameter affecting photocatalysis [10,35]. The maximum output wavelength of the UV lamp in our study was 365 nm, and the RO concentrate had an absorbance of 0.049 cm^{-1} at 365 nm (Table 1). Based on the Beer-Lambert law and the distance from the reactor surface to the sheet, the light transmittance was 82.1%. Therefore, after passing through the RO concentrate in the batch reactor, the light intensity at the sheet surface decreased by 17.9%, which is called the screening effect [36]. This effect contributed to the decrease in the photocatalytic degradation efficiency. Furthermore, the RO concentrate was light brown, and the color of the composite sheet surface changed from white to light brown after treatment of the RO concentrate. This color change could be caused by adsorption of coexisting materials on the sheet surface, and this could negatively affect the performance of TiO$_2$ photocatalysis through occupation of the active sites on the surface of TiO$_2$ [37].

The TiO$_2$ photocatalysis could be inhibited via the scavenger effect by coexisting ions [10,38–40]. The ions Cl$^-$ and HCO$_3^-$ have been found to inhibit photocatalysis through the hydroxyl radical and valence band hole scavenging [10,40]. Rioja et al., reported a marked deactivation effect caused by added salts for two tested acidic drugs [41]. Furthermore, Tokumura et al., suggested that coexisting matter could mitigate the generation of hydroxyl radicals through direct reactions with holes in the valence band and electrons in the conduction band of the photocatalyst [37]. As reported by Song et al., Cl$^-$ can cause agglomeration of TiO$_2$ particles in a slurry by suppressing the stabilizing effect of electrostatic repulsion, reducing the effective contact surface between the photocatalyst and the pollutants [42]. In this research, the TiO$_2$/zeolite composite sheet was used instead of TiO$_2$ powder. Therefore, even if the RO concentrate contained 316 mg/L Cl$^-$ (Table 1), agglomeration of TiO$_2$ and its associated issues should be eliminated. However, the mechanism for this should be investigated in future research.

Organic matter in secondary effluent also competes with target pharmaceuticals during photocatalysis [43]. When the F9/P25 composite sheet was used to degrade raw RO concentrate without crotamiton spiking, approximately 15% of the initial TOC (10.1 mgC/L) was degraded after 24 h of UV irradiation. The TOC concentration in the RO concentrate was 10.1 mgC/L, and the TOC concentration for the 10 mg/L crotamiton solution in ultrapure water was 7.67 mgC/L theoretically. The coexisting organic matter may compete with crotamiton for consumption of the oxidizing agent during photocatalysis by the F9/P25 composite sheet.

Mineralization during photocatalytic degradation was evaluated by plotting TOC/TOC$_0$ against time at an initial crotamiton concentration of 10 mg/L (Figure 4). The TOC concentration provided by the residual crotamiton was also determined by performing stoichiometric calculations. During the photocatalytic degradation of crotamiton by the F9/P25 composite sheet in both ultrapure water and RO concentrate, the solution TOC did not obviously decrease with the degradation of crotamiton. This result implied that crotamiton was degraded step by step and that the intermediate compounds accumulated at the same time. Kuo et al., investigated the photocatalytic mineralization of methamphetamine in a UVA/TiO$_2$ system and found that TOC disappeared more slowly than methamphetamine because the methamphetamine intermediates took some time to be mineralized [44]. A more detailed discussion on the degradation intermediates is given in Section 3.4.

3.3. Adsorption and Photocatalytic Degradation of Crotamiton by the HSZ-385/P25 Composite Sheet

In our previous research, we found that P25 was effective for photocatalytic degradation of crotamiton [31]. The HSZ-385/P25 composite sheet prepared in the present study combines adsorption and photocatalysis processes, which makes it possible to regenerate the adsorbent during treatment.

To clarify the adsorption and degradation performance of crotamiton by the HSZ-385/P25 composite sheet, the mass of crotamiton in the composite sheet ($M_{in\ sheet}$) was determined together with the mass of crotamiton in the aqueous phase ($M_{in\ water}$). The mass of crotamiton in the system

($M_{in\ system}$) was calculated as the sum of the residual mass of crotamiton in both the aqueous phase and in the sheet, which was the mass of undecomposed crotamiton remaining in the system.

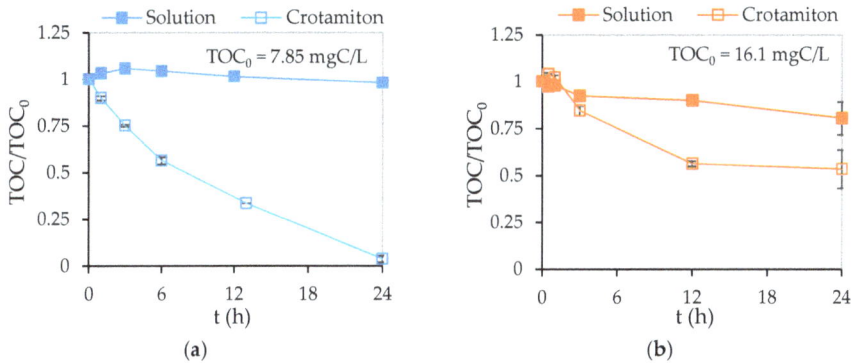

Figure 4. Removal of TOC in the solutions and the TOC derived from the residual crotamiton during the treatment by the F9/P25 composite sheet with ultraviolet irradiation in ultrapure water (**a**) and RO concentrate (**b**). Results are means ± standard deviations (*n* = 2).

The masses of crotamiton in different phases were plotted against time (Figure 5) for solutions with initial crotamiton concentrations of 10 mg/L in ultrapure water (Figure 5a), 10 mg/L in RO concentrate (Figure 5b), and 120 µg/L in RO concentrate (Figure 5c). Although similar trends were observed for $M_{in\ water}$ in ultrapure water and RO concentrate (Figure 5a,b) with an initial crotamiton concentrations of 10 mg/L, the trends for $M_{in\ sheet}$ were very different. The highest value of $M_{in\ sheet}$ (ca. 0.12 mg) was observed after 3 hr treatment of crotamiton in the ultrapure water, and this then decreased with time (Figure 5a); at most, only 23% of the initial crotamiton was accumulated in the sheet, and all the crotamiton was eventually degraded by photocatalysis.

In removing of crotamiton from the RO concentrate, much more crotamiton (0.34 mg) was accumulated in the composite sheet after 6 h treatment (Figure 5b). After that, the mass of crotamiton in the sheet gradually decreased, and finally 0.25 mg remained in the sheet at 24 hr. The higher removal rate obtained with adsorption compared with photocatalysis led to the accumulation and long retention time of crotamiton in the composite sheet. In the treatment of both RO concentrate and ultrapure water, crotamiton could be effectively removed from the aqueous phase, thus purifying the water. The adsorption process was not greatly affected by the water matrix, but inhibition of photocatalysis resulted in low crotamiton degradation in the RO concentrate when using the HSZ-385/P25 composite sheet.

Removing of crotamiton from the RO concentrate with an initial crotamiton concentration of 120 µg/L was investigated (Figure 5c). The $M_{in\ sheet}$ values were maintained at a low level throughout the treatment, and the maximum accumulation of crotamiton in the sheet was only 6.7% of the initial crotamiton mass in the aqueous phase. A rapid decrease was observed in $M_{in\ system}$, showing that rapid decomposition of crotamiton occurred with the low initial crotamiton concentration. Inhibition of the degradation process with high initial crotamiton concentrations may be attributed to competition from intermediates produced by crotamiton degradation [37]. Jang et al., found that the target material (trichloroethylene) saturated the composite catalyst surface and reduced photon efficiency, leading to photocatalyst deactivation [45]. Kuo et al., showed that the degradation rates of codeine and methamphetamine increased with increasing initial concentration (100–250 µg/L) [44,46]. With the initial concentration at microgram per liter levels, the degradation rate may not be limited by the availability of catalytic sites but by contaminant concentration.

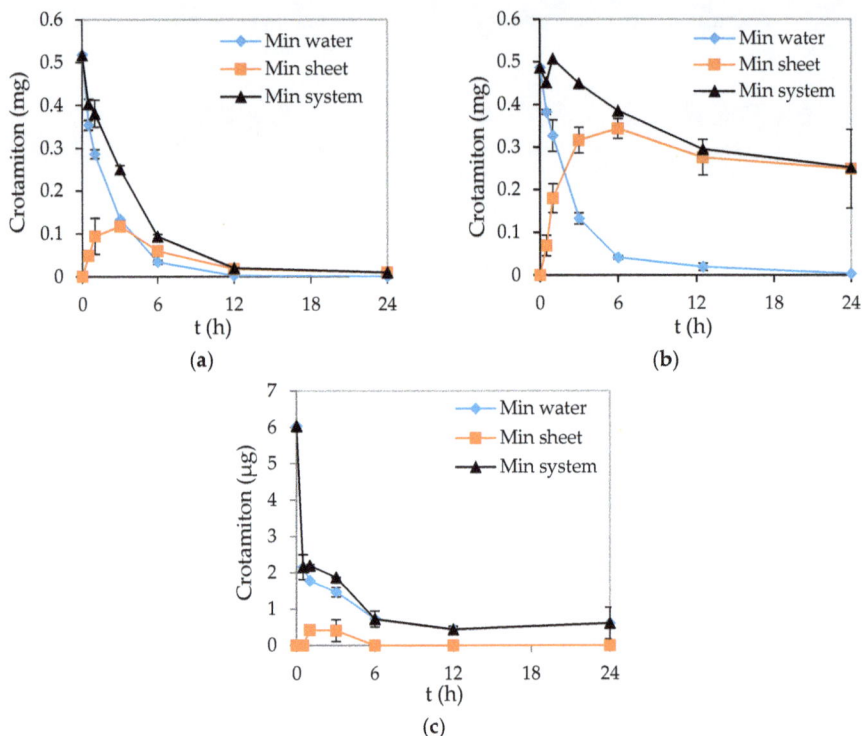

Figure 5. Removal of crotamiton using the HSZ-385/P25 composite sheet with ultraviolet irradiation for solutions of (**a**) 10 mg/L crotamiton in ultrapure water; (**b**) 10 mg/L crotamiton in RO concentrate; and (**c**) 120 µg/L crotamiton in RO concentrate. Results are means ± standard deviations (*n* = 2).

The TOC/TOC$_0$ ratios plotted against time in the HSZ-385/P25 composite sheet experiment are shown in Figure 6. With removal of crotamiton in the ultrapure water by the HSZ-385/P25 composite sheet, TOC was gradually removed and the removal efficiency reached up to 84% after 24 hr (Figure 6a), whereas the TOC removal efficiency was stable at ca. 51% after 6 hr of crotamiton treatment in the RO concentrate by the HSZ-385/P25 composite sheet (Figure 6b). The photocatalytic degradation of crotamiton was significantly inhibited by the other components in the RO concentrate. Furthermore, the low TOC removal by individual adsorption or degradation for the organic matter in the original RO concentrate is another important reason. After 24 hr treatment by the HSZ-385/P25 composite sheet, the majority of the crotamiton was removed, which was similar to that in the experiment using the F9/P25 composite sheet. A rather lower TOC/TOC$_0$ ratio was observed in the treatment using the HSZ-385/P25 composite sheet compared with the F9/P25 composite sheet. It has been assumed that accumulation of degradation intermediates in the experiment using the F9/P25 composite sheet resulted in the high residual TOC concentration in the aqueous phase (Figure 4a). The much lower TOC/TOC$_0$ in the treatment using the HSZ-385/P25 composite sheet indicated other TOC derived from degradation intermediates in the aqueous phase has been removed because of the function of the HSZ-385 in the composite sheet (Figure 6a). That is to say, not only crotamiton in the solution but also the degradation intermediates of crotamiton were removed by the HSZ-385 in the composite sheet.

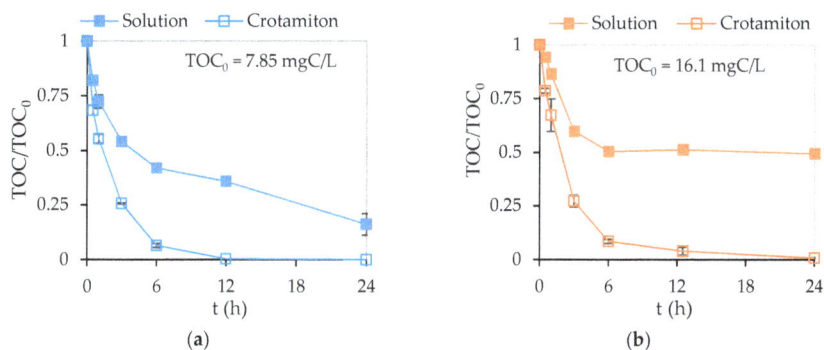

Figure 6. Removal of TOC in the solutions and the TOC derived from the residual crotamiton during the treatment by the HSZ-385/P25 composite sheet with ultraviolet irradiation in ultrapure water (**a**) and RO concentrate (**b**). Results are means ± standard deviations (n = 2).

3.4. Behavior of the Degradation Intermediates during Photocatalysis

The degradation intermediates were characterized following the methods described in our previous study [31]. We proposed that P25-catalyzed photodegradation of crotamiton could initially occur via hydroxylation of the aromatic ring, the double bond of the propenyl group, or the ethyl group. These reactions formed intermediates that we labeled as P189, P217, and P219. The peak areas of the intermediates were measured in the selected ion recording mode of LC/MS/MS.

With the degradation of crotamiton, the intermediates gradually accumulated and reached their highest levels in the treatment using the F9/P25 composite sheet (Figure 7). Intermediate P219, which was produced by hydroxylation of the aromatic ring of crotamiton, was the most noticeable degradation intermediate. The peak area of P219 was higher than those for P189 and P217. After 24 hr of treatment, most of the P189 and P217 had disappeared. In contrast, the peak area of P219 was still high after 24 hr of treatment with the F9/P25 composite sheet. This result corresponded well to the high TOC/TOC_0 level revealed in Figure 4a, validating the assumption of the accumulation of degradation intermediates during the treatment by the F9/P25 composite sheet.

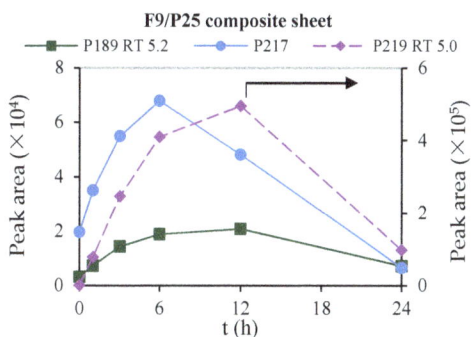

Figure 7. Changes in the peak areas of the major intermediates over time in ultrapure water when crotamiton was photocatalytically degraded using the F9/P25 composite sheet (C_0 = 10 mg/L). The squares are P189 (retention time 5.2 min) and the circles are P217. The diamonds with a dashed line are P219 (retention time 5.0 min, secondary y-axis).

Treatment with the HSZ-385/P25 composite sheet (Figure 8) was compared with that using the F9/P25 composite sheet. The peak areas for the three intermediates obtained with the HSZ-385/P25 composite sheet were clearly lower than those obtained with the F9/P25 composite sheet throughout the treatment, especially for intermediate P219. After 24 hr of treatment with the HSZ-385/P25 composite sheet, the majority of all three intermediates had disappeared. The lower peak areas of degradation intermediates as well as the lower TOC/TOC$_0$ ratios (Figure 6a) confirmed that the HSZ-385/P25 composite sheet captured the degradation intermediates.

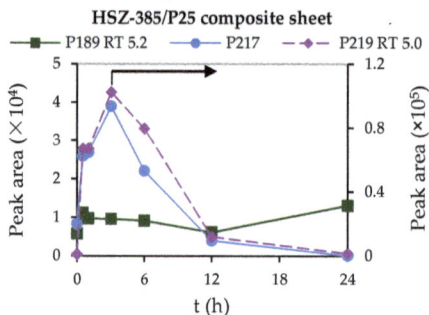

Figure 8. Changes in the peak areas of the major intermediates over time in ultrapure water when crotamiton was photocatalytically degraded using the HSZ-385/P25 composite sheet ($C_0 = 10$ mg/L). The squares are P189 (retention time 5.2 min) and the circles are P217. The diamonds with a dashed line are P219 (retention time 5.0 min, secondary y-axis).

Moreover, changes in the peak areas for the major intermediates in the HSZ-385/P25 composite sheet were evaluated through desorption treatment for the composite sheet after the adsorption and photocatalysis experiment. The methanol solution with the composite sheet after ultrasonic treatment contained crotamiton, as well as large amounts of degradation intermediates, retained in the composite sheet (Figure 9). The peak area for P219 was higher in the composite sheet during treatment than that in the aqueous phase when using the HSZ-385/P25 composite sheet shown in Figure 8. Even if the efficiency of desorption of the intermediates from the sheet could not be confirmed without the standard of every detected intermediate, the high peak areas for the intermediates provided direct evidence of the adsorption of intermediates on the composite sheet.

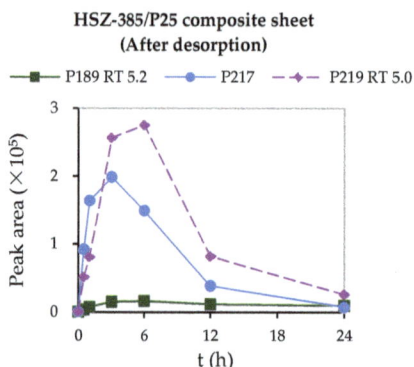

Figure 9. Changes in the areas of the peaks for the major intermediates over time in the HSZ-385/P25 composite sheet after desorption when crotamiton was photocatalytically degraded using the HSZ-385/P25 composite sheet ($C_0 = 10$ mg/L). The squares are P189 (retention time 5.2 min), the circles are P217, and the diamonds with a dashed line are P219 (retention time 5.0 min).

In conclusion, the HSZ-385/P25 composite sheet is effective not only for removing crotamiton, but also for capturing degradation intermediates produced by photocatalysis. This method mitigates the negative impact of harmful degradation intermediates produced by advanced oxidation processes.

3.5. Cyclic Use of the HSZ-385/P25 Composite Sheet

To effectively apply the HSZ-385/P25 composite sheet to remove crotamiton from RO concentrate in practical treatment processes, it is essential that the composite sheet can remove pollutants even after several cycles of reuse. The efficiency of removing crotamiton from RO concentrate achieved by the HSZ-385/P25 composite sheet after 24 hr ultraviolet irradiation in three circles of reuse were all over 95% (Figure 10). The crotamiton amount in the composite sheet after three cycles of reuse was 0.40 mg, which was ca. 27% of the total amount of three cycles of treated crotamiton (C_0 = 10 mg/L, V = 50 mL), demonstrating continuous crotamiton photocatalytic degradation. The TOC removal efficiency slightly decreased with an increase in the cycles of reuse. It can be concluded that the HSZ-385/P25 composite sheet is feasible for the cyclic removal of crotamiton in RO concentrate.

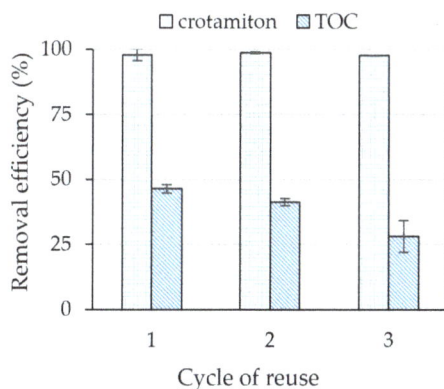

Figure 10. The removal efficiency of crotamiton and TOC in RO concentrate using the HSZ-385/P25 composite sheet after cyclic use. Results are means ± standard deviations (n = 2).

Some other composite materials used as adsorbents and photocatalysts for treating organic pollutants are summarized in Table 2. No previous publications applied composite materials for the treatment of pollutants in RO concentrate. A few cycles of pollutant removal were carried out in studies using TiO$_2$–coconut shell powder composite [47], polyacrylic acid-grafted-carboxylic graphene/titanium nanotube composite [48], multi-walled carbon nanotubes/Fe$_3$O$_4$ composites [49], multi-walled carbon nanotube/TiO$_2$ composites [50] and nitrogen-doped-TiO$_2$/activated carbon composite [51] similar to that in this study. The stability of the photocatalyst and reusability of these materials were confirmed, thus making them promising cost-effective water purification materials.

Except for reusability, some of the composites were designed to promote degradation capability through improving visible light utilization, photon yield, and so on. The presence of MoS$_2$ in the TiO$_2$-MoS$_2$-reduced graphene oxide composite worked as a co-catalyst to reduce electron-hole pairs, and improved the photocatalytic performance of TiO$_2$ for BPA removal [52]. The nitrogen-doped-TiO$_2$/activated carbon composite was synthesized to regenerate spent powdered activated carbon using solar photocatalysis for cost-effective application in wastewater treatment [51]. The graphene/TiO$_2$/ZSM-5 composite material showed higher stability, stronger absorption of visible light, and lower band gap value [53].

Table 2. Composite materials synthesized for removing organic pollutants.

Composite	Target Pollutant	Water Matrix	Reference
TiO_2–coconut shell powder composite	Carbamazepine, clofibric acid, and triclosan	Ultrapure water	[47]
Polyacrylic acid-grafted-carboxylic graphene/titanium nanotube composite	Enrofloxacin	Distilled water and simulated poultry farm effluent	[48]
Multi-walled carbon nanotubes/Fe_3O_4 composites	Bisphenol A	Doubly-distilled deionized water	[49]
Multi-walled carbon nanotubes/TiO_2 nanocomposite	Tetracycline	Pharmaceutical wastewater	[50]
Nitrogen-doped-TiO_2/activated carbon composite	Bisphenol-A, sulfamethazine, and clofibric acid	Ultrapure water	[51]
TiO_2-MoS_2-reduced graphene oxide composite	Bisphenol A	Not mentioned	[52]
Graphene/TiO_2/ZSM-5 composites	Oxytetracycline	Deionized water	[53]

4. Conclusions

TiO_2/zeolite composite sheets were synthesized and used to remove crotamiton from RO concentrate. Crotamiton is effectively adsorbed by the HSZ-385/P25 composite sheet without obvious inhibition by other components of the RO concentrate. The photocatalytic decomposition of crotamiton in the RO concentrate is significantly inhibited by the water matrix at high initial concentrations of crotamiton, whereas rapid decomposition occurs at low initial concentrations. When the HSZ-385/P25 composite sheet is used with UV irradiation for the removal of crotamiton from RO concentrate, crotamiton is removed by adsorption and photocatalysis. The inhibition of photocatalytic degradation by other components resulted in crotamiton remaining in the composite sheet. The degradation intermediates are captured by the HSZ-385/P25 composite sheet, and this capture provides a way to mitigate the potential negative impact of intermediates from advanced oxidation processes. In addition, the HSZ-385/P25 can continually remove crotamiton from RO concentrate with repeated uses.

Supplementary Materials: The following are available online at http://www.mdpi.com/2076-3417/7/8/778/s1, Figure S1: The experimental set-up of the adsorption and photocatalytic degradation of crotamiton using the TiO_2/zeolite composite sheet. The symbol Φ refers to the diameter of the vial; Figure S2: Removal of crotamiton from ultrapure water by adsorption using F9 powder, HSZ-385 powder and the F9/P25 composite sheet ($C_0 = 10$ mg/L, $V = 50$ mL). The dosage for the powder adsorbent was 0.1 g/L. The F9/P25 composite sheet was 2×5.5 cm^2 and submerged at a depth of 4 cm. The composite sheet contained 4 mg F9/cm^2.

Acknowledgments: This work was supported by the Japan Society for the Promotion of Science Grants-in-Aid for Scientific Research Grant Number 16H02372. We appreciate the assistance of Suntae Lee with sampling of the RO concentrate. We thank Gabrielle David, from Edanz Group (www.edanzediting.com/ac) and Dennis Murphy of the United Graduate School of Agricultural Sciences, Ehime University, for editing a draft of this manuscript.

Author Contributions: Qun Xiang, Taku Fujiwara and Shuji Fukahori conceived and designed the experiments; Qun Xiang performed the experiments, analyzed the data and wrote the paper; Taku Fujiwara supervised the research; Shuji Fukahori contributed to the preparation and characterization of the composite sheet; Naoyuki Yamashita and Hiroaki Tanaka contributed to the sampling of the RO concentrate and provided advice on the experiments. All authors have read and approved the final manuscript.

Conflicts of Interest: The authors declare no conflict of interest.

References

1. Joo, S.H.; Tansel, B. Novel technologies for reverse osmosis concentrate treatment: A review. *J. Environ. Manag.* **2015**, *150*, 322–335. [CrossRef] [PubMed]
2. Umar, M.; Roddick, F.; Fan, L. Recent advancements in the treatment of municipal wastewater reverse osmosis concentrate—An overview. *Crit. Rev. Environ. Sci. Technol.* **2015**, *45*, 193–248. [CrossRef]

3. Umar, M.; Roddick, F.; Fan, L. Comparison of coagulation efficiency of aluminium and ferric-based coagulants as pre-treatment for UVC/H_2O_2 treatment of wastewater RO concentrate. *Chem. Eng. J.* **2016**, *284*, 841–849. [CrossRef]

4. Umar, M.; Roddick, F.; Fan, L. Assessing the potential of a UV-based AOP for treating high-salinity municipal wastewater reverse osmosis concentrate. *Water Sci. Technol.* **2013**, *68*, 1994–1999. [CrossRef] [PubMed]

5. Westerhoff, P.; Moon, H.; Minakata, D.; Crittenden, J. Oxidation of organics in retentates from reverse osmosis wastewater reuse facilities. *Water Res.* **2009**, *43*, 3992–3998. [CrossRef] [PubMed]

6. Rodriguez-Mozaz, S.; Ricart, M.; Köck-Schulmeyer, M.; Guasch, H.; Bonnineau, C.; Proia, L.; de Alda, M.L.; Sabater, S.; Barceló, D. Pharmaceuticals and pesticides in reclaimed water: Efficiency assessment of a microfiltration-reverse osmosis (MF-RO) pilot plant. *J. Hazard. Mater.* **2015**, *282*, 165–173. [CrossRef] [PubMed]

7. Tang, F.; Hu, H.Y.; Wu, Q.Y.; Tang, X.; Sun, Y.X.; Shi, X.L.; Huang, J.J. Effects of chemical agent injections on genotoxicity of wastewater in a microfiltration-reverse osmosis membrane process for wastewater reuse. *J. Hazard. Mater.* **2013**, *260*, 231–237. [CrossRef] [PubMed]

8. Subramani, A.; Jacangelo, J.G. Treatment technologies for reverse osmosis concentrate volume minimization: A review. *Sep. Purif. Technol.* **2014**, *122*, 472–489. [CrossRef]

9. Yang, Y.; Ok, Y.S.; Kim, K.-H.; Kwon, E.E.; Tsang, Y.F. Occurrences and removal of pharmaceuticals and personal care products (PPCPs) in drinking water and water/sewage treatment plants: A review. *Sci. Total Environ.* **2017**, *596–597*, 303–320. [CrossRef] [PubMed]

10. Chong, M.N.; Jin, B.; Chow, C.W.K.; Saint, C. Recent developments in photocatalytic water treatment technology: A review. *Water Res.* **2010**, *44*, 2997–3027. [CrossRef] [PubMed]

11. Saha, S.; Wang, J.M.; Pal, A. Nano silver impregnation on commercial TiO_2 and a comparative photocatalytic account to degrade malachite green. *Sep. Purif. Technol.* **2012**, *89*, 147–159. [CrossRef]

12. Salaeh, S.; Perisic, D.J.; Biosic, M.; Kusic, H.; Babic, S.; Stangar, U.L.; Dionysiou, D.D.; Bozic, A.L. Diclofenac removal by simulated solar assisted photocatalysis using TiO_2-based zeolite catalyst; mechanisms, pathways and environmental aspects. *Chem. Eng. J.* **2016**, *304*, 289–302. [CrossRef]

13. Haynes, V.N.; Ward, J.E.; Russell, B.J.; Agrios, A.G. Photocatalytic effects of titanium dioxide nanoparticles on aquatic organisms—Current knowledge and suggestions for future research. *Aquat. Toxicol.* **2017**, *185*, 138–148. [CrossRef] [PubMed]

14. Chen, X.; Fujiwara, T.; Fukahori, S.; Ishigaki, T. Factors affecting the adsorptive removal of bisphenol A in landfill leachate by high silica Y-type zeolite. *Environ. Sci. Pollut. Res.* **2015**, *22*, 2788–2799. [CrossRef] [PubMed]

15. Yang, R.T. *Adsorbents: Fundamentals and Applications*; John Wiley & Sons, Inc.: Hoboken, NJ, USA, 2003; ISBN 0471297410.

16. Kyzas, G.Z.; Fu, J.; Lazaridis, N.K.; Bikiaris, D.N.; Matis, K.A. New approaches on the removal of pharmaceuticals from wastewaters with adsorbent materials. *J. Mol. Liq.* **2015**, *209*, 87–93. [CrossRef]

17. Anastopoulos, I.; Bhatnagar, A.; Hameed, B.H.; Ok, Y.S.; Omirou, M. A review on waste-derived adsorbents from sugar industry for pollutant removal in water and wastewater. *J. Mol. Liq.* **2017**, *240*, 179–188. [CrossRef]

18. Kyzas, G.Z.; Matis, K.A. Nanoadsorbents for pollutants removal: A review. *J. Mol. Liq.* **2015**, *203*, 159–168. [CrossRef]

19. Wei, X.; Wang, Y.; Hernández-Maldonado, A.J.; Chen, Z. Guidelines for rational design of high-performance absorbents: A case study of zeolite adsorbents for emerging pollutants in water. *Green Energy Environ.* **2017**, in press. [CrossRef]

20. Fukahori, S.; Fujiwara, T.; Funamizu, N.; Matsukawa, K.; Ito, R. Adsorptive removal of sulfonamide antibiotics in livestock urine using the high-silica zeolite HSZ-385. *Water Sci. Technol.* **2013**, *67*, 319–325. [CrossRef] [PubMed]

21. Wu, H.; Wu, X.-L.; Wang, Z.-M.; Aoki, H.; Kutsuna, S.; Jimura, K.; Hayashi, S. Anchoring titanium dioxide on carbon spheres for high-performance visible light photocatalysis. *Appl. Catal. B Environ.* **2017**, *207*, 255–266. [CrossRef]

22. Yap, P.-S.; Cheah, Y.-L.; Srinivasan, M.; Lim, T.-T. Bimodal N-doped P25-TiO_2/AC composite: Preparation, characterization, physical stability, and synergistic adsorptive-solar photocatalytic removal of sulfamethazine. *Appl. Catal. A Gen.* **2012**, *427–428*, 125–136. [CrossRef]

23. Fukahori, S.; Fujiwara, T. Modeling of sulfonamide antibiotic removal by TiO_2/high-silica zeolite HSZ-385 composite. *J. Hazard. Mater.* **2014**, *272*, 1–9. [CrossRef] [PubMed]

24. Li, M.; Lu, B.; Ke, Q.F.; Guo, Y.J.; Guo, Y.P. Synergetic effect between adsorption and photodegradation on nanostructured TiO_2/activated carbon fiber felt porous composites for toluene removal. *J. Hazard. Mater.* **2017**, *333*, 88–98. [CrossRef] [PubMed]

25. Fukahori, S.; Ichiura, H.; Kitaoka, T.; Tanaka, H. Photocatalytic decomposition of bisphenol A in water using composite TiO_2-zeolite sheets prepared by a papermaking technique. *Environ. Sci. Technol.* **2003**, *37*, 1048–1051. [CrossRef] [PubMed]

26. Fukahori, S.; Ichiura, H.; Kitaoka, T.; Tanaka, H. Capturing of bisphenol A photodecomposition intermediates by composite TiO_2-zeolite sheets. *Appl. Catal. B Environ.* **2003**, *46*, 453–462. [CrossRef]

27. Nakada, N.; Tanishima, T.; Shinohara, H.; Kiri, K.; Takada, H. Pharmaceutical chemicals and endocrine disrupters in municipal wastewater in Tokyo and their removal during activated sludge treatment. *Water Res.* **2006**, *40*, 3297–3303. [CrossRef] [PubMed]

28. Nakada, N.; Yasojima, M.; Okayasu, Y.; Komori, K.; Suzuki, Y. Mass balance analysis of triclosan, diethyltoluamide, crotamiton and carbamazepine in sewage treatment plants. *Water Sci. Technol.* **2010**, *61*, 1739–1747. [CrossRef] [PubMed]

29. Tamura, I.; Yasuda, Y.; Kagota, K.; Yoneda, S.; Nakada, N. Ecotoxicology and environmental safety contribution of pharmaceuticals and personal care products (PPCPs) to whole toxicity of water samples collected in effluent-dominated urban streams. *Ecotoxicol. Environ. Saf.* **2017**, *144*, 338–350. [CrossRef] [PubMed]

30. Japanese Industrial Standards Committee. Pulps-Preparation of Laboratory Sheets for Physical Testing-Conventional Sheet-Former Method, JIS P8222: 2015. Available online: http://www.jisc.go.jp/app/jis/general/GnrJISSearch.html (accessed on 26 July 2017).

31. Fukahori, S.; Fujiwara, T.; Ito, R.; Funamizu, N. Photocatalytic decomposition of crotamiton over aqueous TiO_2 suspensions: Determination of intermediates and the reaction pathway. *Chemosphere* **2012**, *89*, 213–220. [CrossRef] [PubMed]

32. Bui, T.X.; Choi, H. Influence of ionic strength, anions, cations, and natural organic matter on the adsorption of pharmaceuticals to silica. *Chemosphere* **2010**, *80*, 681–686. [CrossRef] [PubMed]

33. Umar, M.; Roddick, F.; Fan, L. Effect of coagulation on treatment of municipal wastewater reverse osmosis concentrate by UVC/H_2O_2. *J. Hazard. Mater.* **2014**, *266*, 10–18. [CrossRef] [PubMed]

34. Fukahori, S.; Fujiwara, T.; Ito, R.; Funamizu, N. pH-Dependent adsorption of sulfa drugs on high silica zeolite: Modeling and kinetic study. *Desalination* **2011**, *275*, 237–242. [CrossRef]

35. Egerton, T. UV-absorption—The primary process in photocatalysis and some practical consequences. *Molecules* **2014**, *19*, 18192–18214. [CrossRef] [PubMed]

36. Tsydenova, O.; Batoev, V.; Batoeva, A. Solar-enhanced advanced oxidation processes for water treatment: Simultaneous removal of pathogens and chemical pollutants. *Int. J. Environ. Res. Public Health* **2015**, *12*, 9542–9561. [CrossRef] [PubMed]

37. Tokumura, M.; Sugawara, A.; Raknuzzaman, M.; Habibullah-Al-Mamun, M.; Masunaga, S. Comprehensive study on effects of water matrices on removal of pharmaceuticals by three different kinds of advanced oxidation processes. *Chemosphere* **2016**, *159*, 317–325. [CrossRef] [PubMed]

38. Kudlek, E.; Dudziak, M.; Bohdziewicz, J. Influence of inorganic ions and organic substances on the degradation of pharmaceutical compound in water matrix. *Water* **2016**, *8*, 532. [CrossRef]

39. Zhou, T.; Lim, T.T.; Chin, S.S.; Fane, A.G. Treatment of organics in reverse osmosis concentrate from a municipal wastewater reclamation plant: Feasibility test of advanced oxidation processes with/without pretreatment. *Chem. Eng. J.* **2011**, *166*, 932–939. [CrossRef]

40. Burns, R.A.; Crittenden, J.C.; Hand, D.W.; Selzer, V.H.; Sutter, L.L.; Salman, S.R. Effect of inorganic ions in heterogeneous photocatalysis of TCE. *J. Environ. Eng.* **1999**, *125*, 77–85. [CrossRef]

41. Rioja, N.; Benguria, P.; Peñas, F.J.; Zorita, S. Competitive removal of pharmaceuticals from environmental waters by adsorption and photocatalytic degradation. *Environ. Sci. Pollut. Res.* **2014**, *21*, 11168–11177. [CrossRef] [PubMed]

42. Song, L.; Zhu, B.; Gray, S.; Duke, M.; Muthukumaran, S. Hybrid processes combining photocatalysis and ceramic membrane filtration for degradation of humic acids in saline water. *Membranes* **2016**, *6*. [CrossRef] [PubMed]

43. Ito, M.; Fukahori, S.; Fujiwara, T. Adsorptive removal and photocatalytic decomposition of sulfamethazine in secondary effluent using TiO_2-zeolite composites. *Environ. Sci. Pollut. Res.* **2014**, *21*, 834–842. [CrossRef] [PubMed]

44. Kuo, C.-S.; Lin, C.-F.; Hong, P.-K.A. Photocatalytic degradation of methamphetamine by UV/TiO_2-Kinetics, intermediates, and products. *Water Res.* **2015**, *74*, 1–9. [CrossRef] [PubMed]

45. Jang, D.; Ahn, C.; Choi, J.; Kim, J.; Kim, J.; Joo, J. Enhanced removal of trichloroethylene in water using nano-ZnO/polybutadiene rubber composites. *Catalysts* **2016**, *6*, 152. [CrossRef]

46. Kuo, C.S.; Lin, C.F.; Hong, P.K.A. Photocatalytic mineralization of codeine by UV-A/TiO_2-Kinetics, intermediates, and pathways. *J. Hazard. Mater.* **2016**, *301*, 137–144. [CrossRef] [PubMed]

47. Khraisheh, M.; Kim, J.; Campos, L.; Al-Muhtaseb, A.H.; Al-Hawari, A.; Al Ghouti, M.; Walker, G.M. Removal of pharmaceutical and personal care products (PPCPs) pollutants from water by novel TiO_2-coconut shell powder (TCNSP) composite. *J. Ind. Eng. Chem.* **2014**, *20*, 979–987. [CrossRef]

48. Anirudhan, T.S.; Shainy, F.; Christa, J. Synthesis and characterization of polyacrylic acid-grafted-carboxylic graphene/titanium nanotube composite for the effective removal of enrofloxacin from aqueous solutions: Adsorption and photocatalytic degradation studies. *J. Hazard. Mater.* **2017**, *324*, 117–130. [CrossRef] [PubMed]

49. Huang, Y.; Xu, W.; Hu, L.; Zeng, J.; He, C.; Tan, X.; He, Z. Combined adsorption and catalytic ozonation for removal of endocrine disrupting compounds over MWCNTs/Fe_3O_4 composites. *Catal. Today* **2017**, in press.

50. Ahmadi, M.; Motlagh, H.R.; Jaafarzadeh, N.; Mostoufi, A.; Saeedi, R.; Barzegar, G.; Jorfi, S. Enhanced photocatalytic degradation of tetracycline and real pharmaceutical wastewater using MWCNT/TiO_2 nano-composite. *J. Environ. Manag.* **2017**, *186*, 55–63. [CrossRef] [PubMed]

51. Yap, P.-S.; Lim, T.-T. Solar regeneration of powdered activated carbon impregnated with visible-light responsive photocatalyst: Factors affecting performances and predictive model. *Water Res.* **2012**, *46*, 3054–3064. [CrossRef] [PubMed]

52. Luo, L.; Li, J.; Dai, J.; Xia, L.; Barrow, C.J.; Wang, H.; Jegatheesan, J.; Yang, M. Bisphenol A removal on TiO_2-MoS_2-reduced graphene oxide composite by adsorption and photocatalysis. *Process Saf. Environ. Prot.* **2017**, in press.

53. Hu, X.-Y.; Zhou, K.; Chen, B.-Y.; Chang, C.-T. Graphene/TiO_2/ZSM-5 composites synthesized by mixture design were used for photocatalytic degradation of oxytetracycline under visible light: Mechanism and biotoxicity. *Appl. Surf. Sci.* **2016**, *362*, 329–334. [CrossRef]

applied sciences

MDPI

Article

Longitudinal Removal of Bisphenol-A and Nonylphenols from Pretreated Domestic Wastewater by Tropical Horizontal Sub-SurfaceConstructed Wetlands

Andrés Toro-Vélez [1,2,*], Carlos Madera-Parra [3], Miguel Peña-Varón [1], Hector García-Hernández [4], Wen Yee Lee [5], Shane Walker [6] and Piet Lens [4]

[1] Grupo de Saneamiento Ambiental, Instituto Cinara, Unversidad del Valle, Cali 100-00, Colombia; miguel.pena@correounivalle.edu.co

[2] Doctorado en Ciencias Ambientales, Universidad del Cauca, Popayán 190001, Colombia

[3] Escuela EIDENAR-Facultad de Ingeniería, Universidad del Valle, Cali 100-00, Colombia; carlos.a.madera@correounivalle.edu.co

[4] UNESCO-IHE Institute for Water Education, 2611 AX Delft, The Netherlands; h.garcia@un-ihe.org (H.G.-H.); p.lens@un-ihe.org (P.L.)

[5] Department of Chemistry, University of Texas at El Paso, El Paso, TX 79968, USA; wylee@utep.edu

[6] Department of Civil Engineering, University of Texas at El Paso, El Paso, TX 79968, USA; wswalker2@utep.edu

* Correspondence: andres.toro@correounivalle.edu.co

Received: 30 June 2017; Accepted: 20 July 2017; Published: 15 August 2017

Abstract: Bisphenol A (BPA) and nonylphenols (NPs), with a high potential to cause endocrine disruption, have been identified at levels of nanograms per liter and even micrograms per liter in effluents from wastewater treatment plants. Constructed wetlands (CWs) are a cost-effective wastewater treatment alternative due to the low operational cost, reduced energy consumption, and lower sludge production, and have shown promising performance for treating these compounds. A CW pilot study was undertaken todetermine its potential to remove BPA and NP from municipal wastewater. Three CWs were used: the first CW was planted with *Heliconia* sp., a second CW was planted with *Phragmites* sp., and the third CW was an unplanted control. The removal efficiency of the *Heliconia*-CW was 73 ± 19% for BPA and 63 ± 20% for NP, which was more efficient than the *Phragmites*-CW (BPA 70 ± 28% and NP 52 ± 23%) and the unplanted-CW (BPA 62 ± 33% and NP 25 ± 37%). The higher capacity of the *Heliconia*-CW for BPA and NP removal suggests that a native plant from the tropics can contribute to a better performance of CW for removing these compounds.

Keywords: municipal wastewater; constructed wetlands; Bisphenol A; nonylphenol; biodegradation; tropical environment

1. Introduction

Exposure to trace concentrations of certain synthetic and natural chemicals compounds, e.g., pharmaceuticals and personal care products (PPCPs), may induce negative environmental and health effects. The United States Geological Survey found 13 compounds related to organic wastewater contaminants in samples from the water supplies that ranged from 0.0120 to 0.480 $\mu g \cdot L^{-1}$, and concentrations of pharmaceutical and personal care compounds ranged from 0.0037 to 0.0576 $\mu g \cdot L^{-1}$ [1]. The main source of these compounds in the water cycle is their discharge by sewage systems. Several synthetic compounds with a high potential to cause endocrine disruption at low concentrations (e.g., micrograms per liter, $\mu g \cdot L^{-1}$, or even nanograms per liter, $ng \cdot L^{-1}$) have been identified to be present in the effluents of wastewater treatment plants (WWTPs) [2,3]. Furthermore, due to their chemical and recalcitrant characteristics, some of these PPCP compounds pass through conventional wastewater treatment processes without undergoing any transformation, resulting in their direct discharge to the receiving waters [4,5].

Compounds such as Bisphenol A (BPA) are widely used in industrial processes as a primary raw material in the manufacturing of many products such as plastics for engineering applications (e.g., epoxy resins and polycarbonate plastics), electronic devices, food cans, bottles, and dental sealants [6]. There is evidence that relates BPA appearance directly to adverse reproductive and carcinogenic effects in mice with a dose of $25 \text{ ng} \cdot \text{kg}^{-1}$ per day and $1 \mu g \cdot \text{kg}^{-1}$ per day, respectively [7]. Nonylphenols (NPs) are used in the manufacturing of anionic detergents, lubricants, agrochemicals, tanneries, and lubricant oil additives. The main source of NP in municipal wastewater is due to the intermediate degradation products of soaps and detergents. NPs are found to be endocrine disruptive compounds (EDCs), and have effects in the reproductive system of some mammals, including the reduction in testis and ovaries weight, and the appearance of an irregular estrous cycle [8].

Different treatment systems for the reduction of EDCs from wastewater are being evaluated, such as membrane bioreactors, activated sludge systems, ozonation, photocatalysis, and sequencing batch reactors. Most of these processes require a high economic investment with high environmental costs related to their operation and maintenance which makes their implementation difficult in developing countries. Some studies focus on sustainable wastewater treatment by decreasing electricity consumption and mitigating its greenhouse gas footprint [9]. In this sense, constructed wetlands (CWs) are natural wastewater treatment systems that offer cost-effective treatment for small to medium-sized systems [10]. Recently, the removal of pharmaceuticals in horizontal and vertical subsurface flow constructed wetlandswas evaluated [9] with removal efficiencies for ibuprofen of 51–54% in winter and 85–96% in summer and for carbamazepine of 24–36% in winter and 48% in summer.

The behavior of EDC removal in CW wastewater treatment systems is not fully understood. Data display a high variation in removal efficiency, ranging from 20% to 99%, depending on the chemical compound characteristics, plant type, flow conditions (regime), and geographic location [11]. Particularly under tropical climate conditions, some research has observedthat CW are capable of removing phenolic compounds in a range from 60% to 77% [12]. However, more research is required to understand the potential of CWs in the removal of EDCs, specifically regarding the effect of variables such as plant type and hydraulic retention time. Pilot-scale research is essential to enable extrapolation to full-scale CW design for effective removal of EDCs.

2. Materials and Methods

2.1. Location and Description of Horizontal Sub-Surface Constructed Wetlands (HSSF-CWs)

This research was performed at a test site ($3°43'50''$ N and $76°16'20''$ O) located approximately 1.1 km away from the urban area of Ginebra, Colombia, a small city of Valle del Cauca, located approximately 50 km northwest of Cali (Figure 1a). The test site is a research and technology transfer station of the domestic wastewater treatment and reuse research center of the Universidad del Valle (Cali, Colombia).

The study was carried out in a module consisting of three pilot scale horizontal subsurface flow constructed wetlands (HSSF-CWs), as shown in Figure 1b. The influent to the CWs comes from the effluent of an anaerobic pond as primary treatment of the domestic wastewater of the city. Each CW was designed to treat a flow of $3.5 \text{ m}^3 \cdot \text{day}^{-1}$ with an effective volume of 6.35 m^3 ($9 \times 3 \times 0.6$ m and 40% porosity). This design corresponded to a nominal hydraulic retention time of 1.8 days, a nominal surface loading rate of $0.13 \text{ m} \cdot \text{day}^{-1}$. One CW was planted with *Heliconias* sp. (a native flowering and marketable plant) and a second CW was planted with *Phragmites* sp. (a native perennial wetland grass). The third CW (in between the other two CWs) was a control without plants.

(a)

(b)

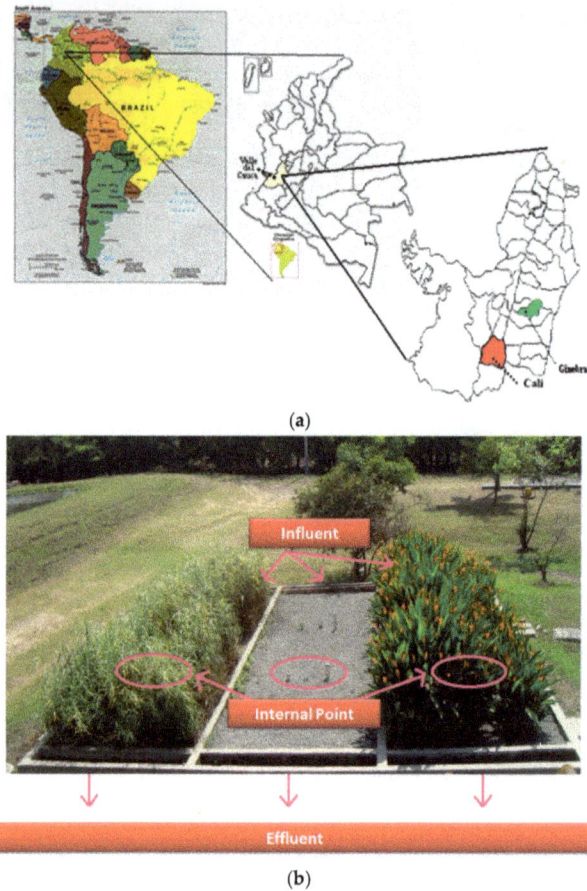

Figure 1. (a) Location of the pilot scale; (b) pilot scale horizontal subsurface flow constructed wetlands used in this study.

The EDC removal efficiency throughout each HSSF-CW was measured and two sampling points were provided to quantify the longitudinal concentration changes in BPA and NP concentration. The first was an internal sampling point, located at six meters from the inlet, which represents two-thirds of the length of the CW. These points were labeled H1 for *Heliconias*-CW, P1 for *Phragmites*-CW and C1 for control-CW. The other sample point was located at the outlet and was labeled H2, P2, and C2 respectively. The main response variable was the removal efficiency of BPA and NP. One sample was collected once per week over a seven-week period at each sampling point, including the beginning of the first week (i.e., eight samples were collected at each sample point). Unfortunately, the third effluent *Phragmites* sample was broken in shipping, and the entire fourth set of samples was lost in shipping.

2.2. Endocrine Disruptive Compounds

Bisphenol A was spiked into the wastewater to ensure detection, whereas NP was detected at higher concentrations naturally in the wastewater samples and did not required an external injection. For BPA spiking, a feeding system was built to supplement the anaerobic pond effluent with an average mass loading of 2.08 mg·day^{-1} BPA (i.e., a daily average concentration of 0.59 µg·L^{-1} BPA in the influent to the CW units). The feeding system consisted of a 50 L storage tank with piping to the CWs. From the BPA stock solution (200 mg·L^{-1}, prepared with analytical grade (>98%) BPA from Sigma-Aldrich (St. Louis, MO, USA) and stored at 4 °C), a 25 mg·L^{-1} solution was prepared separately using distilled water. Every day, at the same time, the storage tank was filled with 50 L of anaerobic pond effluent and 83 mL of 25 mg·L^{-1} BPA, (after completing the research, it was observed that spiking the influent with BPA was unnecessary).

Samples from the CWs were collected in 100 mL pre-cleaned amber bottles, labeled and stored in a dark room at <4 °C until shipment to the University of Texas (UTEP), El Paso, USA. To minimize the effect of microbiological degradation, each sample was centrifuged and filtered through 0.45 and 0.12 µm cellulose membrane filter before the shipping to UTEP. The shipping time was between 1.5 and 2 days. Upon arrival, the samples were kept in a refrigerator at 4 °C until chemical analysis.

To determine the EDC concentrations, samples were analyzed at UTEP using stir bar sorptive extraction (SBSE) with in-line thermal-desorption and gas chromatography/mass spectrometry (TDU-GC–MS). Briefly, (i) twenty milliliters of the filteredsample were transferred to a 20-mL screw cap vial, (ii) sodium carbonate (200 mg) was added to adjust the pHto 11.5, and (iii) acetic acid anhydride (200 µL) was added as the derivatization reagent. A pre-conditioned stir bar (three hours at 300 °C in a flow of nitrogen) was placed in each vial, and the samples were stirred at 1000 rpm for four hours. After the extraction, the stir bar (Gerstel, Linthicum, MD, USA) was removed with forceps, rinsed with purified water, and dried with lint-free tissue paper. The stir bar was thermally desorbed in a thermal desorption (TDU) system at the sample introduction inlet of a GC-MS system (Agilent, Santa Clara, CA, USA).

2.3. Influent Concentration of HSSF-CW

Table 1 shows the physical and chemical composition of the influent supplied to the HSSF-CW units, including the EDC concentrations. The influent composition varied throughout the study period due to variations in the background concentrations present in the domestic wastewater.

Table 1. Composition of the influent supplied to the horizontal subsurface flow constructed wetlands (HSSF-CWs).

Statistical Results	Parameters					
	BPA (µg·L^{-1})	NPs (µg·L^{-1})	DOC * (mg·L^{-1})	COD * (mg·L^{-1})	COD$_f$ * (mg·L^{-1})	TSS * (mg·L^{-1})
$\hat{y}$	8.80	1671	17.6	252	134	63.7
σ	6.40	838	4.23	48.6	28.8	26.2
C.V.	0.73	0.50	0.24	0.19	0.21	0.41

Notes: DOC: Dissolved Organic Carbon; COD: Chemical Oxygen demand; COD$_f$: Filtered COD; TSS: Total Suspended Solids; $\hat{y}$: mean value; σ: standard deviation; and C.V.: Coefficient of variation. * These parameters were measured in accordance with the Standard methods 21th Ed.

Of the three locations sampled in the CWs, the influent had the lowest redox potential (ORP) value (-420 ± 189 mV), which was understandable given that the influent was from an anaerobic pond. Likewise, the influent dissolved oxygen (DO) concentration was low (<0.15 ± 0.3 mg·L^{-1}). The unplanted (control) CW had higher ORP and DO (-253 mV and 0.6 mg·L^{-1}, respectively) than the influent, but the effluents of the planted CWs (*Phragmites* and *Heliconia*) had the highest average ORP values (-158 and -127 mV, respectively) and DO concentrations (0.9 and 0.8 mg·L^{-1}, respectively). The higher ORP and DO in the planted CWs is likely due to oxygen translocation by plants through its roots system [13].

2.4. Statistical Analysis

Data were recorded and analyzedwith Microsoft Excel, and Minitab 15 software was used for the Friedman Two-Way Analysis of Variance by Ranks combined with a non-parametric post hoc analysis (Wilcoxon signed-rank test).

3. Results and Discussion

3.1. Longitudinal Removal of BPA in HSSF-CWs

Figure 2 shows the influent and effluent BPA concentrations for the (a) *Heliconia*-CW, (b) unplanted (control), and (c) *Phragmites*-CW, as well as the longitudinal removal efficiencies for each (parts (d), (e), and (f), respectively). A significant reduction in the BPA concentrations was observed at the internal sampling points H1, C1 and P1—each at two-thirds of the horizontal length of the CW—compared to the influent (Figure 2). This implies that BPA was transformed, sorbed, or consumed in the first two-thirds of each CW. From this point until the effluent discharge point, the BPA concentrations were fairly consistent, and decreased only marginally. The differences in the partial and total BPA removal for the internal sample point and effluent were established by a Wilcoxon signed rank test in each CW and displayed significant differences for the *Heliconia*-CW and unplanted-CW. Regarding the *Phragmites*-CW, the final third of the CW did not contribute to the improvement of the total BPA removal.

From the literature, the main removal mechanism of BPA in a CW is sorption [14–16], due to a higher octanol–water partition coefficient of the hydrophobic BPA molecule (log K_{ow} is 3.4), as well as a large surface area in the CW, which enhances sorption onto the biofilm, suspended solids, support media, and rhizosphere [17]. The sorption capacity of a compound can be better expressed in terms of the organic carbon partition coefficient (K_{oc} or K_d) related to the quantity of the sorbed compound in the solid phase with respect to the concentration in the aqueous phase [18]. This is important because BPA has two hydroxyphenyl groups, which tend to promote sorption in soils (or support media) and sediments [19]. The research of this manuscript did not include BPA measurements on the sediments or support media, but based on other research of BPA partitioning in wastewater sediment [20], a log K_d (L·kg^{-1}) value of 4.37 was calculated for BPA at the average concentration reported in Table 1. If the log K_d < 2.47, sorption can be neglected, while for values higher than 4.0, sorption onto the solid phase is one of the major removal processes [18,20]. The calculated log K_d value of BPA suggests that it is likely that sorption of BPA was one of the significant removal processes in each of the three CW types investigated.

Sorption onto the support media or onto the biofilm and sediments on the media implies a longer EDC residence time for BPA in the CW. This may favor bioremediation by increasing exposure to plant uptake or microbial degradation. Indeed, phytoremediation can also be an important removal pathway, where the EDC may possibly be degraded by phytostimulation or rhizodegradation, phytodegradation, phythoextraction, sequestration or volatilization [13,21]. In some cases, the plants growing in the CW may play a significant role in the direct uptake of many organic pollutants from wastewater. For instance, the presence of *Phragmites australis* in a CW improved the removal efficiency of BPA compared to an unplanted CW [12,22,23]. In a study by Dodgen et al. [24], the plant uptake of BPA, 4-NP, diclofenac (DCL), and naproxen (NPX) during the hydroponic cultivation of *Lactuca sativa* (lettuce) and *Brassica oleracea* (collard) was investigated, and EDC accumulation was observed in both plant species with a trend in descending order of BPA > NP > DCL > NPX [24].

The Friedman Test indicates significant differences (*p*-value: 0.03) for the removal of BPA (Figure 2) at the intermediate sampling point (H1, C1, and P1). The post hoc test revealedthat P1 obtained the highestaverage removal efficiency (64.3%) compared with C1 (55.2%) and H1 (61.4%). Also, significant differences (*p*-value: 0.015) were observed for the average effluent BPA removal (H2, C2, and P2), and the post hoc test showed that the *Heliconia*-CW had the highest average BPA removal efficiency (73.3%) compared withthe unplanted-CW (62.2%) and *Phragmites*-CW (70.2%), as shown in Table 2.

Both planted CWs had greater average effluent removal efficiencies than the unplanted control, and both planted CWs had greater effluent removal efficiencies than the unplanted-CW for all but one of the sampling events; thus, it is assumed that plant vegetation had a role in BPA removal. The specific removal mechanism by the plants was not investigated.

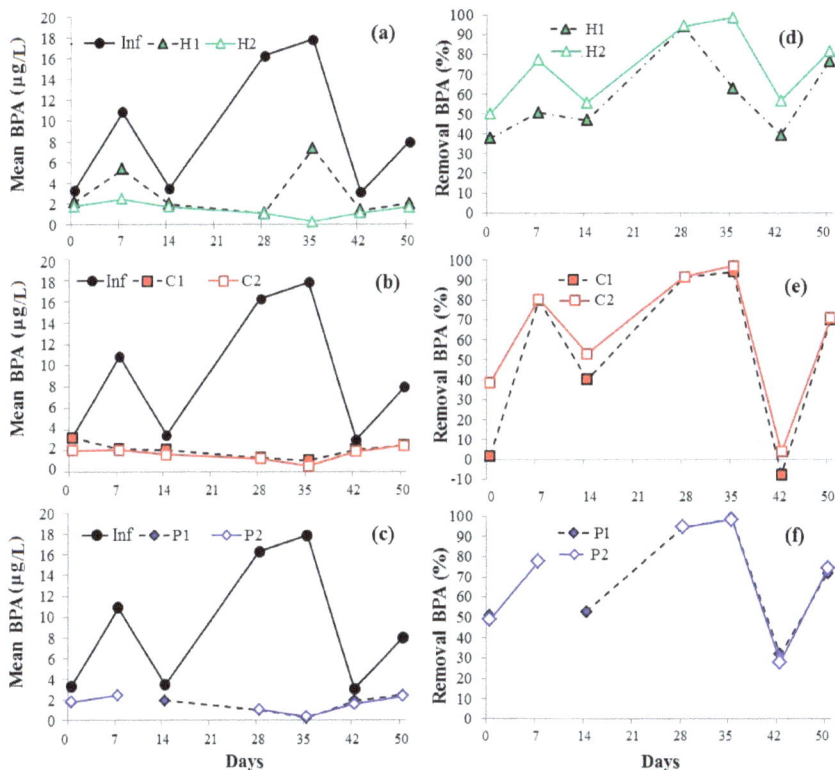

Figure 2. Longitudinal BPA concentrations in CWs of (**a**) *Heliconia* (H); (**b**) unplanted control (C), and (**c**) *Phragmites* (P), for influent (Inf), two-thirds internal (1), and effluent (2) sample points. Removal efficiency of BPA in (**d**) *Heliconia*; (**e**) unplanted control; and (**f**) *Phragmites*.

Table 2. Average effluent BPA removal for *Heliconia*-CW, unplanted-CW and *Phragmites*-CW.

Removal Efficiencies (%)	BPA		
	Heliconia	*Unplanted*	*Phragmites*
$\hat{y}$	73.3%	62.2%	70.2%
Maximum	98.6%	97.1%	98.3%
Minimum	50.0%	3.7%	27.9%
σ	19.6%	33.1%	27.1%
C.V.	0.27	0.53	0.39
n *	7	7	6

* The third *Phragmites* sample was broken in shipping. Notes: $\hat{y}$: mean value; σ: standard deviation; C.V.: coefficient of variation; and n number of samples.

3.2. Longitudinal Removal of NP in HSSF-CWs

The average total NP concentration in the influent was (1671 ± 838) µg·L^{-1}. A reduction of the NP concentration was observed at the internal sampling point (two-thirds of the length) of each wetland (H1, P1, and C1), with average concentrations of (724 ± 453), (1150 ± 499) and (1041 ± 446) µg·L^{-1}, respectively. The lowest final effluent concentrations were obtained for the *Heliconia*-CW (629 ± 318 µg·L^{-1}) and *Phragmites*-CW (736 ± 284 µg·L^{-1}), showing more NP removal in the final third of the CWs. The NP removal efficiencies were less than those of BPA. A Wilcoxon signed rank test was used to establish significance of differences between the internal sampling point and effluent concentrations in each CW. Although the plantedCWs had a higher average NP removal efficiency in the effluent than the internal sample point, these results are not statistically different.

The unplanted-CW displayed a different behavior, showing an increment of the NP concentration in the effluent (1103 ± 538 µg·L^{-1}) compared with its internal point. NP desorption was observed in the final third of the CW, despite NPs having high K_{ow} values (log K_{ow} 3.80 to 4.77). This negative removal efficiency in CW was also reported in other research [18] for seven different PPCPs, attributed to an initial retention and sorption, but subsequent release during passage of the wastewater through the medium of the CW.

The *Heliconia*-CW showed the highest average effluent NP removal efficiency of 62.8 ± 20.1%, while the *Phragmites*-CW and unplanted-CW had a removal efficiency of 25.3 ± 37.1% and 52.1 ± 23.2%, respectively (Table 3). These results confirm a statistically significant difference between planted and unplanted CWs with p-values of 0.042 for the internal sample point and 0.03 for the effluent. The posthoc test corroborated this conclusion, confirming that the *Heliconia*-CW showed a higher NP removal efficiency than the *Phragmites*-CW and the unplanted-CW.

Table 3. Average effluent NP removal for *Heliconia*-CW, unplanted-CW, and *Phragmites*-CW.

Removal Efficiencies (%)	NP		
	Heliconia	*Unplanted*	*Phragmites*
$\hat{y}$	62.8%	25.3%	52.1%
Maximum	90.0%	83.7%	80.2%
Minimum	28.0%	−12.3%	20.4%
σ	20.1%	37.1%	23.2%
C.V.	0.32	1.46	0.4
n *	7	7	6

* The third *Phragmites*-CW sample was broken in shipping. Notes: $\hat{y}$: mean value; σ: standard deviation; C.V.: coefficient of variation; and n number of samples.

3.3. EDC Removal Rate Against Mass Loading Rates

The total mass removed of each EDC compound was plotted against its total inlet mass loading rate (Figure 3). The removal rate of BPA in the *Heliconia*-CW, unplanted-CW and *Phragmites*-CW increased as the mass loading rate increased as well. BPA removal efficiency in all three CW types investigated was not sensitive to the mass loading rate and was almost completely removed in all CWs investigated (Figure 3a). However, NP removal decreased at a high mass loading rate. Moreover, NP was poorly removed in the unplanted-CW both at low and high mass loading rates (Figure 3b).

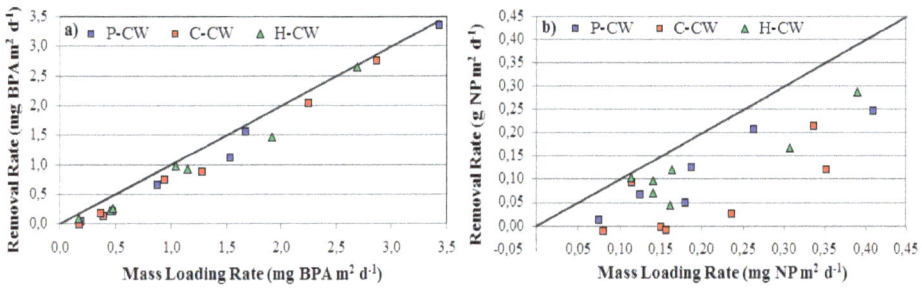

Figure 3. Endocrine disruptive compound(EDC) removal rate for H-CW, C-CW, and P-CW against their mass loading rates (**a**) BPA and (**b**) NP. Continuous line represents 100% removal.

3.4. Overall Performance of Each HSSF-CW in EDC Removal

Four scenarios (quadrants) were identified (Figure 4) according to the weekly effluent removal efficiencies of BPA and NP. Zone I is the worst scenario in which both compounds were removed with less than 50% efficiency. In Zone II, BPA removal was greater than 50%, but NP was less than 50%. Zone III shows a BPA removal efficiency less than 50% and NP greater than 50%. Finally, in Zone IV, both compounds were removed with efficiencies greater than 50%. The best CW removal efficiencies were observed for the *Heliconia*-CW, in which six of seven (85%) of the data points are in zone IV, compared with the unplanted-CW with more than five of seven (71%) of the data points in zone II. Overall, the BPA and NP removal efficiencies were in the following descending order: *Heliconia*-CW > *Phragmites*-CW > unplanted-CW. This performance is likely due to sorption in the HSSF-CW onto support media. Also, the rhizosphere of planted CWs likely generates benefits such as increasing DO concentrations and releasing organic exudates that serve as nutrient sources for the growth of microorganisms [25].

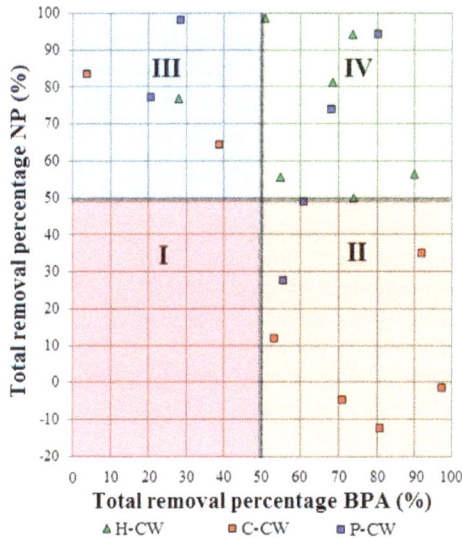

Figure 4. Quadrant chart for NP removal efficiency versus BPA removal efficiency for the *Heliconia*-CW (H), *Phragmites*-CW (P) and unplanted-CW (C).

This study showed that BPA and NP can beeffectively removed by planted HSSF CW under tropical conditions, with removal being more efficient withthe native and marketable plant *Heliconia* sp. in the CW.

4. Conclusions

The *Heliconia*-CW removed BPA (73%) and NP (62%) more efficiently than the *Phragmites*-CW (70% and 52%, respectively) and unplanted-CW (62% and 25%). The *Heliconia*-CW showed an improvement in BPA removal in the last third of the length of the wetland (p-value: 0.015). In contrast, the last third section of the *Phragmites*-CW and unplanted-CW did not contribute to additional BPA removal. Desorption of NP in the unplanted-CW was observed in the last third of the lengthof the CW, suggesting that a sorption–desorption equilibrium can be reached. This situation was not found for *Phragmites* sp. and *Heliconias* sp., corroborating that plants had a positive influence in the CW.

Acknowledgments: This study was carried out under the framework of UNESCO-IHE Partnership Research Fund (UpaRF), Natural System for Wastewater Treatment and Reuse: Technology Adaptations and Implementations in Developing Countries (NATSYS Project). The authors express special gratitude to ACUAVALLE (local water agency) and "Ricclisa: Programa para el Fortalecimiento de la Red Interinstitucional de Cambio Climático y Seguridad Alimentaria RC0853-2012 supported by Colciencias" for their contribution to this research.

Author Contributions: Andrés Toro-Vélez, Miguel Peña-Varón, Hector García-Hernández and Carlos Madera-Parra conceived and designed the experiments; Andrés Toro-Vélez performed the experiments; Andrés Toro-Vélez, analyzed the data; Wen Yee Lee and Shane Walker contributed reagents/materials/analysis tools; Andrés Toro-Vélez, Piet Lens and Shane Walker wrote the paper.

Conflicts of Interest: The authors declare no conflict of interest.

References

1. Zimmerman, M.J. *Occurrence of Organic Wastewater Contaminants, Pharmaceuticals, and Personal Care Products in Selected Water Supplies, 2005. Cape Cod, Massachusetts, June 2004*; Open-File Report 2005-1206; U.S. Geological Survey: Denver, CO, USA, 2005; p. 16.
2. Luo, Y.; Guo, W.; Ngo, H.H.; Nghiem, L.D.; Hai, F.I.; Zhang, J.; Wang, X.C. A review on the occurrence of micropollutants in the aquatic environment and their fate and removal during wastewater treatment. *Sci. Total Environ.* **2014**, *473*, 619–641. [CrossRef] [PubMed]
3. Terzić, S.; Senta, I.; Ahel, M.; Gros, M.; Petrović, M.; Barcelo, D.; Jabucar, D. Occurrence and fate of emerging wastewater contaminants in Western Balkan Region. *Sci. Total Environ.* **2008**, *399*, 66–77. [CrossRef] [PubMed]
4. Kasprzyk, B.; Dinsdale, R.M.; Guwy, A.J. The occurrence of pharmaceuticals, personal care products, endocrine disruptors and illicit drugs in surface water in South Wales, UK. *Water Res.* **2008**, *42*, 3498–3518. [CrossRef] [PubMed]
5. Suárez, S.; Carballa, M.; Omil, F.; Lema, J.M. How are pharmaceutical and personal care products (PPCPs) removed from urban wastewaters? *Rev. Environ. Sci. Bio/Technol.* **2008**, *7*, 125–138. [CrossRef]
6. Huang, Y.Q.; Wong, C.K.C.; Zheng, J.S.; Bouwman, H.; Barra, R.; Wahlström, B.; Wong, M.H. Bisphenol A (BPA) in China: A review of sources, environmental levels, and potential human health impacts. *Environ. Int.* **2012**, *42*, 91–99. [CrossRef] [PubMed]
7. Newbold, R.R.; Jefferson, W.N.; Padilla-Banks, E. Prenatal exposure to bisphenol at environmentally relevant doses adversely affects the murine female reproductive tract later in life. *Environ. Health Perspect.* **2009**, *117*, 879–885. [CrossRef] [PubMed]
8. United Nations Environment Programme and World Health Organization. *State of the Science of Endocrine Disrupting Chemicals–2012*; Bergman, A., Heindel, J.J., Jobling, S., Kidd, K.A., Zoeller, R.T., Eds.; WHO Press: Geneva, Switzerland, 2013.
9. Yan, P.; Qin, R.C.; Guo, J.S.; Yu, Q.; Li, Z.; Chen, Y.P.; Shen, Y.; Fang, F. Net-Zero-Energy Model for Sustainable Wastewater Treatment. *Environ. Sci. Technol.* **2017**, *51*, 1017–1023. [CrossRef] [PubMed]
10. Nahlik, A.M.; Mitsch, W.J. Tropical treatment wetlands dominated by free-floating macrophytes for water quality improvement in Costa Rica. *Ecol. Eng.* **2006**, *28*, 246–257. [CrossRef]

11. Hijosa, M.; Matamoros, V.; Sidrach, R.; Martín, J.; Bécares, E.; Bayona, J. Comprehensive Assessment of the Design Configuration of Constructed Wetlands for the Removal of Pharmaceuticals and Personal Care Products from Urban Wastewaters. *Water Res.* **2010**, *44*, 3669–3678. [CrossRef] [PubMed]

12. Abira, M.A.; van Bruggen, J.J.A.; Denny, P. A Potential of a tropical subsurface constructed wetland to remove phenol from pre-treated pulp and papermill wastewater. *Water Sci. Technol.* **2005**, *51*, 173–176. [PubMed]

13. Zhang, D.; Gersberg, R.M.; Ng, W.J.; Tan, S.K. Removal of pharmaceuticals and personal care products in aquatic plant-based systems: A review. *Environ. Pollut.* **2014**, *184*, 620–639. [CrossRef] [PubMed]

14. Fountoulakis, M.S.; Terzakis, S.; Kalogerakis, N.; Manios, T. Removal of polycyclic aromatic hydrocarbons and linear alkylbenzene sulfonates from domestic wastewater in pilot constructed wetlands and a gravel filter. *Ecol. Eng.* **2009**, *35*, 1702–1709. [CrossRef]

15. Matamoros, V.; Bayona, J.M. Elimination of pharmaceuticals and personal care products in subsurface flow constructed wetlands. *Environ. Sci. Technol.* **2006**, *40*, 5811–5816. [CrossRef] [PubMed]

16. Dordio, A.V.; Teimão, J.; Ramalho, I.; Carvalho, A.J.P.; Candeias, A.J.E. Selection of a support matrix for the removal of some phenoxyacetic compounds in constructed wetlands systems. *Sci. Total Environ.* **2007**, *380*, 237–246. [CrossRef] [PubMed]

17. Avila, C.; Reyes, C.; Bayona, J.M.; García, J. Emerging organic contaminant removal depending on primary treatment and operational strategy in horizontal subsurface flow constructed wetlands: Influence of redox. *Water Res.* **2013**, *47*, 315–325. [CrossRef] [PubMed]

18. Verlicchi, P.; Galletti, A.; Petrovic, M.; Barceló, D.; Al Aukidy, M.; Zambello, E. Removal of selected pharmaceuticals from domestic wastewater in an activated sludge system followed by a horizontal subsurface flow bed-analysis of their respective contributions. *Sci. Total Environ.* **2013**, *454–455*, 411–425. [CrossRef] [PubMed]

19. Imfeld, G.; Braeckevelt, M.; Kuschk, P.; Richnow, H. Monitoring and Assessing Processes of Organic Chemicals Removal in Constructed Wetlands. *Chemosphere* **2009**, *74*, 349–362. [CrossRef] [PubMed]

20. Clara, M.; Strenn, B.; Saracevic, E.; Kreuzinger, N. Adsorption of bisphenol-A, 17 beta-estradiole and 17 alpha-ethinylestradiole to sewage sludge. *Chemosphere* **2004**, *56*, 843–851. [CrossRef] [PubMed]

21. Zhang, B.Y.; Zheng, J.S.; Sharp, R.G. Phytoremediation in Engineered Wetlands: Mechanisms and Applications. *Procedia Environ. Sci.* **2010**, *2*, 1315–1325. [CrossRef]

22. Li, Y.; Zhu, G.; Ng, W.J.; Tan, S.K. A review on removing pharmaceutical contaminants from wastewater by constructed wetlands: Design, performance and mechanism. *Sci. Total Environ.* **2013**, *468–469*, 908–932. [CrossRef] [PubMed]

23. Toyama, T.; Yusuke, S.; Daisuke, I.; Kazunari, S.; Young, C.; Shintaro, K.; Michihiko, I. Biodegradation of Bisphenol A and Bisphenol F in the Rhizosphere Sediment of Phragmites Australis. *J. Biosci. Bioeng.* **2009**, *108*, 147–150. [CrossRef] [PubMed]

24. Dodgen, L.K.; Li, J.; Parker, D.; Gan, J.J. Uptake and accumulation of four PPCP/EDCs in two leafy vegetables. *Environ. Pollut.* **2013**, *182*, 150–156. [CrossRef] [PubMed]

25. Song, H.L.; Nakano, K.; Taniguchi, T.; Nomura, M.; Nishimura, O. Estrogen Removal from Treated Municipal Effluent in Small-Scale Constructed Wetland with Different Depth. *Bioresour. Technol.* **2009**, *100*, 2945–2951. [CrossRef] [PubMed]

applied sciences

MDPI

Article

An Innovative Dual-Column System for Heavy Metallic Ion Sorption by Natural Zeolite

Amanda L. Ciosek * and Grace K. Luk

Department of Civil Engineering, Faculty of Engineering and Architectural Science, Ryerson University, Toronto, ON M5B 2K3, Canada; gluk@ryerson.ca
* Correspondence: amanda.alaica@ryerson.ca; Tel.: +1-647-444-7201

Received: 7 July 2017; Accepted: 2 August 2017; Published: 5 August 2017

Abstract: This study investigates the design and performance of a novel sorption system containing natural zeolite. The apparatus consists of packed, fixed-bed, dual-columns with custom automated controls and sampling chambers, connected in series and stock fed by a metering pump at a controlled adjustable distribution. The purpose of the system is to remove heavy metallic ions predominately found in acid mine drainage, including lead (Pb^{2+}), copper (Cu^{2+}), iron (Fe^{3+}), nickel (Ni^{2+}) and zinc (Zn^{2+}), combined in equal equivalence to form an acidified total 10 meq/L aqueous solution. Reported trends on the zeolite's preference to these heavy metallic ions is established in the system breakthrough curve, as $Pb^{2+} \gg Fe^{3+} > Cu^{2+} > Zn^{2+} \gg Ni^{2+}$. Within a 3-h contact period, Pb^{2+} is completely removed from both columns. Insufficient Ni^{2+} removal is achieved by either column with the promptest breakthrough attained, as zeolite demonstrates the least affinity towards it; however, a 48.97% removal is observed in the cumulative collection at the completion of the analysis period. The empty bed contact times for the first and second columns are 20 and 30 min, respectively; indicating a higher bed capacity at breakthrough and a lower usage rate of the zeolite mineral in the second column. This sorption system experimentally demonstrates the potential for industrial wastewater treatment technology development.

Keywords: zeolite; sorption; packed fixed-bed columns; heavy metallic ions; automated sampling design

1. Introduction

Acid mine drainage (AMD) generated by industrial mines contains highly acidic wastewater and toxic heavy metallic ions (HMIs). These HMIs are a serious threat to human health and the environment [1–5], with their high solubility [6], as well as non-biodegradable and bio-accumulative properties [7,8]. If mines are abandoned, there is a risk of AMD contamination to both surface and groundwater, causing catastrophic damage to the ecosystem [4,9]. Therefore, the HMIs must be removed by advanced treatment methods prior to discharge [10–12], to abide by the effluent maximum allowable limits [7,8]. Due to the high site-to-site variability present in mines, mitigation feasibility can become a challenge [3], and a simple, resilient and cost-effective strategy must be developed [1].

Various industrial wastewater treatment methods include precipitation, electro-chemical remediation, oxidation and hydrolysis, neutralization, ion-exchange and solvent extraction, ion-exchange and precipitation, titration, bio-sorption, adsorption, reverse osmosis [4,9], and ultrafiltration [13]. Among these, sorption is a simple but promising treatment method [5,14], based on demonstrated industrial viability and effectiveness, cost efficiency and environmental sustainability [15,16]. The removal of HMIs is attributed to the mechanisms of both adsorption (on the surface of the sorbents' micropores) and ion-exchange (through the sorbents' framework pores and channels) [17]; and culminate in the unified process of sorption [18,19].

Zeolites hold great potential as naturally occurring minerals [1], and are recognized as effective materials for the removal of HMIs from industrial wastewater by sorption [20]. They have attracted a lot of researchers' interests [11], due to its coexisting molecular sieve action, ion-exchange and catalytic properties [14,21]. The mineral zeolite is a hydro-aluminosilicate with a crystalline structure comprised of SiO_4 and AlO_4 tetrahedras linked by oxygen atoms, which form an open, homogeneous microporous three-dimensional framework creating voids and channels [21]. Clinoptilolite, a globally abundant and well-documented form of zeolite [13,20,22,23], is used in this research. Two of the most significant properties of zeolite are its high cation exchange capacity [23], and its selectivity of certain metals [24]. In addition to wide deposit distribution and low exploitation costs [7,25], zeolite is considered as a strong candidate for the removal of wastewater contaminants [26], especially those containing high HMIs from mine waste.

Currently, there is limited comprehension of the sorption by zeolite for waste with multi-component HMIs [24,27], in order to fully benefit from using it in tertiary treatment processes [1]. In particular, the composition of AMD waste contains numerous contaminants, which include HMIs and other pollutants, and the presence of these in solution affect the overall removal potential [3,18]. Research is needed on the simultaneous sorption of the multi-metallic components that are prevalent in industrial effluent, and to quantify uptake interference of these HMIs in combination [16]. Helfferich [18] (p. 201) points out that for multi-component systems, the exchange rate may vary for the various counter-ions of HMIs in solution, with the possibility that the concentrations of certain species in either the sorbent or solution may fluctuate prior to attaining its balanced state. The performance of columns or fixed-bed reactors (FBR) is convenient for industrial scale applications [7], which requires less investment and operational costs, and is more economically feasible than its discontinuous batch mode counterparts [8]. FBR columns have demonstrated performance efficiency in treating large volumes, and are frequently implemented in sorption studies. However, the removal of multi-component solutions of HMIs is complex due to ion competition, different affinity sequences, and zeolite selectivity. Its operations are affected by equilibrium (isotherm and capacity), kinetic (diffusion and convection coefficients) and hydraulic (liquid holdup, geometric analogies and mal-distribution) factors [28]. In practice, the influence of operative conditions on the overall system performance is not experimentally verified [17,29,30], but they are extremely important to large-scale development. Although the FBR system is highly valuable, its analysis is unpredictably multi-faceted [28,31] and even more so with the presence of numerous interfering ions. Complications due to ion competition and solute-surface interactions [4], as well as the unique affinity sequences and sorbent material selectivity [25], have been reported.

The authors have conducted a four-phase research project, consisting of the analysis of: (1) the effects of preliminary parameters and operative conditions (particle size, sorbent-to-sorbate dosage, influent concentration, contact time, set-temperature, heat pre-treatment), (2) heavy metallic ions component system combinations and selectivity order [32], (3) kinetic modelling trends [33], and finally in this current study, (4) the design of a packed fixed-bed, dual-column sorption system. The first three phases are conducted in batch mode, to which reveal a key trend among the HMIs selected as lead (Pb^{2+}) >> iron (Fe^{3+}) > copper (Cu^{2+}) > zinc (Zn^{2+}) > nickel (Ni^{2+}). The findings of these preliminary phases have established a platform for the design of the sorption system in continuous mode, presented in this paper.

Existing column experimental designs involve various limitations, including:

1. The evaluation of predominantly single- or dual-component HMI system combinations;
2. The implementation of primarily slender column aspect ratios (i.e., bed depth/particle diameter, column height/diameter), causing a challenge to eventual scale-up design;
3. The use of inconsistent and/or vague sorbent compaction techniques, and;
4. The application of simple, idealized flow patterns (i.e., set single and continuous flow rate).

The objective of this paper is to develop a novel dual-column sorption system to overcome some of these shortcomings. Important design factors such as the zeolite compaction, column dimensions and aspect ratios, flow control, sampling and analytical procedure, will be taken into consideration. The exclusivity of this prototype is attributed to an automated, variable-flow configuration with a custom sampling technique. In contrast to most previous single-component sorption set-ups, this study evaluates the simultaneous sorption process by natural zeolite of the five most commonly occurring HMIs found in AMD, including Pb^{2+}, Cu^{2+}, Fe^{3+}, Ni^{2+} and Zn^{2+}. This paper will demonstrate the effectiveness and the removal efficiency in a continuous-flow FBR system over a 3-h duration from the dual columns, providing insights on HMIs selectivity and treatment system breakthroughs. It is envisaged that this research will provide much-needed information to the wastewater treatment industry for the design and implementation of innovative sorption technologies.

2. Experimental Design

2.1. Packed Fixed-Bed Column Design Considerations

When the concentration of the effluent reaches 5%–10% of the influent, this point on a typical S-shaped breakthrough curve is commonly referred to as the 'breakthrough point' or 'breakpoint' (BP). The point of column exhaustion (EP) is when the effluent reaches maximum capacity to 90%–95% of its influent value [12,34]. The efficiency of the column performance is related to the bed capacity at breakthrough and at exhaustion, represented by the following relationship [12,34]:

$$\eta = \frac{C_{BP}}{C_{EP}} \tag{1}$$

where η is the column efficiency (degree of saturation), C_{BP} is the breakthrough capacity of the bed (in meq/g), and C_{EP} is the maximum capacity at exhaustion of the bed, defined by the total amount of HMI ions bound in the zeolite (or exchanged in the packed fixed-bed) (in meq/g).

The breakthrough capacity and equilibrium capacity are further expressed in Equations (2) and (3), respectively [12,34]:

$$C_{BP} = \frac{\int_0^{V_{BP}} (C_0 - C)dV}{\rho HA} = \frac{C_0 \cdot V_{BP}}{m} = \frac{\eta_{BP}}{m} \tag{2}$$

$$C_{EP} = \frac{\int_0^{V_{EP}} (C_0 - C)dV}{\rho HA} = \frac{\int_0^{V_{EP}} (C_0 - C)dV}{m} = \frac{\eta_{EP}}{m} \tag{3}$$

where V_{BP} is the effluent volume collected up to breakthrough point (BP) (in L), V_{EP} is the effluent volume at which the exhaustion point (EP) is reached in the zeolite bed (in L), C_0 is the influent concentration (in meq/L), C is the effluent concentration (in meq/L), ρ is the packing density of the bed (in g/cm^3), H is the bed depth (in cm), A is the bed cross-sectional area (in cm^2), m is the zeolite mass (in g); where η_{BP} and η_{EP} is the total amount of HMI ions removed up to BP and EP (in meq), respectively.

Empty bed contact time (EBCT) is the residence time (in min) a fluid element is in contact with the bed, and is related to the systems' removal kinetics [35]. This is represented by the relationship between the bed depth (H) in the column and the feed solution velocity (v) [25,34], as given by:

$$EBCT = \frac{H}{v} = \frac{H}{(Q/A)} = \frac{d^2 \pi H}{4Q} \tag{4}$$

Research conducted by Vukojevic Medvidovic et al. [34] demonstrates that the breakthrough curve results reveal that the flow through the column determines the EBCT; with the same initial concentration, the increase in flow rate decreases the contact time and increases the mass transfer zone (MTZ) height. The MTZ is the restricted area where the exchange process occurs, and is defined as the zeolite layer height between the equilibrium zone and the unused bed zone [34]; where the effluent concentration varies from 5% to 95% of the influent concentration [25]. As the HMI solution is fed through the packed fixed-bed, the MTZ moves in the direction of flow and eventually reaches the exit [12,34]. Peric et al. [36] distinctly demonstrates the importance of the column bed depth on the removal of lead from aqueous solutions. The results show that as the bed depth increases, a delay in breakthrough and exhaustion occurs, with an increase of the MTZ height. The higher the bed depth, the longer the service time at various breakthrough points due to the increase in binding sites on the sorbent material (zeolite mineral) [9]. Adequate wetting of the zeolite, and ideal contact time between the zeolite and solution interface are important for mass transfer and equilibrium conditions based on the selection of the flow rate and particle size. To minimize possible wall and axial dispersion effects in the fixed-bed column, the bed depth-to-particle diameter ratio (H/d_p) must be greater than 20. At a higher H/d_p ratio, the breakthrough point appears later and the curve is steeper.

The usage rate (v_U, in g/L) determines the rate at which the sorbent would be exhausted and how often it must be replaced or regenerated, and is expressed in the following relationship [4,25]:

$$v_U = \frac{m_Z}{V_{BP}} \tag{5}$$

where m_Z is the zeolite mass in the bed (in g) and V_{BP} is the volume of the effluent treated at breakthrough (BP) (in L) [25]. Inglezakis [28] states that it is extremely difficult to model multi-component system interactions, as numerous time-consuming data are required and the process involves significant mathematical complexity. Breakthrough and exhaustion thresholds of specific HMIs within a fixed-bed are important for experimental specific conditions. In order to optimize the liquid-solid contact time and removal capacity, it is necessary to develop these relationships, between EBCT and usage rate [28].

2.2. Natural Zeolite Mineral

This study employs a natural zeolite mineral sample composed primarily of 85%–95% clinoptilolite (CAS No. 12173-10-3) and is sourced from a deposit located in Preston, Idaho [37]. The natural zeolite sample specifications are provided in Table 1, where typical elemental analysis indicates the presence of various elements, including Na^+, Ca^{2+}, Mg^{2+} and K^+, as well as lead, copper, iron, and zinc. No significant concentrations of toxic trace elements are present in its composition, nor are trace metal elements water soluble. The low-clay content unique to this sample ensures good hydraulic conductivity, low dust content, and a harder and more resistant structure [37]. The zeolite mineral sample is applied in its natural state, without any chemical modifications, to minimize all associated costs and environmental impacts of this study. As suggested by the laboratory-scale packed bed system investigations by Inglezakis et al. [38], a zeolite fraction of 0.8 mm to 1 mm nominal diameter is recommended for the columns to ensure the full exploitation of the material but also to prevent considerable pressure drop during the analysis period. Therefore, the zeolite sample used for this research is obtained from standard mechanical mesh sieves set to a range of 0.841 mm to 1.19 mm (standard mesh $-16 + 20$) [27], resulting in a geometric mean diameter of 1.00 mm [39]. The sieved zeolite is exposed to a cleaning cycle, which involves rinsing in deionized distilled water to remove residual debris and dust, and drying at $80 \pm 3°C$ for 24 h to remove any residual moisture [30].

Table 1. Natural zeolite specifications [32,37].

Chemical Composition	Mineral Component	85%–95% Clinoptilolite (non-crystalline silica opaline balance)		
	Cation Exchange Capacity (CEC)	180–220 meq/100 g (as ammonium, -N) (high)		
	Maximum Water Retention	>55 wt % (hydrophilic)		
	Overall Surface Area	24.9 m²/g (large)		
	Bulk Density	approx. 55–60 lb/f³		
	Hardness	Moh's No. 4 (high)		
	pH	7–8.64		
	Colour	Pale Green		
MSDS Composition Information		**Chemical**	**wt %**	**CAS No.**
		Clinoptilolite	90–97	12173-10-3
		Water	3–10	7732-18-5
Analytical Rock Data		SiO_2		67.4%
		Al_2O_3		10.6%
		MgO		0.45%
		K_2O		4.19%
		MnO		<0.01%
		CaO		2.23%
		TiO_2		0.27%
		Fe_2O_3		1.70%
		Na_2O		0.59%
		P_2O_5		0.10%
		Loss-On-Ignition (LOI) 925 °C		11.40%
Major Cation Range		Ca		1.60%–2.0%
		K		2.93%–3.47%
		Na		<0.5%

2.3. Heavy Metallic Ion Solution

Due to a greater presence in various Ontario mine waste streams as presented by Wilson [40] and the strict limitations required by the Canadian Government [41], this study focuses on the presence of the heavy metallic ions of Pb^{2+}, Cu^{2+}, Fe^{3+}, Ni^{2+} and Zn^{2+} [32,42]. The cations present in the sorbent have valences that differ from those in solution. Consequently, as the dilution increases, the selectivity of the adsorbent for the ion with a higher valence also increases. Accordingly, comparative analysis of various metal ions should be conducted at the same normality and temperature, in order to minimize the changes observed in isotherm configuration with dilution [43]. The synthetic metallic ion solutions are prepared from analytical grade nitrate salts in deionized distilled water, namely $Pb(NO_3)_2$ (CAS No. 10099-74-8), $Cu(NO_3)_2 \cdot 3H_2O$ (CAS No. 10031-43-3), $Fe(NO_3)_3 \cdot 9H_2O$ (CAS No. 7782-61-8), $Ni(NO_3)_2 \cdot 6H_2O$ (CAS No. 13478-00-7), and $Zn(NO_3)_2 \cdot 6H_2O$ (CAS No. 10196-18-6), respectively, and combined equally to maintain a total normality of 0.01 N (10 meq/L). The five metals in a multi-component system of 2.0 meq/L per metal correspond to concentrations of approximately 207 mg/L for Pb^{2+}, 64 mg/L for Cu^{2+}, 37 mg/L for Fe^{3+}, 59 mg/L for Ni^{2+}, and 65 mg/L for Zn^{2+}, respectively. The NO_3^- anions in the aqueous solution do not influence the ion-exchange process, since they do not form any metal-anion complexes and do not hydrolyze in solution [13,44].

The uptake of multiple HMIs from aqueous solutions on natural zeolite is a complex process consisting of predominately ion-exchange and adsorption. At high initial concentrations, this process could be accompanied by precipitation and the metal ion hydroxo-complexes formed can be sorbed on zeolite surface sites that encompass different sorption affinity [13]. Research has demonstrated [26,43,45] that the sorbate solution acidity level affects the uptake of metals, and this is particularly the case for the HMIs that have low preference by zeolite. Such factors include the metal ion speciation and natural stability, as well as the electro-kinetic properties of zeolite in aqueous solutions. At a low pH level, the hydrogen cation (H^+) is considered as a competitive ion to the HMI during the ion-exchange process [43]; evidently, the process is preferred at higher pH levels, which should be lower than the minimum pH of precipitation [10]. The pH level of the effluent solution decreases, depending on the metal removed. Therefore, the pH range under which sorption takes place should be specified [11].

The Canada-Wide Survey of Acid Mine Drainage [40] reports a seasonal average of a majority of the mines surveyed to have documented pH values ranging from 2 to 5. This present study is conducted in the conservative end of this range, with the influent stock acidified to a pH level of around 2 with concentrated nitric (HNO_3) acid (CAS No. 7697-37-2) [46], to prevent precipitation of the metal ions [14,43]. With this low pH, however, the H^+ ion competition is significant, and so the removals obtained are on the conservative side and are lower than would normally be expected in field installations.

2.4. Analytical Procedure and Quality Control

There are various atomic spectrometry techniques, which include Flame Atomic Adsorption (AA), Graphite Furnace AA, and Inductively Coupled Plasma Mass Spectrometry (ICP-MS). In particular, the Inductively Coupled Plasma—Atomic Emission Spectroscopy (ICP-AES) technique permits the complete atomization of the elements in a sample. This feature minimizes the potential for chemical interferences. It is considered as a true multi-element technique with exceptional sample throughput, and with a very wide range of analytical signal intensity [47]. Therefore, the HMIs are analyzed in their aqueous phase by ICP-AES technology (Optima 7300 DV, Part No. N0770796, Serial No. 077C8071802, Firmware Version 1.0.1.0079; Perkin Elmer Inc.; Waltham, MA, USA); with corresponding WinLab32 Software (Version 4.0.0.0305). The analyte primary wavelengths of each HMI element targeted were selected as 327.393 for Cu, 238.204 for Fe, 231.604 for Ni, 220.353 for Pb, and 206.200 for Zn, respectively; on the basis that these wavelengths have the strongest emission and provide the best quantifiable detection limits (QDL). Analysis is conducted with a plasma setting in radial view (to concentrations of greater than 1 mg/L), with QDLs of 0.05 µg/mL for Cu, Fe, Ni, and Zn, and 0.10 µg/mL for Pb. The spectrometer settings involve auto sampling of 45 s normal time at a rate of 1.5 mL/min, and a processing setting of 3 to 5 points per peak with 2 point spectral corrections. The calibration curve is generated 'through zero' by applying a stock blank and a multi-element Quality Control Standard 4 with 1, 10, 50, 90, and 100 mg/L concentrations (as per Standard Methods Part 3000) [46]. In comparison to the 'linear calculated intercept' calibration method, only a 0.15 mg/L (or 0.66%) maximum discrepancy is observed among all average concentration values, demonstrating an accurate overall calibration. In addition, the median 50 mg/L calibration standard is applied as a check parameter, with the intent to ensure a higher accuracy of all experimental measurements. Triplicate readings and their mean concentrations in calibration units are generated in mg/L by the corresponding WinLab32 Software. The sorbed amount of HMI is calculated from the difference between the starting concentration and its concentration in the 0.45 µm filtered samples' supernatant.

During every ICP-AES analytical session, several quality control methods are applied, and evaluated by three check parameters to assess the calibration quality [48]. Firstly, the percent relative standard deviation (%RSD) reports an average of 0.433%, which is well within the <3% limit recommended. The triplicate concentration of the median standard has an average value of 49.26 mg/L, and is within 5% of the known value. Finally, the correlation coefficient of each HMI analyte primary wavelength reports an average of 0.999977, which is very close to unity. Therefore, the data is relatively accurate, highly reproducible, and the experimental replicates are reliable based on the calibration relationship established.

The multi-component stock is created by diluting the respective HMI nitrate salts of three 1-L stock solutions, acidified to a pH of 2.0 ± 0.1, and then re-combined. These 3 stocks (denoted as X, Y, Z) are diluted by one 50% step to be within the 0–100 mg/L calibration range, analyzing each separately and combined (denoted as M). The consistency in stock preparation is demonstrated in Table 2. The average diluted concentrations of the X, Y, and Z influent stocks for Cu^{2+}, Fe^{3+}, Ni^{2+}, Pb^{2+}, and Zn^{2+} are 70.15, 39.25, 61.18, 216.02, and 66.54 mg/L, respectively. The diluted concentration of the M influent stock for Cu^{2+}, Fe^{3+}, Ni^{2+}, Pb^{2+}, and Zn^{2+} are 70.22, 39.21, 60.48, 213.22, and 65.98 mg/L, respectively. A maximum difference of 2.80 mg/L, equivalent to 1.3%, is detected for the Pb^{2+} stock. Also, the corresponding HMI concentrations in mg/L are comparable to the theoretically expected values, based on the selected total 10 meq/L initial concentration; only a 0.05% difference between the average of all initial concentrations of the theoretical and combined M stock is detected. Overall, this demonstrates that strong quality control has been implemented.

Table 2. Inductively Coupled Plasma—Atomic Emission Spectroscopy (ICP-AES) generated multi-component stock concentration.

Sample ID	Analyte	Int (Corr)	RSD (Corr Int)	Conc (Calib) (mg/L)
M-X	Cu 327.393	188,070.71	0.27	34.26
	Fe 238.204	79,641.94	0.39	19.18
	Ni 231.604	41,071.22	0.50	29.81
	Pb 220.353	32,330.16	0.52	105.28
	Zn 206.200	55,015.91	0.38	32.31
M-Y	Cu 327.393	186,885.03	0.74	34.04
	Fe 238.204	79,083.95	0.90	19.04
	Ni 231.604	40,721.53	0.48	29.55
	Pb 220.353	31,973.87	0.31	104.12
	Zn 206.200	54,758.60	1.09	32.16
M-Z	Cu 327.393	202,742.71	0.91	36.93
	Fe 238.204	85,771.53	1.02	20.65
	Ni 231.604	44,652.28	3.73	32.41
	Pb 220.353	35,199.76	3.77	114.63
	Zn 206.200	60,176.84	4.12	35.34
MM	Cu 327.393	192,776.82	0.63	35.11
	Fe 238.204	81,419.80	0.80	19.60
	Ni 231.604	41,667.28	0.38	30.24
	Pb 220.353	32,738.45	0.40	106.61
	Zn 206.200	56,170.16	0.94	32.99

2.5. Sorption System Design

Based on qualitative observations, the uptake of counter-ions in a continuous column system is favoured by various factors, including: a strong preference of the zeolite for the HMI counter-ions in solution, low concentration of HMI counter-ions, small and uniform particle size, high volume capacity and low degree of cross-linking, elevated temperature and low flow rate, as well as a high column height or aspect ratio [18] (p. 427). With this is mind, the apparatus development considers an extensive material and equipment selection process, with numerous stages of optimization in order to maintain flow continuity and repeatability. The final design was adopted in consideration of the following factors:

- Zeolite Compaction Technique

  ○ Regulated Layers of Dry Mass
  ○ Systematic Tampered Compaction

- Column Dimensions

  ○ Modular Design
  ○ Internal Diameter (1 in)
  ○ Sorption Column Height (1 ft)

- Flow Configuration

  ○ Upflow Distribution
  ○ Dual-Column Series Connection
  ○ Methodical Flow Rate Variability

- Pump Type

  ○ Diaphragm Metering

- Sampling Method

  - ○ Automated Mode Controls
  - ○ Customized Sampling Chambers
  - ○ Modes' Interchange in Five (5)-minute Intervals

- Analysis Period

  - ○ Three (3)-hour Contact Period

Based on these critical parameters, the sorption system design is finalized. Figure 1 is a schematic representation of the constructed prototype, detailing the flow paths through the system. The fundamental components include:

- HMI Multi-Component Influent Stock
- Metering Pump
- Silicon Tubing and Polyvinyl chloride (PVC) Connections
- Check Valves
- Automatable Solenoid Valves (symbol S)
- Packed Fixed-Bed Sorption Columns
- Custom Sampling Chambers
- Sampling Ports
- Effluent Collection Basin

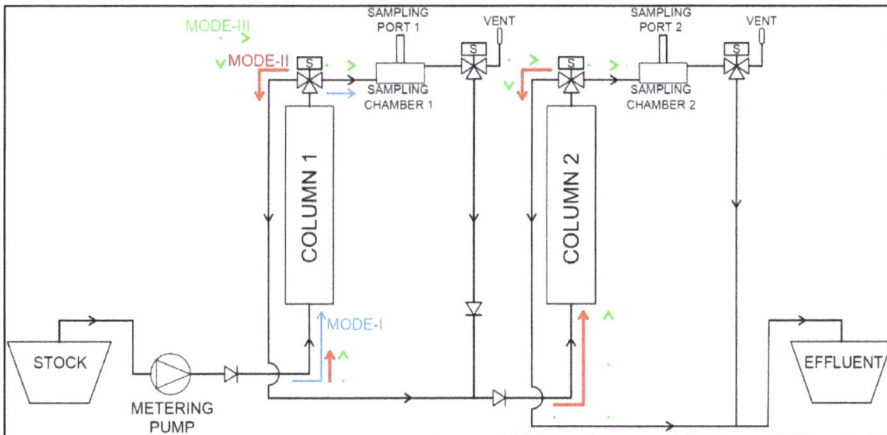

Figure 1. Schematic representation of automated sorption system prototype flow path layout.

2.5.1. Column Dimensions

The column is made of a circular section of clear PVC SCH-40 pipe (Part No. r4-1000; Fabco Plastics; Maple, ONT, Canada), 30.48 cm in height with 2.61 cm internal diameter. In order to minimize potential effects of wall and axial dispersion in the columns, the bed depth-to-particle diameter ratio should be kept greater than 20 [36]. Using the average nominal zeolite diameter of 1.00 mm as a reference, this ratio works out to be over 300 for the design. The cleaned and dry zeolite particles are added to the column at nine layers applied at 20-mL or 16.9 g amounts. Each layer is compacted with medium force, pounding six times with a customized PVC plunger of a diameter equal to the internal diameter of the column; such that the column height and the zeolite bed depth are equivalent. Inert plastic mesh with a smaller size than the minimum zeolite particle gradation of 0.841 mm is used to contain the zeolite material, and to permit sample flow through the columns. This mesh is set at each end of the column, within the two halves of a PVC SCH-80 socket union fitting incorporating a viton o-ring (Part No. 897010; Fabco Plastics; Maple, ONT, Canada), connected to a nominal 1 × 1/2-inch PVC SCH-40 reducer bushing (Part No. 438130 (slip × FPT); Fabco Plastics; Maple, ONT, Canada). All components are connected to 1/4-inch silicon tubing and with corresponding adapters and nipples fittings.

2.5.2. Flow Rate and Configuration

As a dual-stage system, the two columns are connected in a serial-flow arrangement, such that the first column receives the original stock with a higher HMI concentration and the second column receives the effluent from the first column. The upflow configuration ensures an overall better quality of flow, with a low liquid hold-up and a good stock feed distribution across the column cross sectional area. In contrast, for the downflow mode, an increase in pressure drop and flooding of the column bed is more probable [28]. Consequently, the stock is fed in an upflow direction to ensure proper and thorough distribution to the column beds and to minimize the need for backwashing and head loss effects.

A critical parameter in the design process is the flow rate. Existing research demonstrates that lower flow rates result in high detention times in the column, which is needed due to the relatively slow uptake rate of zeolite [24,27]. The HMI solution volume element is in contact with a given zeolite bed layer for only a limited period of time. Consequently, equilibrium is not usually achieved and thereby results in a lower overall uptake of HMIs from the influent stock solution. Preliminary testing involved a peristaltic pump, using the corresponding silicon rubber tubing. Significant back pressure was observed and the capacity of the peristaltic pump was hindered. Consequently, the required flow rate was unachievable; the rotational speed and strength decreased for the feed to completely traverse through the entire system. Subsequently a diaphragm-type metering pump (No. 950218125-C Plus, max 45-LPD, 80-psi, 125-AC, 50/60-Hz; PULSAtron; Punta Gorda, FL, USA) is employed in the final design, which mechanically facilitates the desired stock feeding rate. Based on the 45-LPD (31.25 mL/min) capacity of the metering pump, preliminary flow rate testing of the pump set to 100% stroke (mechanically pumped volume) and 50% rate established an initial, repeatable, point-of-reference flow rate of 6.36 ± 0.32 mL/min. This stroke-rate setting is maintained and is comparable to the lower end of the 6–18 mL/min range recommended by Inglezakis et al. [24,27], to provide sufficient detention time in the system.

2.5.3. Sampling Method

Another critical component to the design is the sampling method, and how to maintain continuous flow through the system while sampling the effluent of both columns. Due to the relatively slow feeding rate, the time to collect the desired sample volume for dilution and ICP-AES analysis would require residual sample volume and minutes of valuable contact time. Three-way solenoid valves (No. 00457979, 0124-C, 1/8-FKM-PP, NPT-1/4, max 145-psi, 24-V, 60-Hz, 8-W, 38-mL; burkert; Ingelfingen, Germany) are implemented to ensure that while a sample volume is collected at the desired sampling time, both columns would still be fed continuously. The MODE valves and custom fabricated rotating 30-mL sampling chambers are attached to the top exit of each sorption column, with accessible sampling ports. A second three-way solenoid VENT valve is included at the exit of each sampling chamber to introduce an air vent to assist in rapid sample extraction by preventing vacuum pockets within the sample chamber and discharge tubing. A multi-turn valve is included at the exit of the vent for the first column to introduce minor back pressure similar to that of the second column, so as not to alter the flow characteristics through the first column. Check valves are placed at critical locations throughout the hydraulic circuit to prevent back flows.

2.5.4. System Modes of Operation

The sorption system presented in this paper is comprised of three distinct modes of operation that are controlled by the MODE valves for each column:

- **Mode-I**

  ○   Sorption System Activation
  ○   Fill Sorption Column 1 and Sample Chamber 1

- **Mode-II**

  ○   Flow Circulation through Sorption Columns 1 (C1) and 2 (C2)
  ○   Detour of flow to Sampling Chambers 1 (SC1) and 2 (SC2)
  ○   VENT Valve Activation for Sample Collection

- **Mode-III**

  ○   Flow Rate Division
  ○   Concurrently 'Pulse' Fill Sampling Chambers 1 and 2

Figure 2 presents the arrangement of the prototype components, including an adjustable bi-stable timer which determines the time division modulation of the MODE and VENT solenoid valves.

Activating the process in Mode-I, the fluid element is mechanically pumped from the acidified 3-L multi-component influent stock. Once the pump is turned on, the inlet tubing is primed with the influent stock and passes the column check valve at the system inlet. The fluid element passes through the mesh-union fitting and reaches the base of the first column (C1), and traverses up through the sample chamber entry solenoid valve to the first sampling chamber. Once the 30-mL sample chamber is filled, the fluid element begins to drip at its exit against the multi-turn valve, which is an indication to switch the sample chamber entry solenoid MODE valves to Mode-II using the automated mode controls.

In Mode-II, the fluid element by-passes the first sampling chamber (SC1), continues to traverse through column 1 (C1) and begins to fill column 2 (C2). The fluid element does not cross-circuit back towards the exits of first column, due to the additional check valves connection located at the entry of the second column. While the fluid element traverses up both columns C1 and C2, the sampling chamber exit solenoid VENT valve is switched from closed to open. The sampling port tube is uncapped, twisted using the custom rotating handle and inverted to draw a 30-mL sample. The VENT valve is then turned off (closed from atmosphere). It is important to note that the inlet-outlet offset of the sampling chambers guarantees a highly repeatable sample volume. It is designed to minimize cross-contamination, for when the chamber is rotated from the vertical upward (sample collection in Mode-III) to downward (sample dispense in Mode-II and VENT) position, the chamber contents are completely void.

Once C2 is filled, both MODE valves of the sampling chambers are switched from Mode-II to Mode-III. The fluid element now simultaneously traverses through C1 and C2, while filling SC1 and SC2, dividing the flow rate and maintaining a continuous flow through the system. Once both sampling chambers are filled, the MODE valves are switched from Mode-III back to Mode-II, such that the fluid element by-passes the sample chambers and only traverses through the columns. At this time, the VENT valves are switched from closed to open, and the samples are taken from the sampling ports of both SC1 and SC2. Once both samples are collected, the VENT valves are closed and the MODE valves are once again switched back to pulse in Mode-III until SC1 and SC2 are filled. This sequence is repeated at approximate 5-min increments between Mode-II and Mode-III, for a total analysis period of just over 3-h. The prototype is secured to a sturdy, level frame that includes supporting clamps for the packed fixed-bed columns and a removable sampling chamber lock mechanism for maintenance accessibility.

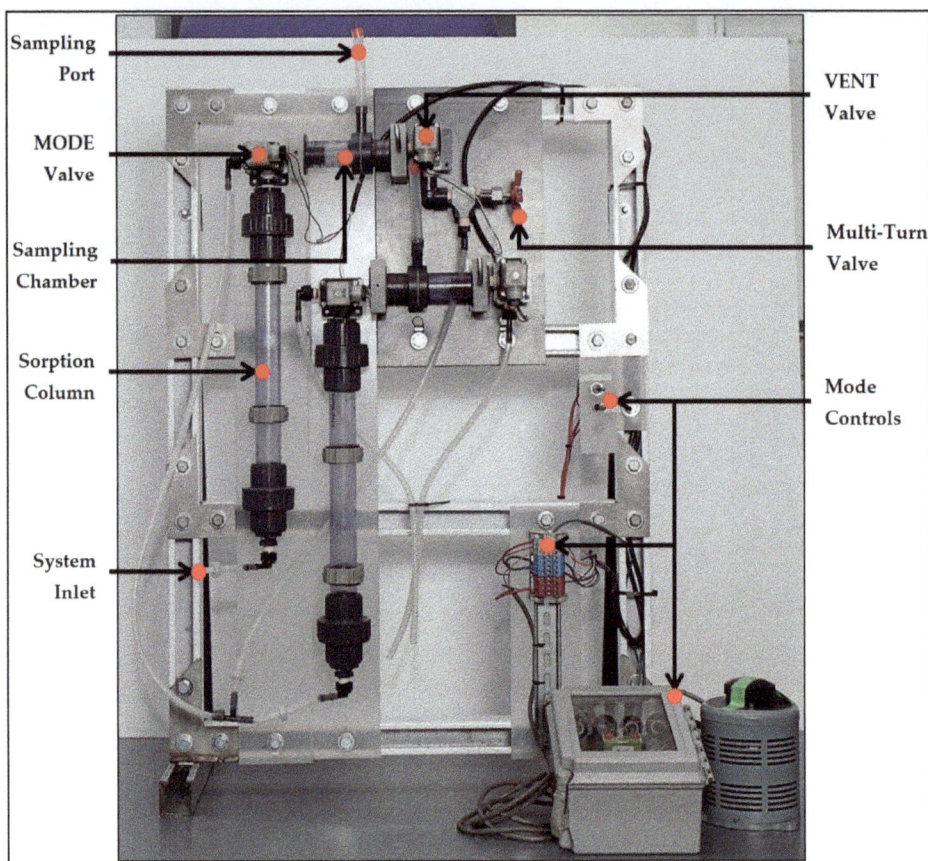

Figure 2. Image of automated sorption system prototype design.

It is important to note that the sample chambers are fabricated to a 30-mL capacity, to ensure that the 25-mL required volume is attained, to be filtered for dilution in preparation of ICP-AES analysis. This influences the time to collect the sample volume, based on the selected pump flow rate of this study. Also, the spacing of the prototype components influences the tubing connection lengths. The dual-column sorption system design presented in Figures 1 and 2 provides the opportunity to analyse higher flow rates and/or prolonged sample collection in Mode-III in future research endeavours.

3. Results and Discussion

3.1. Preliminary Batch Mode Results

Detailed analysis on the selected HMIs of this study was conducted by Ciosek and Luk [32,33] in batch-mode configuration, consisting of a synthetic nitrate salt solution at 10 meq/L total concentration, acidified to a pH of 2 by concentration HNO_3 acid, with a zeolite dosage of 4 g per 100-mL HMI solution. The aqueous solution is agitated within a contact period 180 min by means of a triple-eccentric drive orbital shaker operating at 400 r/min set to 22°C. The five (5) HMIs were methodically combined in single-, dual-, triple-, and multi-component systems. Elemental analysis by ICP-AES concludes that after 3 contact h, a total HMI uptake of 0.0986 meq/g is achieved the multi-component system.

The percent removal of Pb^{2+}, Cu^{2+}, Fe^{3+}, Ni^{2+} and Zn^{2+} are 94.0%, 21.9%, 56.2%, 9.10%, and 16.5%, respectively. The zeolite's preference among the HMIs is demonstrated by the selectivity series, which is established as $Pb^{2+} \gg Fe^{3+} > Cu^{2+} > Zn^{2+} > Ni^{2+}$. One of the objectives of this current study is to investigate how these HMIs interact and affect the removal uptake in a continuous flow, dual-column settling.

3.2. Automated Column Sorption System

3.2.1. Sampling Sequence and Flow Rate

Table 3 provides the timeline of modes in the system set-up sequence. Once the inlet tubing and check valve are primed, the pump starts to fill the inlet connection cavity. At full flow rate in Mode-I, it requires approximately 8:39 min:s to travel from the base to the top of column 1 (C1). After approximately 3:40 min:s, sample chamber 1 (SC1) is filled, and Mode-II (circulation) is initiated while the first sample (C1-A) is collected. In the continuous flow of Mode-II, and it requires approximately 8:27 min:s for the flow to travel from the base to the top of column 2 (C2). The flow is then switched to Mode-III (pulse), which divides the flow to fill both sample chambers SC1 and SC2. Once the 30-mL volumes are filled, Mode-III is switched back to Mode-II and the samples C1-B and C2-B are collected at 42:50 min:s.

Table 3. Sorption system set-up sequence.

MODE	Function	Flow Description	Time (min:s)
		Primed Inlet to C1 Base	2:26
I	Fill C1	C1 Base to C1 Top	11:05
		C1 Top to SC1 Drip	15:10
	Fill SC1		
		Sample C1-A	18:50
II		C2 Inlet to C2 Base	24:08
	Fill C2	C2 Base to C2 Top	32:35
		C2 Top to SC2 Drip	36:14
III	Fill SC1 and SC2		
II		Sample C1-B and C2-B	42:50
III	Fill SC1 and SC2		48:04
II		Sample C1-1 and C2-1	54:27

Once the system is set-up, there is an orderly switch between Mode-II (circulation) and Mode-III (pulse). Table 4 summarizes this sampling sequence. Altogether, there are twenty-nine 30-mL samples collected throughout the analysis period. During the system set-up, the collection of the first sample (C1-A) is followed by the second column 1 sample (C1-B) and first column 2 sample (C2-B). The orderly sequence begins at the collection of Cx-1 (48:04 min:s), for a total of two samples for each of the thirteen (13) runs. A total waste (TW) sample in the collection basin of the sorption system is also collected half-way through sampling (115:45 min:s) and at the end of the analysis period (195:00 min:s). The final influent stock and total effluent volumes are approximately 1.45-L and 550-mL, respectively.

Table 4. Sorption system sampling sequence.

Sample	MODE	Start Time (min:s)	End Time (min:s)	
			SC1	SC2
C1-A	I	15:10	18:50	-
	II	18:50		
Cx-B	III	36:40	42:42	42:49
	II	42:50		
Cx-1	III	48:04	54:10	54:27
	II	54:27		
Cx-2	III	59:39	66:15	66:34
	II	66:34		
Cx-3	III	71:37	77:55	78:19
	II	78:20		
Cx-4	III	83:24	89:56	89:56
	II	89:57		
Cx-5	III	95:05	101:32	101:45
	II	101:46		
Cx-6	III	106:52	112:53	113:10
	II	113:11		
TW1		115:45		
Cx-7	III	118:10	123:45	123:56
	II	123:57		
Cx-8	III	129:00	136:07	136:19
	II	136:20		
Cx-9	III	141:22	147:29	147:34
	II	147:35		
Cx-10	III	152:40	158:35	158:45
	II	158:46		
Cx-11	III	163:45	169:40	169:45
	II	169:46		
Cx-12	III	174:50	181:06	181:04
	II	181:10		
Cx-13	III	186:11	192:08	192:36
	II		PUMP OFF	
TW2		195:00		

The flow patterns are continuous and methodically kept consistent throughout the analysis period. Once samples C1-B and C2-B are collected, an average time of 6:26 min:s passes to switch from Mode-III to Mode-II, and 5:05 min:s from Mode-II to Mode-III. When the flow is divided in Mode-III, the average sampling acquisition time of 6:19 min:s is required to fill the 30-mL chambers, which is then collected for the filtering and dilution of the 25-mL sub-sample. The adjustable bi-stable timer at an approximate 50% duty setting automatically toggles the pulsing in Mode-III, to maintain a relatively consistent division of flow between the two columns, creating partial diversion to the two sampling chambers. This is demonstrated in relation to the start and end times of the sampling sequence.

It is important to note that the first sorption column (C1) receives a continuous inlet flow rate, as observed by the Mode-I filling rate of 8.18 mL/min for SC1 sample C1-A. Immediately after SC1 is filled and before 30-mL collection, the switch to Mode-II diverts the flow to begin filling the second sorption column (C2). Once both columns are filled, the flow is divided in Mode-III at the top outlet of C1, between SC1 and SC2, while maintaining consistent contact throughout the system. Again, during

Mode-III, C1 receives the same inlet flow rate, but the sampling chambers SC1 and SC2 receives this division of flow. It is column C2 that receives a variable flow rate during the analysis period, set by the adjustable division timer. Based on the filling start time (36:40 min:s) of C1-B and C2-B, and the end time of collection (192:36 min:s) for C1-13 and C2-13, a geometric mean flow rate between Mode-II and Mode-III in C2 is established as 5.39 mL/min. With these unique flow rates recognized for both columns, their corresponding EBCTs are established by Equation (4) to yield:

$$EBCT_1 = \frac{d^2\pi H}{4(Q_{C1})} = \frac{(2.61 \text{ cm})^2 \pi (30.48 \text{ cm})}{4(8.18 \text{ cm}^3/\text{min})} \cong 20:00 \text{ min}:s$$

$$EBCT_2 = \frac{d^2\pi H}{4(Q_{C2,AVG})} = \frac{(2.61 \text{ cm})^2 \pi (30.48 \text{ cm})}{4(5.39 \text{ cm}^3/\text{min})} \cong 30:20 \text{ min}:s$$

Due to the relatively slow kinetics of zeolites, long residence times are required. Any solution volume-element in contact with a given zeolite bed layer is for only a limited time period, which is usually insufficient to reach the equilibrium state. The failure of zeolite to attain local equilibrium causes a lower uptake of HMIs from solution [11]. The detention time that the fluid element is in contact with the fixed-bed per sorption column is a result of the flow rate selected in this present study. This trend between the columns provides insight into the overall treatment availability of the zeolite material in this very unique configuration.

3.2.2. Acidity Levels

Natural zeolites are known to raise the pH level in acidic aqueous solutions, which is due to: (1) the ion-exchange of H^+ ions, (2) the binding of H^+ ions to the Lewis basic sites linked to the oxygen atoms in the zeolite framework, and (3) the OH^- ions in solution deriving from hydrolysis of some species present in the zeolite [11]. The pH level of the aqueous solution controls the overall sorption process; adsorption of the HMI at the solid–water interfaces as well as the ion-exchange of cations within the zeolite structure. Stylianou et al. [10] points out that for all minerals, a decrease in the ion-exchange capacity of HMIs occurs for a pH range of 1 to 2. However, very low pH levels may positively influence the sorption process with the hydrolysis of the HMIs in solution [10]. Table 5 presents the pH levels of the effluent for both columns, of equally distributed selected time-step checkpoints of Cx-3, Cx-6, Cx-9, and Cx-13. When the acidified influent stock is combined to a 3-L volume, the average multi-component MM pH level has a value of 1.90. By maintaining a very low initial pH level and the use of highly soluble nitrate salts, the precipitation of the HMIs is avoided. Additional trials verified that the filtered and unfiltered HMI influent stock concentrations are the same, indicating both effective dilution practices and complete solubility.

Table 5. The pH levels of selected sorption column samples.

Sample	pH Level	
	SC1	SC2
C1-A	6.34	-
Cx-3	4.79	6.84
Cx-6	3.99	6.72
TW1	6.05	
Cx-9	3.86	6.33
Cx-13	3.60	5.76
TW2	5.44	

As the sample traverses through the first column C1, the H^+ ions are captured by the zeolite, resulting in an increase in the pH level to 6.34 from the first sample C1-A. There is an interesting observation between the columns' pH levels, which is a direct reflection of the zeolite's removal capacity for both the HMIs of interest and the competitive H⁺ ions in solution. The pH level gradually decreases in both columns, with the levels of C2 being slightly greater than that of C1. The total waste (TW) collects in the effluent basin throughout the analysis period, and its pH level decreases from the half-way check point of 6.05 (115:45 min:s) to 5.44 at the final collection (195:00 min:s). This is a clear indication that the zeolite capacity is becoming exhausted for the competing H⁺ ions, as well as hindering the sorption process of the HMIs.

Research conducted by Vukojevic Medvidovic et al. [34] also demonstrated that the pH values changed during the uptake process, following the opposite shape of the typical breakthrough curves. At breakpoint, a drastic change in the pH value occurred, which corresponded with a rapid Pb^{2+} concentration increase. The maximum pH level is reached at the breakthrough point, due to the absence of HMIs in the effluent. The minimum pH level is reached at the exhaustion point, due to the increase of the concentration of HMIs in the effluent and due to their hydrolysis in solution. After the exhaustion point, the pH level is constant [12]. These findings suggest that the continuous monitoring of pH levels is important and considerably contributes to the prediction of breakthrough and exhaustion points [11,12,34]; in order to monitor the progress of the service life and inevitably regeneration (adsorption/desorption cycles), both of which are very significant for practical industrial applications [34].

Also, the pH level may influence the ionization degree of the sorbate (HMI solution) and the surface property of the sorbent (zeolite mineral) [44]. The structural stability of the sorbent should not be compromised; for once the pH level reaches below 1, the structure of clinoptilolite breaks down in a process termed 'dealumination'. Precipitation should be avoided, for once the ions of interest have precipitated they cannot be sorbed [43]. It should be noted that while low pH levels prevent precipitation, the competitive H⁺ ions present would hinder the sorption of HMIs. Therefore, it is to be expected that future field installations for the treatment of AMD (with typical pH range of 2 to 5 [40]) should potentially demonstrate even higher removal efficiencies. However, care should be taken in the design of industrial applications to incorporate pre-treatment processes to reduce particulates prior to applying the waste to any sorption system, to avoid flow obstruction in the sorption columns.

3.2.3. Hydraulic Conductivity Considerations

Following HMI uptake analysis, the columns are drained to sit overnight. The standard test method for permeability of granular soils (constant head) (ASTM 2434-68) [49] is adopted to determine the variance in the overall hydraulic conductivity between the sorption columns. The hydraulic coefficient of permeability is given by adapting the standard test in the following relationship:

$$k_T = \frac{V_C \cdot L}{A \cdot H_C \cdot T} \tag{6}$$

where k_T (in cm/s) is the coefficient of permeability, V_C (in cm³) is the quantity of water that has discharged from the column and collected, L (in cm) is the column height, A (in cm²) is the column cross-sectional area, H_C (in cm) is the constant head of water on the column or the vertical distance between the feed head level and the column overflow level, and T (in s) is the time required to collect the V_C volume. With a plumb tank clamp support system, the water is fed in upflow mode from its base. The collection volume (V_C) is set to 50-mL by a graduated cylinder, with a column height (zeolite bed depth) and cross-sectional area of 30.48 cm and 5.37 cm², respectively. Based on an 18 °C detected water temperature, the viscosity correction factor of $n_T/n_{20} = 1.0508$ is applied to reveal the hydraulic conductivity of columns C1 and C2 as 4.08×10^{-4} m/s and 3.89×10^{-4} m/s, respectively. With a 4.84% difference to the average between the columns, this demonstrates a consistency in the overall executed compaction method.

3.2.4. Heavy Metallic Ion Concentration Analysis

Table 6 provides the results of the HMI concentrations (C_t) in both sorption columns based on triplicate readings obtained by the ICP-AES software. The percent removal (%R) values are presented with respect to the 2.0 meq/L influent concentrations of each HMI. During the experimental sequence of sample collection, C2-11 is lost due to improper handling when transferring from the sampling port to filtration at 169:45 min:s. However, this sample is of a lower HMI concentration and the overall removal trend has been well-established by the time this sample is collected. The first major observation is that throughout the analysis period, Pb^{2+} is not detected in both column effluents as well as the total waste, indicating a complete removal of the ion. The C1-13 sample for Cu^{2+}, Fe^{3+}, and Zn^{2+} reports a removal of 18.09%, 82.54%, and 10.71%, respectively, from the first column. The dual-column configuration provides a substantial improvement on the removal as observed with the second pass in sample C2-13, to achieve a final removal of Cu^{2+}, Fe^{3+}, and Zn^{2+} of 80.07%, 99.98%, and 51.53%, respectively. This improvement is also attributed to the unique feeding rate and design of the second column C2; the additional EBCT of approximately 10-minutes is available for the sorption process to occur as well as the slightly higher pH levels (and therefore lower presence of competitive H^+ ions). The final total waste (TW2) effluent concentrations report very good removal for all ions except for Ni^{2+}, with a removal of 48.97%. This removal trend is also consistent with the batch analyses conducted by Ciosek and Luk [32,33]. This is significant, as it proves that results from complex experimental batch studies, which are in high abundance, are useful in providing information on the sorption performance (i.e., removal efficiency, selectivity, and kinetics) in industrial applications where the process is run in a continuous flow-feeding configuration. In summary, the results demonstrate for the first time the effectiveness of multiple HMIs sorption by zeolite in a dual-column system with continuous flow.

Table 6. The heavy metallic ion (HMI) concentration (meq/L) and percent removal (%R) in the sorption columns.

Sample		Cu^{2+}		Fe^{3+}		Ni^{2+}		Pb^{2+}		Zn^{2+}	
		meq/L	%R	meq/L	%R	meq/L	%R	meq/L	%R	meq/L	%R
	C1-A	0.000	100.00	0.000	99.98	0.007	99.67 [BP]	0.0003	99.98	0.002	99.92 [BP]
	C1-B	0.129	93.55 [BP]	0.000	99.99	0.911	54.47	0.0006	99.97	0.541	72.96
	C1-1	0.517	74.17	0.000	99.99	1.483	25.85	0.0004	99.98	0.974	51.30
	C1-2	0.938	53.09	0.001	99.93	1.906	4.71	0.0006	99.97	1.320	34.02
	C1-3	1.221	38.93	0.011	99.46	2.116	0.00	0.0003	99.98	1.507	24.66
	C1-4	1.369	31.54	0.030	98.49	2.231	0.00	0.0006	99.97	1.622	18.91
	C1-5	1.431	28.47	0.052	97.42	2.269	0.00	0.0004	99.98	1.671	16.43
C1	C1-6	1.468	26.60	0.072	96.40	2.273	0.00	0.0005	99.97	1.703	14.83
	C1-7	1.584	20.78	0.102	94.90 [BP]	2.316	0.00	0.0005	99.98	1.816	9.20
	C1-8	1.563	21.86	0.118	94.08	2.199	0.00	0.0005	99.98	1.751	12.43
	C1-9	1.543	22.84	0.138	93.11	2.174	0.00	0.0004	99.98	1.730	13.51
	C1-10	1.571	21.44	0.167	91.64	2.134	0.00	0.0004	99.98	1.739	13.04
	C1-11	1.598	20.12	0.209	89.55	2.123	0.00	0.0004	99.98	1.752	12.38
	C1-12	1.604	19.79	0.268	86.59	2.096	0.00	0.0001	100.00	1.750	12.49
	C1-13	1.638	18.09	0.349	82.54	2.130	0.00	0.0002	99.99	1.786	10.71
	C2-B	0.00	100.00	0.0004	99.98	0.002	99.90	0.0003	99.99	0.002	99.88
	C2-1	0.00	100.00	0.0003	99.98	0.002	99.88	0.0003	99.99	0.001	99.94
	C2-2	0.00	100.00	0.0003	99.98	0.012	99.42	0.0004	99.98	0.001	99.95
	C2-3	0.00	100.00	0.0002	99.99	0.046	97.68	0.0006	99.97	0.001	99.96
	C2-4	0.00	100.00	0.0002	99.99	0.131	93.43 [BP]	0.0002	99.99	0.001	99.93
	C2-5	0.00	100.00	0.0002	99.99	0.285	85.75	0.0002	99.99	0.009	99.55
C2	C2-6	0.00	100.00	0.0003	99.98	0.505	74.77	0.0003	99.99	0.049	97.56
	C2-7	0.00	100.00	0.0002	99.99	0.835	58.26	0.0003	99.98	0.155	92.23 [BP]
	C2-8	0.004	99.80	0.0002	99.99	1.163	41.85	0.0003	99.99	0.312	84.39
	C2-9	0.029	98.57	0.0002	99.99	1.444	27.82	0.0004	99.98	0.455	77.24
	C2-10	0.085	95.77 [BP]	0.0003	99.99	1.675	16.26	0.0004	99.98	0.597	70.17
	C2-12	0.289	85.57	0.0002	99.99	2.126	0.00	0.0006	99.97	0.895	55.23
	C2-13	0.399	80.07	0.0003	99.98	2.198	0.00	0.0004	99.98	0.969	51.53
TW1		0.0514	97.43	0.0004	99.98	0.3077	84.61	0.0004	99.98	0.1107	94.46
TW2		0.1659	91.71	0.0057	99.72	1.0207	48.97	0.0000	100.00	0.4088	79.56

Table header: HMI

3.2.5. Breakthrough Curve, Capacity and Usage Rate Analysis

The breakthrough curve is displayed in Figure 3, as a plot of the solute outlet concentration (C_t) from Table 6 normalized to the inlet concentration (C_o) [34]. This normalized ratio trend over the service time of analysis [12,25] at which the sampling chambers (SC1, SC2) are extracted is presented for each of the five (5) HMIs combined in the multi-component solution; for both the first sorption column (C1) and second sorption column (C2). The breakthrough point (BP) and exhaustion point (EP) of each HMI in each column are indicated. The first observation to be had is that the breakthrough curves of the first column C1 do not have a defined S-shape. However, in the second sorption column C2, the curves take on this typical shape. Vukojevic Medvidovic et al. [34] points out that the shape-change may be attributed to an improved solid-solution phase contact for sorption to take place.

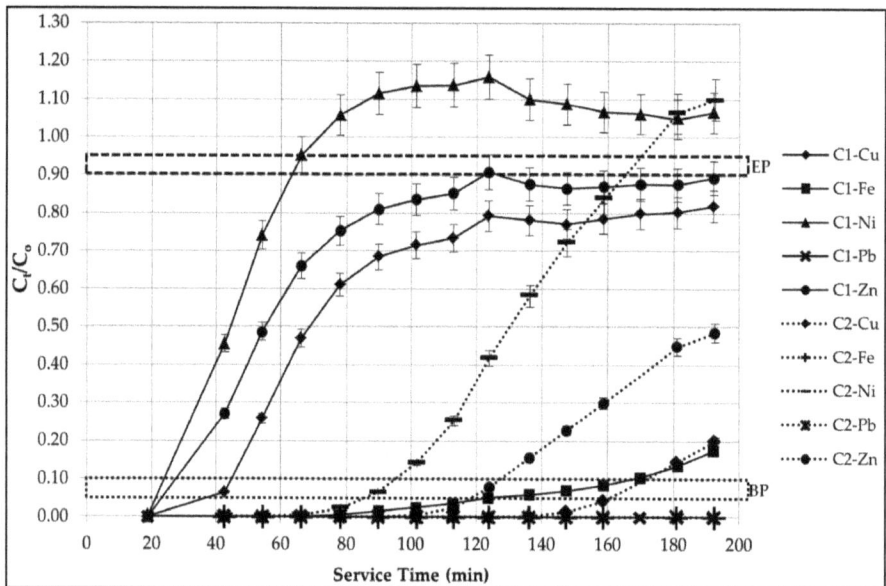

Figure 3. The multi-component system breakthrough curve.

The influent stock concentration of the Ni^{2+} ion is exceeded in the effluent solution, where the normalized C_t/C_o ratio surpasses 1 to reach an approximate maximum of 1.16 at 125 min and 1.10 at 190 min of service time, in sorption columns C1 and C2, respectively. The final ratio readings plateau at the end of service to approximately 5%–10% of the value of 1, given the nature of this experimental investigation. The effluent concentration that overshoots the influent concentration translates to concentration wave extremes inside the column [50]. Nuic et al. [12] investigates the breakthrough curves of Pb^{2+} and Zn^{2+} ions in dual-component solutions by natural zeolite; a similar trend is observed compared to this present study, where the C_t/C_o ratio even reaches a value of 2 for one set of operation conditions. This is attributed to the displacement of the bound Zn^{2+} by the Pb^{2+} from the influent, which is supported by a lower breakthrough capacity and higher exhaustion capacity in favour of the Pb^{2+} ion, specifically [12]. It is important to note that the ion-exchange mechanism that attributes to the sorption process of HMIs transpires through the zeolite's framework of pores and channels. The presence of stronger binding HMIs, such as Pb^{2+}, weaken the chemical bonds between the functional group on the surface of zeolite and the weaker HMIs, such as Ni^{2+} ions [9]. Given that zeolite demonstrates its highest preference towards Pb^{2+}, sorption site availability has reached its threshold, which may cause the leaching out of ions that zeolite holds a lower preference towards during this process. Therefore, careful screening on the selectivity of HMIs by zeolite should be conducted prior to adaptation.

In summary, the major trends observed from the breakthrough curves are as follows:

1. The zeolite holds the greatest preference towards to Pb^{2+} ion, based on its complete removal throughout the analysis period;
2. The zeolite demonstrates the least preference towards the Ni^{2+} ion;

    a. A more sudden breakpoint occurring after just 25 min and 90 min of service time in columns C1 and C2, respectively;
    b. An approximate exhaustion point after just 65 min and 165 min of service time in columns C1 and C2, respectively;

3. The Fe^{3+} ion is removed entirely and sustained throughout the analysis period in C2, and;
4. The removal of both the Cu^{2+} and Zn^{2+} ions begin to plateau at 120 min of service time in C1, acting in parallel and do not reach the lower threshold of the exhaustion point in both columns throughout the analysis period.

The breakthrough curves provide significant information from a perspective of sorption process performance, feasibility and optimization, which are vital for scaling-up the sorption system for industrial applications [28,50].

The overall column performance efficiency and its relationship between breakthrough capacity (C_{BP}) and exhaustion capacity (C_{EP}) of each sorption column are unique to the individual HMIs selected. As observed in the data displayed Table 6 and the trends visualized in Figure 3, the EP is only attained by Ni^{2+}; the remaining four HMIs have yet to reach this point due to the constraints of the 3-h analysis period. Evidently, the optimization of future works would be to prolong the service time in order to quantify the overall columns' performance efficiency. Qualitatively speaking, for both columns' effluent and total waste (TW), the Pb^{2+} ion is completely removed throughout the analysis period, demonstrating the utmost efficiency; neither BP nor EP are attained. On the opposite end of the spectrum, the Ni^{2+} ion reaches exhaustion quite rapidly.

The volume of the effluent treated at breakthrough (V_{BP}) is determined with the use of the mean flow rates of 8.18 mL/min (Q_{C1}) and 5.39 mL/min (Q_{C2}) for the sorption columns C1 and C2, respectively. These flow rates are applied to the time data of Table 4, at which BP (approximately 95% removal or 5% of the 2.0 meq/L influent concentration per HMI) is observed; as indicated by the superscript in Table 6. The 2:26 min:s time required for inlet priming to the base of C1 as well as the 24:08 min:s time observed for solute contact to the base of C2 are deducted from these BP times. As summarized in Table 7, the approximate effluent volumes treated at BP (V_{BP}) are provided for both sorption columns and each HMI selected; based on the zeolite mass in each bed (m) of 152.10 g, Equations (2) and (5) are employed to determine the corresponding breakthrough capacity (C_{BP}) and usage rate (v_U), respectively.

Table 7. System breakthrough point performance.

HMI	Sorption Column 1			Sorption Column 2		
	$V_{BP,1}$ (L)	$C_{BP,1}$ (meq/g)	$v_{U,1}$ (g/L)	$V_{BP,2}$ (L)	$C_{BP,2}$ (meq/g)	$v_{U,2}$ (g/L)
Cu^{2+}	0.3295	0.00433	461.67	0.7257	0.00954	209.59
Fe^{3+}	0.9926	0.01305	153.24	-	-	-
Ni^{2+}	0.1342	0.00176	1133.54	0.3547	0.00466	428.80
Pb^{2+}	-	-	-	-	-	-
Zn^{2+}	0.1342	0.00176	1133.54	0.5380	0.00707	282.71

The Pb^{2+} and Fe^{3+} ions are completely removed in sorption column C2, which essentially transforms the influent stock of a five HMI multi-component solution to a triple-component solution containing Cu^{2+}, Ni^{2+}, and Zn^{2+} ions within the 3-h analysis period. The zeolite does not have to address the competition of the two preferred HMIs, which provides greater sorption site availability for the three HMIs remaining in solution. It is important to develop a relationship between the EBCT and usage rate. The time that the fluid element is in contact with the zeolite bed in sorption column C2 is ten minutes greater than the detention time in C1. This thereby demonstrates a stronger overall treatment availability in C2; reflected in the average 195.2% increase of bed capacity at breakthrough and the average 63.94% decrease in usage rate of the zeolite between the columns; based on the trends observed for the Cu^{2+}, Ni^{2+} and Zn^{2+} ions detected in solution.

Evidently, the columns' usage rate provides significant insight into the operation and management required for this unique sorption system. It has a direct impact on the financial viability of performing either replacement (disposal) or regeneration (on- or off-site), and is affected by factors that include HMI influent concentration, zeolite bed depth, and flow rate. Research into other sorbent materials demonstrates that the order of usage rate is consistent with the sorption capacity [51]. Due to the unique automated variable influent feeding rate and sampling technique proposed in this study, the usage rate and performance efficiency are very complex [28]. However, the major removal trend of Pb2+ >> Fe3+ > Cu2+ > Zn2+ >> Ni2+ is well-established and supports previous results [22,27,32,33,38,42,43], providing significant validation of this design.

The use of natural zeolites as sorbents in industrial wastewater treatment and environmental management is motivated by the non-toxicity of these minerals, their abundant global availability, and economic feasibility. The removal and recovery processes of HMIs from aqueous solutions by natural zeolites take into consideration the regeneration potential of the zeolite bed to be reused in multiple cycles, as well as the use of the recovered metals [52] in applicable industrial applications. Metal processing effluents contain high concentrations of recoverable metals, triggering a movement towards technologies to recover these metals from AMD waste [53]. The removal-regeneration-recovery process has the potential to generate additional revenue streams with the use of metals of value [54]; such as the HMIs investigated in this innovative study.

4. Conclusions

This research has demonstrated the performance of natural zeolite (clinoptilolite) to remove multiple heavy metallic ions (HMIs) commonly found in acid mine drainage. With the design and development of a novel dual-column sorption system, the lead (Pb^{2+}) ion is removed completely and sustained throughout the analysis period. The relationships between empty bed contact time (EBCT), breakthrough capacity, and usage rate are evident. The additional ten minutes of EBCT in the second sorption column contributes to an enhancement in overall removal for Cu^{2+}, Fe^{3+}, and Zn^{2+} by 75.67%, 99.90%, and 45.72%, respectively, from the first sorption column. This improvement is also apparent in the greater breakthrough capacity and lower usage rate in the second column, and visualized in an improved S-shape to the characteristic breakthrough curve. Based on the multi-component influent stock of 10 meq/L total concentration, the second column demonstrates a removal of 99.98%, 99.98%, 80.07%, 51.53%, and 0.00%, for Pb^{2+}, Fe^{3+}, Cu^{2+}, Zn^{2+}, and Ni^{2+}, respectively; and the final cumulative collection of effluent reports a removal of 100.00%, 99.72%, 91.71%, 79.56% and 48.97%, for Pb^{2+}, Fe^{3+}, Cu^{2+}, Zn^{2+}, and Ni^{2+}, respectively at the completion of the analysis period. These HMI sorption removal trends confirm the consistency between batch and continuous mode operations.

The modular design inventively incorporated a 'circulation-pulse' method to distribute the flow, rather than operating on a more commonly implemented fixed flow rate. With the consideration of this unique stock feed method, the findings of the service time and flow rate with respect to the removal trends are both interesting and significant. Forthcoming works in this research project include the advancement of service life and regeneration cycles, with further design development and optimization. The potential for variable flow rate operation and automatic adjustable sampling in a packed fixed-bed dual-column sorption design reveals practicality for treatment applications.

This study has provided greater insight into the immense potential that the natural mineral zeolite holds for the future of industrial wastewater treatment.

Acknowledgments: This research was conducted with the financial support of a Natural Sciences and Engineering Research Council of Canada (NSERC) Discovery Grant to Grace K. Luk.

Author Contributions: Amanda L. Ciosek and Grace K. Luk conceived and designed the experiments; Amanda L. Ciosek constructed the prototype, performed the experiments and analytical simulations, and analyzed the data; Amanda L. Ciosek and Grace K. Luk wrote the paper.

Conflicts of Interest: The authors declare no conflict of interest. The founding sponsors had no role in the design of the study; in the collection, analyses, or interpretation of data; in the writing of the manuscript, and in the decision to publish the results.

References

1. Vaca-Mier, M.; Lopez-Callejas, R.; Gehr, R.; Jimenez-Cisneros, B.E.; Alvarez, P.J.J. Heavy metal removal with mexican clinoptilolite: multi-component ionic exchange. *Wat. Res.* **2001**, *35*, 373–378. [CrossRef]
2. Johnson, D.B.; Hallberg, K.B. Acid mine drainage remediation options—A review. *Sci. Total Environ.* **2005**, *338*, 3–14. [CrossRef] [PubMed]
3. Akcil, A.; Koldas, S. Acid Mine Drainage (AMD): Causes, treatment and case studies. *J. Clean. Prod.* **2006**, *14*, 1139–1145. [CrossRef]
4. Mohan, D.; Chander, S. Removal and recovery of metal ions from acid mine drainage using lignite—A low cost sorbent. *J. Hazard. Mater.* **2006**, *B137*, 1545–1553. [CrossRef] [PubMed]
5. Motsi, T.; Rowson, N.A.; Simmons, M.J.H. Kinetic studies of the removal of heavy metals from acid mine drainage by natural zeolite. *Int. J. Miner. Process.* **2011**, *101*, 42–49. [CrossRef]
6. Baraket, M.A. New trends in removing heavy metals from industrial wastewater. *Arab. J. Chem.* **2011**, *4*, 276–282. [CrossRef]
7. Vukojevic Medvidovic, N.; Peric, J.; Trgo, M.; Nuic, I.; Ugrina, M. Design of fixed bed column for lead removal on natural zeolite based on batch Studies. *Chem. Biochem. Eng. Q.* **2013**, *27*, 21–28.
8. Nuic, I.; Trgo, M.; Vukojevic Medvidovic, N. The application of the packed bed reactor theory to Pb and Zn uptake from the binary solution onto the fixed bed of natural zeolite. *Chem. Eng. J.* **2016**, *295*, 347–357. [CrossRef]
9. Han, R.; Zou, W.; Li, H.; Li, Y.; Shi, J. Copper(II) and lead(II) removal from aqueous solution in fixed-bed columns by manganese oxide coated zeolite. *J. Hazard. Mater.* **2006**, *137*, 934–942. [CrossRef] [PubMed]
10. Stylianou, M.A.; Inglezakis, V.J.; Moustakas, K.G.; Malamis, S.P.; Loizidou, M.D. Removal of Cu(II) in fixed bed and batch reactors using natural zeolite and exfoliated vermiculite as adsorbents. *Desalination* **2007**, *215*, 133–142. [CrossRef]
11. Stylianou, M.A.; Hadjiconstantinou, M.P.; Inglezakis, V.J.; Moustakas, K.G.; Loizidou, M.D. Use of natural clinoptilolite for the removal of lead, copper and zinc in fixed bed column. *J. Hazard. Mater.* **2007**, *143*, 575–581. [CrossRef] [PubMed]
12. Nuic, M.; Trgo, J.; Peric, N.; Vukojevic Medvidovic, N. Analysis of breakthrough curves of Pb and Zn sorption from binary solutions on natural clinoptilolite. *Micropor. Mesopor. Mater.* **2013**, *167*, 55–61. [CrossRef]
13. Peric, J.; Trgo, M.; Vukojevic Medvidovic, N. Removal of zinc, copper and lead by natural zeolite—A comparison of adsorption isotherms. *Water Res.* **2004**, *38*, 1893–1899. [CrossRef] [PubMed]
14. Wang, C.; Li, J.; Sun, X.; Wang, L.; Sun, X. Evaluation of zeolites synthesized from fly ash as potential adsorbents for wastewater containing heavy metals. *J. Environ. Sci.* **2009**, *21*, 127–136. [CrossRef]
15. Anari-Anaraki, M.; Nezamzadeh-Ejhieh, A. Modification of an Iranian clinoptilolite nano-particles by hexadecyltrimethyl ammonium cationic surfactant and dithizone for removal of Pb(II) from aqueous solution. *J. Colloid Interf. Sci.* **2015**, *440*, 272–281. [CrossRef] [PubMed]
16. Borandegi, M.; Nezamzadeh-Ejhieh, A. Enhanced removal efficiency of clinoptilolite nano-particles toward Co(II) from aqueous solution by modification with glutamic acid. *Colloids Surf. A Physicochem. Eng. Asp.* **2015**, *479*, 35–45. [CrossRef]
17. Curkovic, L.; Cerjan-Stefanovic, S.; Filipan, T. Metal ion exchange by natural and modified zeolites. *Water. Res.* **1997**, *31*, 1379–1382. [CrossRef]

18. Helfferich, F. Equilibria; Kinetics; Ion-Exchange in Columns. In *Ion Exchange*; Series in Advanced Chemistry; McGraw-Hill Book Company: New York, NY, USA, 1962; pp. 95–322, 421–506.

19. Inglezakis, V.J.; Poulopoulos, S.G. Chapter 4—Adsorption and Ion-Exchange (Kinetics). In *Adsorption, Ion Exchange and Catalysis—Design of Operations and Environmental Applications*, 1st ed.; Elsevier Science: Amsterdam, The Netherlands, 2006; pp. 262–266. ISBN 13 978-0-444-52783-7.

20. Trgo, M.; Peric, J.; Vukojevic Medvidovic, N. A comparative study of ion exchange kinetics in zinc/lead—Modified zeolite-clinoptilolite systems. *J. Hazard. Mater.* **2006**, *136*, 938–945. [CrossRef] [PubMed]

21. Tsitsishvili, G.V. Perspectives of Natural Zeolite Applications. Occurrence. In *Properties and Utilization of Natural Zeolites—2nd International Conference 1985*; Akademiai Kiado: Budapest, Hungary, 1988; pp. 367–393.

22. Wang, S.; Peng, Y. Natural zeolites as effective adsorbents in water and wastewater treatment. *Chem. Eng. J.* **2010**, *156*, 11–24. [CrossRef]

23. Nezamzadeh-Ejhieh, A.; Shirzadi, A. Enhancement of the photocatalytic activity of Ferrous Oxide by doping onto the nano-clinoptilolite particles towards photodegradation of tetracycline. *Chemosphere* **2014**, *107*, 136–144. [CrossRef] [PubMed]

24. Inglezakis, V.J.; Loizidou, M.D.; Grigoropoulou, H.P. Equilibrium and kinetic ion exchange studies of Pb^{2+}, Cr^{3+}, Fe^{3+} and Cu^{2+} on natural clinoptilolite. *Water Res.* **2002**, *36*, 2784–2792. [CrossRef]

25. Nuic, I.; Trgo, M.; Peric, J.; Vukojevic Medvidovic, N. Uptake of Pb and Zn from a binary solution onto different fixed bed depths of natural zeolite—The BDST model approach. *Clay Miner.* **2015**, *50*, 91–101. [CrossRef]

26. Ersoy, B.; Celik, M.S. Electrokinetic properties of clinoptilolite with mono- and multivalent electrolytes. *Micropor. Mesopor. Mater.* **2002**, *55*, 305–312. [CrossRef]

27. Inglezakis, V.J.; Grigoropoulou, H. Effects of operating conditions on the removal of heavy metals by zeolite in fixed bed reactors. *J. Hazard. Mater.* **2004**, *112*, 37–43. [CrossRef] [PubMed]

28. Inglezakis, V.J. Ion exchange and adsorption fixed bed operations for wastewater treatment—Part I: Modelling fundamentals and hydraulics analysis. *J. Eng. Stud. Res.* **2010**, *16*, 29–41.

29. Erdol Aydin, N.; Nasun Saygili, G. Column experiments to remove copper from wastewaters using natural zeolite. *Int. J. Environ. Waste Manag.* **2009**, *3*, 319–326. [CrossRef]

30. Inglezakis, V.J.; Papadeas, C.D.; Loizidou, M.D.; Grigoropoulou, H.P. Effects of pretreatment on physical and ion exchange properties of natural clinoptilolite. *Environ. Technol.* **2001**, *22*, 75–82. [CrossRef] [PubMed]

31. Inglezakis, V.J. Ion exchange and adsorption fixed bed operations for wastewater treatment—Part II: scale-up and approximate design methods. *J. Eng. Stud. Res.* **2010**, *16*, 42–50.

32. Ciosek, A.L.; Luk, G.K. Lead Removal from mine tailings with multiple metallic ions. *Int. J. Water Wastewater Treat.* **2017**, *3*, 1–9. [CrossRef]

33. Ciosek, A.L.; Luk, G.K. Kinetic modelling of the removal of multiple heavy metallic ions in mine waste by natural zeolite sorption. *Water* **2017**, *9*, 482. [CrossRef]

34. Vukojevic Medvidovic, N.; Peric, J.; Trgo, M. Column performance in lead removal from aqueous solutions by fixed bed of natural zeolite–clinoptilolite. *Sep. Purif. Technol.* **2006**, *49*, 237–244. [CrossRef]

35. Reed, B.E.; Jamil, M.; Thomas, B. Effect of pH, empty bed contact time and hydraulic loading rate on lead removal by granular activated carbon columns. *Water Environ. Res.* **1996**, *68*, 877–882. [CrossRef]

36. Peric, J.; Trgo, M.; Vukojevic Medvidovic, N.; Nuic, I. The Effect of Zeolite Fixed Bed Depth on Lead Removal from Aqueous Solutions. *Sep. Sci. Technol.* **2009**, *44*, 3113–3127. [CrossRef]

37. Bear River Zeolite Co. Inc. Zeolite—Specifications and MSDS. Available online: http://www.bearriverzeolite.com (accessed on 1 September 2012 and 1 April 2017).

38. Inglezakis, V.J.; Hadjiandreou, K.J.; Loizidou, M.D.; Grigoropoulou, H.P. Pretreatment of natural clinoptilolite in a laboratory-scale ion exchange packed bed. *Water Res.* **2001**, *35*, 2161–2166. [CrossRef]

39. Mullin, J. Physical and thermal properties. In *Crystallization*, 4th ed.; Read Educational and Professional Publishing Ltd: Woburn, MA, USA, 2001; pp. 76–77, IBSN 0-7506-4833-3.

40. Wilson, L.J. Canada-Wide Survey of Acid Mine Drainage Characteristics. Project Report 3.22.1—Job No. 50788. Mineral Sciences Laboratories Division Report MSL 94–32 (CR). Ontario Ministry of Northern Development and Mines. Mine Environment Neutral Drainage (MEND) Program. Canada, 1994. Available online: http://mend-nedem.org/wp-content/uploads/2013/01/3.22.1.pdf (accessed on 30 October 2014).

41. Canadian Minister of Justice—Metal Mining Effluent Regulations. Consolidation SOR/2002-222. Justice Laws—Government of Canada. Available online: http://laws-lois.justice.gc.ca (accessed on 1 September 2014).

42. Wingenfelder, U.; Hansen, C.; Furrer, G.; Schulin, R. Removal of Heavy Metals from Mine Waters from Natural Zeolites. *Environ. Sci. Technol.* **2005**, *39*, 4606–4613. [CrossRef] [PubMed]

43. Inglezakis, V.J.; Loizidou, M.D.; Grigoropoulou, H.P. Ion exchange of Pb^{2+}, Cu^{2+}, Fe^{3+}, and Cr^{3+} on natural clinoptilolite: Selectivity determination and influence of acidity on metal uptake. *J. Colloid Interface Sci.* **2003**, *261*, 49–54. [CrossRef]

44. Minceva, M.; Fajgar, R.; Markovska, L.; Meshko, V. Comparative Study of Zn^{2+}, Cd^{2+}, and Pb^{2+} Removal From Water Solution Using Natural Clinoptilolitic Zeolite and Commercial Granulated Activated Carbon: Equilibrium of Adsorption. *Sep. Sci. Technol.* **2008**, *43*, 2117–2143. [CrossRef]

45. Ouki, S.K.; Kavannagh, M. Treatment of metals-contaminated wastewaters by use of natural zeolites. *Water. Sci. Tech.* **1999**, *39*, 115–122. [CrossRef]

46. Rice, E.W.; Baird, R.B.; Eaton, A.D.; Clesceri, L.S. Part 1000-Introduction, Part 3000-METALS. In *Standard Methods for the Examination of Water and Wastewater*, 22nd ed.; The American Public Health Association (APHA): Washington, DC, USA; The American Water Works Association (AWWA): Denver, CO, USA; The Water Environment Federation (WEF): Alexandria, VA, USA, 2012; pp. 1.1–68, 3.1–112, ISSN 978-087553-013-0.

47. Perkin Elmer Inc. *Atomic Spectroscopy—A Guide to Selecting the Appropriate Technique and System: World Leader in AA, ICP-OES, and ICP-MS*; Perkin Elmer Inc.: Waltham, MA, USA, 2011.

48. Perkin Elmer Inc. *WinLab32 for ICP—Instrument Control Software, Version 5.0*; Perkin Elmer Inc.: Waltham, MA, USA, 2010.

49. ASTM D2434–68. Standard Test Method for Permeability of Granular Soils (Constant Head). ASTM International, West Conshohocken, PA. 2000. Available online: www.astm.org (accessed on 1 March 2016).

50. Naja, G.; Volesky, B. Multi-metal biosorption in a fixed-bed flow-through column. *Colloid. Surf.* **2006**, *281*, 194–201. [CrossRef]

51. Othman, M.Z.; Roddick, F.A.; Snow, R. Removal of Dissolved Organic Compounds in Fixed-Bed Columns: Evaluation of Low-Rank Coal Adsorbents. *Water Res.* **2001**, *35*, 2943–2949. [CrossRef]

52. Sprynskyy, M.; Buszewski, B.; Terzyk, A.P.; Namiesnik, J. Study of the selection mechanism of heavy metal (Pb^{2+}, Cu^{2+}, Ni^{2+}, and Cd^{2+}) adsorption on clinoptilolite. *J. Colloid Interface Sci.* **2006**, *304*, 21–28. [CrossRef] [PubMed]

53. Zinck, J. Review of Disposal, Reprocessing and Reuse Options for Acidic Drainage Treatment Sludge. Report 3.42.3. Mine Environment Neutral Drainage (MEND) Program. Mining Association of Canada. CANMET Mining and Mineral Sciences Laboratories. Canada, 2005. Available online: http://mend-nedem.org/wp-content/uploads/2013/01/3.42.3.pdf (accessed on 30 October 2014).

54. Dinardo, O.; Kondos, P.D.; MacKinnon, D.J.; McCready, R.G.L.; Riveros, P.A.; Skaff, M. Study on Metals Recovery/Recycling from Acid Mine Drainage Phase IA: Literature Survey. Report 3.21.1a. Mine Environment Neutral Drainage (MEND) Program. CANMET, Energy, Mines and Resources Canada and WTC, Environment Canada. 1991. Available online: http://mend-nedem.org/wp-content/uploads/2013/01/3.21.1a.pdf (accessed on 30 October 2014).

**applied
sciences**

MDPI

Article

Adsorption of Chromium (VI) on Calcium Phosphate: Mechanisms and Stability Constants of Surface Complexes

**Ahmed Elyahyaoui [1,*], Kawtar Ellouzi [1], Hamzeh Al Zabadi [2], Brahim Razzouki [3],
Saidati Bouhlassa [1], Khalil Azzaoui [4], El Miloud Mejdoubi [4], Othman Hamed [5],
Shehdeh Jodeh [5,*] and Abdellatif Lamhamdi [4,6]**

1   Laboratory of Radiochemistry, Department of Chemistry, Faculty of Sciences, Mohamed V-Agdal,
     B.P 1014 Rabat, Morocco; ellouzi.kawtar@gmail.com (K.E.); bouhlass@fsr.ac.ma (S.B.)
2   Public Health Department, An-Najah National University, P.O. Box 7, Nablus 44830, Palestine;
     halzabadi@gmail.com
3   Laboratory of Spectroscopy, Molecular Modeling, Materials and Environment, Department of Chemistry,
     Faculty of Sciences, Mohamed V, B.P 1014 Rabat, Morocco; bramrwk@gmail.com
4   Laboratory LMSAC, Faculty of Sciences, Mohamed 1st University, P.O. Box 717, Oujda 60000, Morocco;
     k.azzaoui@yahoo.com (K.A.); ee.mejdoubi@gmail.com (E.M.M.); abdellatiflamhamdi@hotmail.com (A.L.)
5   Department of Chemistry, An-Najah National University, P.O. Box 7, Nablus 44830, Palestine;
     ohamed@najah.edu
6   National School of Applied Sciences Al Hoceima, Mohamed 1st University, P.O. Box 717,
     Oujda 60000, Morocco
*   Correspondence: yahyaoui@fsr.ac.ma (A.E.); sjodeh@hotmail.com (S.J.); Tel.: +212-537-77-54-40 (A.E.);
     +970-599-590-498 or +970-923-459-82 (S.J.)

Academic Editor: Faisal Ibney Hai

Received: 4 December 2016; Accepted: 7 February 2017; Published: 28 February 2017

Abstract: The adsorption of chromate on octacalcium phosphate (OCP) was investigated as a function of contact time, surface coverage, and solution pH. The ion exchange method was adapted to establish the interaction mechanism. Stoichiometry exchange of H^+/OH^- was evaluated at a pH range of 3–10, and obtained values ranged between 0.0 and 1.0. The surface complexes formed between chromate and OCP were found to be $> S(\overline{HCrO_4})$ and $> S(\overline{CrO_4})$. The logarithmic stability constant $logK_{1-1}$, and the $logK_{10}$ values of the complexes, were 6.0 in acidic medium and 0.1 in alkaline medium, respectively. At low pH and low surface coverage, the bidentate species $> S(\overline{HCrO_4})_2$ with $logK_{10.5}$ of about 2.9, was favored at a hydration time of less than 150 min. The contribution of an electrostatic effect to the chromium uptake by the OCP sorbent, was also evaluated. The results indicate that the adsorption of chromate on OCP is of an electrostatic nature at a $pH \leq 5.6$, and of a chemical nature at a $pH > 5.6$.

Keywords: octacalcium phosphate; chromium (VI); adsorption; environmental; ion exchange

---

1. Introduction

The hexavalent chromium Cr (VI) generated from various industrial processes, such as metallurgy, dyes, paints, inks, and plastics, is a major global concern, due to its harmful effects on humans and nature [1].

As a result, the presence of this metal cation in nature is well controlled. The maximal concentration level of Cr (VI) allowed in drinking water, as determined by the US-Environmental Protection Agency (EPA), is 0.05 mg/L [2]. Compliance with the EPA's chromium rule requires additional industrial monitoring. Designing a treatment process which reduces the number of chromium ions in industrial effluent to an acceptable level, has become crucial. A number of methods for this purpose have already been developed, among which are reduction and precipitation [3], ion exchange [4], solvent extraction [5], adsorption [6], and electrochemical precipitation [7]. When considering these methods, adsorption is the most promising technique [8]. This process is usually performed using conventional adsorbents, such as silica, zeolites, iron(III) (hydr)oxides, and activated carbon, or nonconventional adsorbents, such as red mud, sewage sludge, and bone char [9–11]. Phosphate materials (synthetics and minerals) have also been used as effective adsorbents for heavy metals from wastewaters and polluted soils. They have an excellent stabilization efficiency for several metal ions. For this reason, they are highly efficient metal adsorbents [12–14], particularly calcium phosphate. This material has a large specific area, high thermal and geochemical stability, low solubility, high ionic exchange capacity, and high stability towards ionization by radiation. For these reasons, it has been used as a backfill material for geological repositories for nuclear waste [15], and as an adsorbent in engineered barriers for environmental restoration [16]. It was also reported that phosphates with an amorphous structure are more efficient adsorbents of lead, uranium, and plutonium [17].

Although several adsorbents have been developed, and the retention of the metals hasbeen extensively studied, the sorption mechanisms are still rather difficult to identify. This could be because several phenomena, such as iso-morphous substitutions, surface complexation, and dissolution–precipitation, can occur simultaneously during the sorption process [18–21]. In addition, there is a lack of data on the adsorption of many metals on amorphous or poorly-crystallized phosphates.

Published studies show that the uptake of hexavalent chromium by calcium phosphate, exhibits typical anionic (such as $HCrO_4^-$ and CrO_4^{2-}) sorption behavior, and that the adsorption decreases by increasing the pH [22]. They also show that Cr (VI) adsorption is favored on phosphate that ispositively charged, at a low to neutral pH level (i.e., high point of zero charge (PZC)) [23]. The findings suggest that the retention of Cr (VI) by phosphates occurs through an electrostatic attraction and via binding to the surface functional groups OH_2^+ and OH^-. In other studies, it was found that hydroxapatite (HAp) and tricalcium phosphate (TCP) composite are able to adsorb a significant amount of chromium (VI), at a pH level of about 5. So, the composites showed a lower PZC than HAp (6.2–8.5) [22–24].

Moreover, there is no systematic understanding of the mechanism of chromium immobilization that involves the protonation/deprotonation of surface hydroxyl groups, and their interaction with the metal oxyanion. The combined effect of both pH and contact time on the adsorption mechanism requires more investigation (Figure 1).

Taking these considerations into account, the present study aims to investigate the complexation of hexavalent chromium with low-crystallized octacalcium phosphate (OCP). To achieve the aim of this study, the removal of Cr (VI) from aqueous solution was studied through batch experiments, as a function of contact time, the amount of adsorbent, and the equilibrium pH of chromate (10^{-4} M) solution. The ion exchange method, which has already been successfully implemented, especially in solvent extraction, has been chosen by the authors, in order to study the behavior of chromium (VI) on the surface of OCP. Another aim of this work is to investigate the surface complexation of OCP.

2. Experiment

2.1. Materials and Methods

Octacalcium phosphate (OCP) was synthesized in our laboratory. Phosphoric acid (99%), Chromium (VI), Calcium hydroxide (99%), Potassium hydroxide (KOH) (99%), and Nitric acid (HNO_3) (99%), were purchased from Sigma Aldrich, and were used in the same form as they were received. High-purity distilled water was used for all of the experiments.

2.2. Synthesis and Characterization of OCP

Calcium phosphate was synthesized using the microwave-hydrothermal method. In this method, phosphoric acid (0.3 M) and calcium hydroxide solution (0.5 M) are used as the starting materials. The preparation method has been described in [22]. The obtained mixture was heated at 150 °C for 1 h, and then irradiated in a microwave oven (800 w) for 5 min. The resulting gel was filtered off and dried overnight in an air oven, at 80 °C. The obtained solid was repeatedly washed with hot distilled water, and was identified as OCP by the associated XRD patterns and FT-IR. The characteristics of the diffraction line observed in the XRD patterns at $2\theta = 4.7°$, is evidence for the formation of OCP [22].

This result was confirmed by the FTIR characteristic peaks of OCP at 1089 and 1033, and by the Ca/P ratio of 1.34, which was close to the theoretical OCP ratio of 1.33 [25,26].

2.3. Surface Properties and Adsorption Experiments

The adsorption experiments were carried out by a batch method. A stock solution of Cr (VI) (10^{-4} M) was prepared from potassium dichromate. The pH level of the solution was measured using a Hanna combined electrode (Hanna pH 210). Nitric acid and KOH were used to adjust the starting acidity of the aqueous solutions of Cr (VI), with a concentration of 10^{-4} M in all cases. Sorption experiments were conducted at room temperature, as a function of the pH, contact time, and amount of adsorbent (m). The supernatants (5.0 mL) were filtered and analyzed for aqueous chromium using the 1,5-diphenylcarbazide (EPA 7196A) spectrophotometry method. The adsorbed chromium was calculated from the difference between the concentrations before and after equilibrium with calcium phosphate. The ratio of Cr (VI) concentrations in solid and aqueous phases led to the distribution coefficient D.

3. Results and Discussion

3.1. Effect of pH on Chromium Adsorption

The logarithmic variation of D with pH at different contact times, for 0.5, 1.0, and 1.5 g/L of calcium phosphate solutions, is plotted in Figure 1.

As shown in Figure 1, all cases exhibit similar behavior; log D increased by increasing the pH, to reach the maximum at pH_{max} of 4.0 to 5.0, before decreasing as the pH continued to increase. The maximum adsorption efficiencies were found to increase with the amount of adsorbent (m). The results also show that the pH_{max} is dependent on the contact time; when the contact time increased from 5 to 150 min, the pH_{max} rose from about 4 to 5.

The variation in the extraction efficiency with the solution's pH could be related to the protonation/deprotonating of both surface groups, and to the acidity of H_2CrO_4 ($pK_{a1} = 0.2$ and $pK_{a2} = 6.5$) [27–29]. According to the chromium speciation pH-diagram, the chromium (VI) was adsorbed as hydrogen chromate ($HCrO_4^-$) at pH $\leq$ 5.0 ($\geq$95%), as chromate (CrO_4^{2-}) at pH $\geq$ 7.6 ($\geq$95%), and as a mixture of these species between pH 5.0 and 7.6. In this case, both electrostatic and chemical sorption mechanisms could occur, and generally, it is not possible to distinguish between these two mechanisms [30]. It was assumed that the maximum adsorption occurred when the combination of a high positive surface charge and a high concentration of anionic chromium species, are achieved. Thus, the uptake of weak acid was maximal at a pH value around its dissociation constant, or near the PZC of surface sorbent materials [31]. In general, anion adsorption is strongly dependent on the pH of the medium, exhibiting the greatest removal in acidic to neutral solution. As has been demonstrated in previous studies, optimal Cr (VI) adsorption occurs at a pH lower than 4 for various sorbents, such as some metallic (oxy)hydroxides [32–34] and natural bio-sorbents, for example, larch bark [35], cooked tea dust [36], papaya seeds [37], raw Bagasse [38], and activated carbon [39]. In the case where iron and aluminum oxides are used as adsorbents, an adsorption efficiency of higher than 80% was reached at a pH of 4 to 6. The adsorption was seen to be highly dependent on the pH of the medium; the Cr (VI) uptake increased by increasing the pH values from 1.0 to 7.0, after which the uptake decreased.

Similar results were reported for the variation of hexavalent chromium adsorption with pH solution, with other adsorbents such as clay minerals [40], oxide-coated sand [40], aluminosilicate [41], zeolite [42], chitosan [43], and hydroxyapatite [44], which all exhibited a maximum uptake at a pH of 5 to 7. It has been suggested that, at a low pH, the adsorption is also low, due to the competition between the metal ions and protons for the adsorption sites. In these cases, the uptake of Cr (VI) followed the ion exchange mechanism. From these results, it could be concluded that, at a pH higher than pH_{max}, Cr (VI) exhibited typical anionic sorption behavior, with adsorption decreasing when the pH was increased. Similar results were reported for the retention of similar anions on oxide surfaces [45,46]. This adsorption pattern is the result of the protonation of surface hydroxyl sites and of Cr (VI) hydrolysis [47]. Thus, at a pH higher than pH_{max}, the retention of Cr (VI) was due to the interaction of $HCrO_4^-$ and/or CrO_4^{2-} with OH surface groups, rather than with OH_2^+ groups, which were predominant at a pH lower than pH_{max} [44]. The significant influence on the adsorption of CrO_4^{2-} was found at the slope of the pH adsorption edges. Distinct pH regions with different slopes characterized the various adsorbed species. This result reflected the change in the sorption mechanism [48].

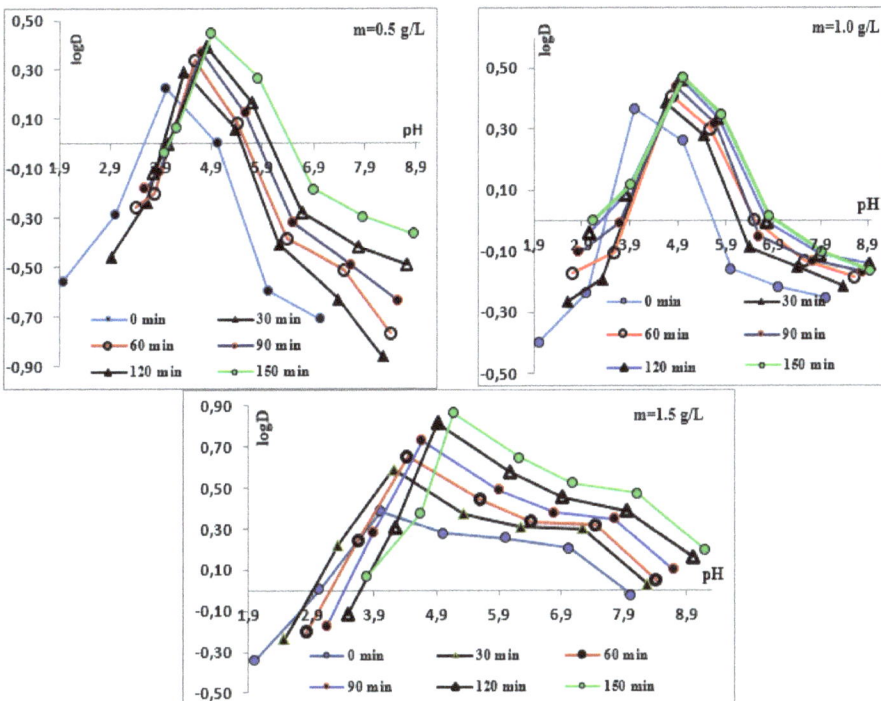

Figure 1. Log(D) versus pH for different amounts of OCP and different contact times.

3.2. Effect of Contact Time

The effect of contact time on the adsorption efficiency was examined, and the results are shown in Figure 2. At a pH lower than pH_{max}, a distinct difference was observed for adsorption envelopes in relation to the sorbent dose, or sorbet/sorbent ratio (surface coverage). At a sorbent dose of 0.5 and 1.0 g/L, the pH adsorption edges followed a similar trend for $t \geq 15$ min. At a pH of 50% of adsorption (pH_{50}), a negligible variation occurred at pH 3.9–4.1. When the sorbent dose was 1.5 g/L, the adsorption process was dependent on the hydration time. Therefore, the adsorption envelopes shifted to higher pH values, resulting in an increase in pH_{50} from 2.9 to 3.9, as the time increased from 0 to 150 min. The mechanism for the oxyanion adsorption was dependent on the surface coverage and hydration of the surface sites, involving the formation of distinct surface complexes. This was in agreement with the spectroscopic study results, which indicated that chromium (VI) resulted in the formation of both monodentate and bidentate surface complexes on iron oxides [49,50]. The proportion of these species was dependent on the metallic ion concentration. Recently, through the use of a spectroscopic study, it has been shown that the monodentate chromate complexes on ferrihydrite were predominant at a low surface coverage and a pH $\geq$ 6.5. In contrast, bidentate surface complexes were formed at a high surface coverage and pH $\leq$ 6 [51].

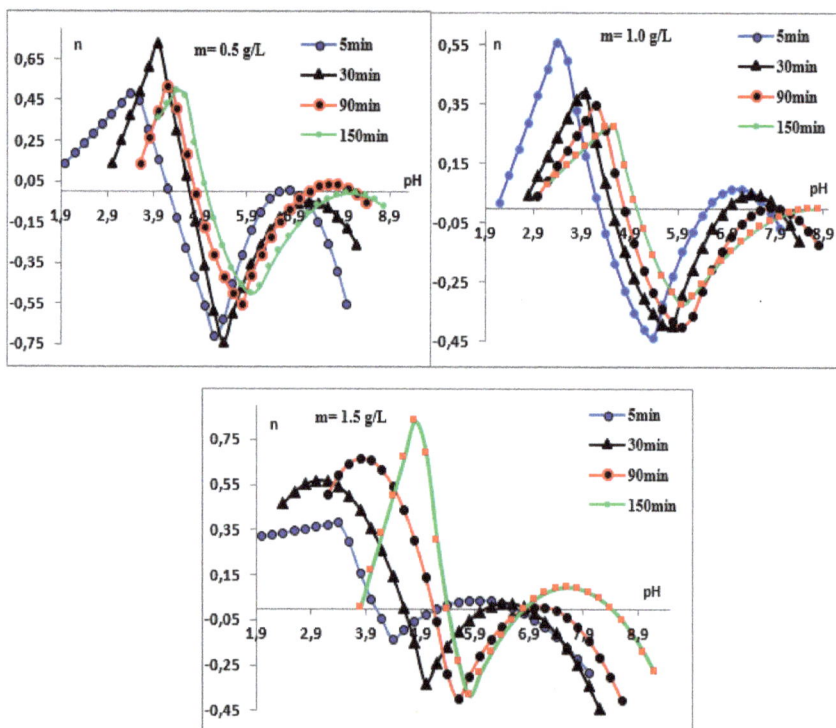

Figure 2. Variation of n = f(pH) obtained at various contact time and at different OCP sorbent amounts of m = 0.5, 1.0 and 1.5 g/L: 5 min ($\bullet$); 30 min ($\blacktriangle$); 90 min ($\bullet$); and 150 min ($\blacksquare$).

Taking into account the previous results, it could be deduced that, at a low adsorbent dose, monodentate surface complexes prevailed for all of the examined experimental conditions. In contrast, at a higher adsorbent dose, the bidentate surface complexes became predominant at a low hydration level of sorbent materials at pH $\leq$ pH_{max}, and were subsequently converted to monodentate species, when hydration equilibrium was reached.

At a pH higher than pH_{max}, adsorption envelopes become more alkaline. The shift to this alkalinity increased by increasing the hydration time. Due to electrostatic repulsion with the negative surface charges, the chromium uptake initially involved the adsorption of a $HCrO_4^-$ anion, which was followed by the subsequent slower, and less important, uptake of CrO_4^{2-} in the alkaline region. These electrostatic factors could influence both the kinetics and equilibrium of chromate ions, as observed for the adsorption of a similar oxyanion on ferrihydrite [52]. When the contact time increased, the adsorption of the chromate anion also increased, and the Log(D) = f(pH) curves show a pronounced difference in the slopes at $t = 150$ min. As previously discussed, this phenomenon could be due to a change in the adsorbed fraction of the Cr (VI)-predominant surface complexes.

3.3. Chromium (VI) Adsorption Reactions and Stability Constants

The acid-based properties of calcium phosphate were described by the protonation and deprotonation reactions of the phosphate surface functional groups, as shown in Equations (1) and (2), $> \overline{SOH}$ [53,54]:

$$> \overline{SOH} + H^+ \leftrightarrow > \overline{S(OH_2)^+} \quad K^+ \tag{1}$$

$$> \overline{SOH} \leftrightarrow > \overline{SO^-} + H^+ \quad K^- \tag{2}$$

K^+ and K^- were the surface stability constants, and the on lined species referred to the solid phase.

The adsorption reaction of chromium (VI) on calcium phosphate can be expressed as follows:

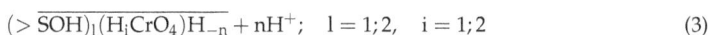

$$(> \overline{SOH})_l \overline{(H_i CrO_4)H_{-n}} + nH^+; \quad l = 1;2, \quad i = 1;2 \tag{3}$$

where n is the number of protons, which ranges from -1 to $+1$.

The initial ionization of $H_2 CrO_4$ is relatively strong, so $HCrO^-_4$ was the main species found at pH > 5, and i = 1 at equilibrium (Equation (3)), under these conditions.

The symbol H_{-n}, stands for both hydrogen atoms (n < 0) and for OH groups (n > 0). Taking into account that $H_2 O \equiv H_{-1} + H_1$, the surface complexes $> \overline{(SOH)_1 (H_i CrO_4)H_{-n}}$ noted thereafter C_{ln}, represented a general formulation of species, differentiated by water composition. So $> \overline{(SOH)_1 (H_2 CrO_4)H_{-n}}$ could be $> \overline{(SOH)_1 (HCrO_4)H_{-n+1}} + H_2 O$ or $> \overline{(SOH)_1 (CrO_4)H_{2-n}}$ $+ 2H_2 O$, and even $> \overline{(S)_1 (H_2 CrO_4)H_{-n-1}}$ or $> \overline{(S)_1 (HCrO_4)H_{-n-l+1}} + H_2 O$ or $> \overline{(S)_1 (H_2 CrO_4)H_{-n-l+2}} + 2H_2 O$.

The surface complexation constant for the relationship (3) is:

$$K_{ln} = \frac{\left[(> \overline{SOH})_l \overline{(H_i CrO_4)H_{-n}} \right] [H^+]^n}{[> \overline{SOH}]^l \left[H_i CrO_4^{(i-2)+} \right]} \tag{4}$$

The distribution coefficient being:

$$D = \frac{\left[(> \overline{SOH})_l \overline{(H_i CrO_4)H_{-n}} \right]}{\left[H_i CrO_4^{(i-2)+} \right]} \tag{5}$$

where $[H_i CrO_4^{(i-2)+}]$ represents the equilibrium concentration of Cr (VI) in solution, it was obtained that:

$$\log D = \log K_{ln} + \log m + npH \tag{6}$$

where $> \overline{SOH} = m$, was the concentration of sorbent used in g/L.

Assuming that the first approximation is the essential formation of mononuclear (l = 1), the surface complexes of Equation (6) become:

$$\log D = \log K_{1n} + \log m + npH \tag{7}$$

The variation of the distribution coefficient with pH allowed us to define the nature of the C_{1n}-adsorbed species. The values of (l,n) were obtained according to:

$$\left(\frac{\delta \log D}{\delta pH} \right)_m = n \tag{8}$$

So, the surface complexes of Cr (VI) with calcium phosphate could be well described from the experimental data $\log D = f(pH)$, shown in Figure 1. The analysis of the obtained results showed that the plots of $\log D = f(pH)$ were linear at various pH ranges. The straight lines of the slope correspond to the mean values of n, and varied between -1 and 1. The value of the surface complexes $(1,n)$, involved in this case were $(1,0)$, $(1,1)$, $(1,2)$, and $(1,-1)$. In this case, the co-precipitation/adsorption process of chromium (VI), and the species distribution diagram as a function of pH, are needed. It is worth noting that the interaction of Cr (VI) with the OCP surface could be described as an $n = f(pH)$ variation. For this purpose, fitting of the data into a polynomial equation was carried out for the three pH regions. The obtained results show that, when in acidic solution (pH < 5), a second-degree equation fitted ($R^2 > 99\%$) the ascendant curve, whereas in low acidic to alkaline media, a cubic polynomial fitted ($R^2 > 99\%$) two distinct segments of the descendant curve, at pH regions of about 4 to 6 and 6 to 10.

3.3.1. Surface Complexes and Effect of pH and Contact Time on H_3O^+/OH^- Exchange

The variations of $n = f(pH)$ are illustrated in Figure 2. As shown in Figure 2, the protonation/deprotonating reaction followed a similar trend, with respect to the pH value. In all cases, the maximum H_3O^+ and OH^- exchange of $|n| = 0.8$ occurred at a pH range of 4.0–6.0. When considering the obtained results, it becomes evident that the H_3O^+/OH^- stoichiometry was not an integer, as might be expected from the theoretical single reaction. Similar results were observed for other adsorbed elements on iron (oxy) hydroxides [55,56].

It was assumed that the chromium adsorption occurred by different reactions, and consequently, resulted in a combination of at least two predominant surface complexes. In the case $l = 1$, the predominant complexes would be different for the sorbent amounts of 0.5 and 1.0 g/L, while a different adsorption behavior was observed for 1.50 g/L. Indeed, a low value of around 0.5 was found at low pH and at $t < 150$ min, indicating that C_{10} is not the predominant species under these conditions.

Taking into account these considerations, the uptake of Cr (VI), characterized by $n = 0.5$, and the general adsorption reaction, can be expressed by:

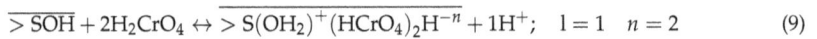

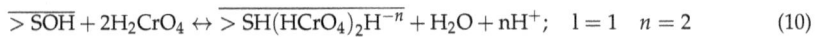

$$> \overline{SOH} + 2H_2CrO_4 \leftrightarrow > \overline{S(OH_2)^+(HCrO_4)_2H^{-n}} + 1H^+; \quad l = 1 \quad n = 2 \tag{9}$$

$$> \overline{SOH} + 2H_2CrO_4 \leftrightarrow > \overline{SH(HCrO_4)_2H^{-n}} + H_2O + nH^+; \quad l = 1 \quad n = 2 \tag{10}$$

The adsorption constant $K_{10.5}$ for equilibrium (10) was given by:

$$K'_{1n} = \frac{\left[> \overline{SH(HCrO_4)_2H^{-n}}\right][H^+]^n}{[> \overline{SOH}][H_2CrO_4]^2} = \frac{D}{[> \overline{SOH}]} \frac{[H^+]^n}{[H_2CrO_4]} \tag{11}$$

Based on this adsorption mechanism, the following relationship could be obtained:

$$\log(D(D+1)) = \log K'_{1n} + \log m + \log[Cr]_0 + n\,pH \tag{12}$$

For $[Cr]_0 = 10^{-4}$ M, which was the chromium analytical concentration used, the equation becomes:

$$\log(D(D+1)) = \log K'_{1n} + \log m - 4 + n\,pH \tag{13}$$

At a pH range of 2.5 to 5.0, the relation $\log(D(D+1)) = f(pH)$ exhibited a linear variation, with slopes increasing from 0.5 to 1.0, as the time increased from 0 to 150 min.

Accordingly, C_{11}, $C_{10.5}$ and C_{10} were the predominant surface species in the case of $l = 1$. A non-protonated C_{11} ($> \overline{SOH(H_2CrO_4)H_{-1}} \equiv > \overline{S(HCrO_4)H_{-1}} \equiv > \overline{SCrO_4}$) complex was always formed at low acidity (pH ~4.4), combined with the protonated C_{10} ($> \overline{SOH(H_2CrO_4)H_0} \equiv > \overline{S(HCrO_4)H_0} \equiv > \overline{SHCrO_4}$), complex during hydration equilibrium conditions. Nevertheless, when this equilibrium was not reached at the 1.5 g/L sorbent amount, the C_{10} complex was not favored, and disappeared, to the benefit of the bidentate: $C_{10.5}$ ($> \overline{SOH(H_2CrO_4)_2H_{-1}} \equiv > \overline{SH(HCrO_4)_2H_{-1}} \equiv > \overline{S(HCrO_4)_2}$) species. As a result, in this condition, $C_{10.5}$ and C_{11} were the prevailing complexes. The $C_{10.5}$ surface complex can also be formulated as the $> \overline{S(H_2Cr_2O_7)}$ species. However, the bichromate surface complexes were not found at 0.5 and 1.0 g/L sorbent doses and were neglected, as previously reported [57].

The results show that, as the sorbent amount increased, the number of surface sites also increased. At high-hydration equilibrium, the hydrogen chromate anions ($HCrO_4^-$) and water molecules were competing for the active surface sites. At a low-hydration level of the sorbent material's surface site, the $HCrO_4^-$ ions displaced the H_2O molecules, in the hydration sphere of the C_{11} complex. Bidentate surface species were then formed at a high sorbent dose and short contact time. These results were in accordance with previous studies, suggesting that high-chemisorbed water molecules prevented the bidentate complexes from forming, as noted for the complexation of chromium with ferrihydrite [48]. A similar substitution of water molecules by $HCrO_4^-$ anions could take place at a higher chromium concentration, even at hydration equilibrium, explaining the formation of bidentate surface species when there is a high surface coverage, observed for the adsorption of Cr (VI) or a similar anion on ferrihydrite [29].

It should be noted that the $C_{10} \equiv > \overline{S(HCrO_4)H_0}$ complex could also be expressed as an outer-sphere $> \overline{SOH_2^+ - HCrO_4^-}$ species, since we could not distinguish between the different complexes' formulations or structures, based on one H_2O molecule. Generally, when at a low pH value, the formation of such outer-sphere Cr (VI) surface complexes is favored, similar to that previously obtained on amorphous aluminum oxides [58], and supported by the formation of SOH_2^+ in acidic media, as was indicated in a titration experiment [59].

At a pH higher than 4.5, n varied between -1 and 0, in all investigated conditions. The predominant surface species formed in these conditions, were C_{10} and C_{1-1}. Thus, Cr (VI) was adsorbed via the formation of $> \overline{S(OH)(HCrO_4)H_0}$ (C_{10}) and $> \overline{S(OH)(HCrO_4)H_1}$ (C_{1-1}) complexes, that could be expressed as $> \overline{SCrO_4}$ and $> \overline{SHCrO_4}$, respectively. As the contact time increased, the optimal pH formation of protonated $C_{1-1} \equiv > \overline{SHCrO_4} \equiv > \overline{S(OH_2^+)(HCrO_4^-)}$ and the un-protonated $C_{10} \equiv > \overline{SCrO_4} \equiv > \overline{S(OH)(HCrO_4)}$ species, shifted from 5 to 6 and 7 to 8, respectively. The adsorption of Cr (VI) increased with t, and at a higher hydration time, the adsorption reaction was likely to comply to OH^- surface exchange. Thus, for a pH lower than PZC, the negative surface charge of the phosphate material reacts in acidic media to form $>S(OH_2)^+$, which adsorbs chromium as $HCrO_4^-$. Maximal OH^- exchange was observed at a pH range of 5.2–6.0, which approximately coincided with the pH range of zero charge and the iso-electric points of unloaded material. Whilst higher than PZC, the repulsion between chromium anions and negatively sorbent surface charge, increased with pH. The contribution of the columbic effect to the overall uptake of Cr (VI) could be a more important process. Nevertheless, at a pH value around the iso-electric point, the electrostatic repulsion reached a minimum value, and then the intrinsic process (with n = 0) became the major adsorption mechanism. As shown from the obtained results, the pH value became higher with contact time, reaching 8.2 at $t \approx 150$ min.

Consequently, the overall adsorption equilibrium could be obtained by the intrinsic reaction with 1 mole of H^+ (n = -1), or 1 mole of OH^- (n = 1) exchange reactions per mole of $HCrO_4^-$.

3.3.2. Equilibrium Constants

In the Log(D)= f(pH), Figure 1 plots the various contact times and sorbent amounts, and the straight lines show slopes ranging from -0.4 to 0.6. In the cases when the surface complexes are mononuclear (l = 1), apparent equilibrium constants, $K_{ap} = K_{1n}$ or K'_{1n}, were obtained from the origin ordinates A = $LogK_{1n} + Log(m)$ or $logK'_{1n} + Log(m) - 4$. From the obtained results, it can be that the variations of $logK_{1n} = f(n)$ (Figure 3) and $logK'_{1n} = f(n)$ (not shown), were linear under all experimental conditions.

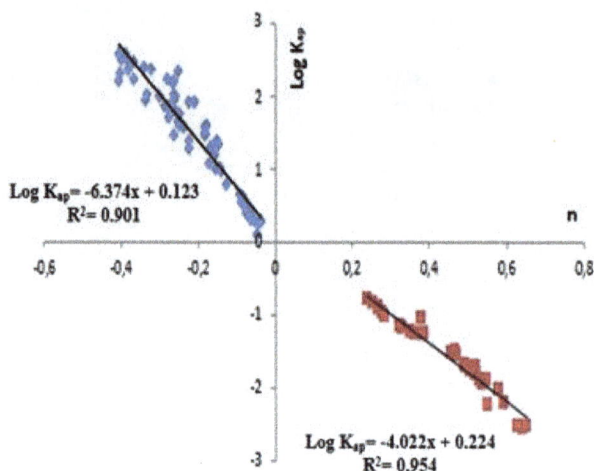

Figure 3. Variations of log Kap = f(n).

Taking into account that $|n|$ also represented the fraction (x) of the predominant $1H^+$ or $1OH^-$ exchange reactions contributing to the overall adsorption equilibrium, the apparent constants (K_{ap}) of the overall equilibrium were given by the expressions shown in Equations (14) and (15):

$$K_{ap} = (K_{10})^{(1-|n|)} (K_{1\pm1})^{|n|} \tag{14}$$

$$\log K_{ap} = (1 - |n|)\log K_{10} + |n|\log K_{1\pm1} \tag{15}$$

with $K_{1\pm1} = K_{11}$ or K_{1-1}.

As an example, the overall partition equilibrium prevailing for pH > 5.5, and involving successive exchange reactions $n = 0$ and -1, can be summarized in the following reactions:

$$(1-x)(> \overline{SOH} + HCrO_4^- \leftrightarrow > \overline{S(OH)(HCrO_4)H_0} \equiv > \overline{S(CrO_4)} + H_2O): \quad (K_{10})^{(1-x)}, n = 0 \tag{16}$$

Overall adsorption reactions are:

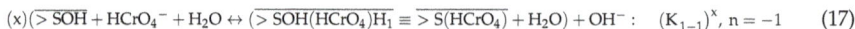

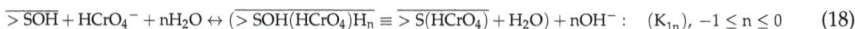

$$(x)(> \overline{SOH} + HCrO_4^- + H_2O \leftrightarrow (> \overline{SOH(HCrO_4)H_1} \equiv > \overline{S(HCrO_4)} + H_2O) + OH^- : \quad (K_{1-1})^x, n = -1 \tag{17}$$

$$> \overline{SOH} + HCrO_4^- + nH_2O \leftrightarrow (> \overline{SOH(HCrO_4)H_n} \equiv > \overline{S(HCrO_4)} + H_2O) + nOH^- : \quad (K_{1n}), -1 \leq n \leq 0 \tag{18}$$

Where x is the molar fraction of the sorption mechanism with 1 OH^- exchange. The apparent constant, $K_{ap} = K_{1n}$, is then given by:

$$\log K_{1n} = (1 - x)\log K_{10} + x\log K_{11} = (1 - |n|)\log K_{10} + |n| \log K_{10} \tag{19}$$

As shown above, the K_{ap} value could be determined experimentally, and was variable, depending on the surface charge and pH. Therefore, the constants K_{10} and $K_{1\pm1}$ were the intersect and the slope of $\log K_{ap} = f(n)$, respectively, whereas K_{1-1} was the opposite of the slope. From this information, the intrinsic constants K_{10}, for H^+ (formation of $> \overline{S(OH_2^+)(HCrO_4^-)H_0} \equiv > \overline{S(HCrO_4)}$) and OH^- (formation of $> \overline{S(OH)(HCrO_4)H_0} \equiv > \overline{S(CrO_4)}$) adsorption reaction exchange, were determined, with corresponding logarithmic values of $\log K_{10} = 0.2$ and 0.1, respectively. The intrinsic constants ($K_{int} = K_{10}$) were obtained in general, by extrapolating the apparent constants to a zero surface charge [57,58].

As discussed previously, the variations of $\log K_{ap} = f(n)$ could be due to the contribution of the electrostatic effect (K_{col}), related to K_{ap} [60,61] according to:

$$K_{int} = K_{ap}K_{col} = K_{ap}\exp(-nF\Psi_0/RT) \tag{20}$$

$$K_{ap} = K_{int}(K_{col})^{-1} = (K_{10})^{(1-|n|)}(K_{1\pm1})^{|n|} \tag{21}$$

In the particular case of $|n| = 1$, Equation (22) was obtained:

$$K_{ap} = K_{int}(K_{col})^{-1} = K_{1\pm1} \tag{22}$$

where ψ_0 was the surface potential, and R, T, and F were the gas constant, absolute temperature, and Faraday constant, respectively.

Since $K_{int} = K_{10}$, the columbic effect was determined according to

$$\log K_{col} = \log K_{10} - \log K_{1\pm1} \tag{23}$$

While $\log K'_{ap}$ for equilibrium (10) was:

$$\log K_{col} = \log K'_{10} - \log K_{10.5} \tag{24}$$

The origin ordinate of the $\log K'_{1n} = f(n)$ plots resulted in $\log K'_{10} = 5.3$. These results are summarized in Table 1.

Table 1. Surface complexation constants for Cr (VI) sorption onto Octacalcium phosphate.

Species	n	Adsorption Reaction	$\log K_{1n}$	$\log K_{col}$
		Acidic medium (pH < 5)		
$> \overline{SOH_2^+ - HCrO_4^-}$	0	$(> \overline{SOH}) + H_2CrO_4 \leftrightarrow > \overline{SOH_2^+ - HCrO_4^-}$	0.1	
$> \overline{SHCrO_4}$	0	$(> \overline{SOH}) + H_2CrO_4 \leftrightarrow > \overline{SHCrO_4} + H_2O$	0.1	
$> \overline{SCrO_4^-}$	+1	$(> \overline{SOH}) + H_2CrO_4 \leftrightarrow > \overline{SCrO_4^-} + H_2O + 1H^+$	−4.0	4.1
$> \overline{S(HCrO_4)_2}$	0.5	$(> \overline{SOH}) + 2H_2CrO_4 \leftrightarrow > \overline{SH(SCrO_4)_2H^-} + H_2O + 1H^+$	3.0	2.4
		Lower acidic to alkaline medium (pH > 5)		
$> \overline{SHCrO_4^-}$	0	$(> \overline{SOH}) + HCrO_4^- \leftrightarrow > \overline{SCrO_4^-} + H_2O$	0.2	
$> \overline{SHCrO_4}$	−1	$(> \overline{SOH}) + HCrO_4^- \leftrightarrow > \overline{SHCrO_4} + 1OH^-$	−6.7	6.1

It is worth noting that the interaction of chromate anions with phosphate materials was of an essentially electrostatic nature in acidic media (pH < 5), and of a chemical character at a lower acidity (pH > 5) to alkaline solution. Two surface species were always formed when hydration equilibrium was reached; deprotonated $(> \overline{SCrO_4})$ and protonated $(> \overline{SHCrO_4})$ complexes, which were more stable in near neutral, than acidic, solution. This was consistent with modeling adsorption data, indicating a mixture of both monodentate and bidentate chromate surface complexes on goethite [62,63].

4. Conclusions

The adsorption of hexavalent chromium on OCP material was thoroughly investigated. One goal of this study was to develop a method for studying the surface complexation of OCP. The obtained distribution coefficient (D) was dependent on the contact time, pH, and surface coverage. The treatment of Log(D) = f(pH) experimental data were used to evaluate H^+/OH^- exchange stoichiometry in adsorption reactions, and the results were used to specify the predominant Cr (VI) surface species. At hydration equilibrium, protonated hexavalent chromium formed $> \overline{SHCrO_4}$ and unprotonated $> \overline{SCrO_4}$ complexes, under all explored conditions. When the hydration equilibrium was not reached at a low surface coverage, the protonated species disappeared, to the benefit of the bidentate $(> \overline{S(HCrO_4)_2})$ complex. The stability constants were $\log K_{10} = 0.123$, for $> \overline{SHCrO_4}$, which could be formulated as $> S(OH_2^+)(HCrO_4^-)$ and $\log K_{11} = -4.0$ for $> \overline{SCrO_4}$, in acidic media.

In alkaline media, the log $K_{1\text{-}0}$ = 0.2 for $> \overline{SCrO_4})$ and log $K_{1\text{-}1}$ = −6.4 for $> \overline{SHCrO_4}$. Whilst, for the bidentate $> S(HCrO_4)_2$ surface species, the log$K_{10.5}$ = 2.9. The electrostatic effect was evaluated for the predominant adsorption reactions. The obtained results suggested that Cr (VI) adsorption on OCP was of an electrostatic nature in acidic solutions, and of a chemical nature in lower acidic to alkaline solutions. The results could have practical and promising applications in the fields of environmental health, for the removal of hazardous chromium from industrial wastewater before dumping it into the environment (water and soil). This could enhance the environmental risk management process and will play a major role in preventing future coastal contamination.

Abbreviations

OCP	Octacalcium Phosphate ($Ca_8H_2(PO_4)_{6.5}H_2O$)
XRD	X-ray Diffraction
FTIR	Fourier transform infrared spectroscopy
D	distribution coefficient (the ratio of concentration of adsorbed Cr (VI) to its concentration in aqueous phase)
m	OCP sorbent amounts in g/L
t	contact time.
K^+, K	surface stability of active site > SOH
l	number of functional surface group involved in adsorption reaction
H n	hydrogen atoms (n < 0) or OH groups (n > 0)
K_{ln}	adsorption constant
Ψ_0	surface potential
R	universal gas constant (8.31 J/mol·K)
T	temperature (K)

References

1. Dhal, B.; Thatoi, H.N.; Das, N.N.; Pandey, B.D. Chemical and microbial remediation of hexavalent chromium from contaminated soil and mining/metallurgical solid waste: A review. *Hazard J. Mater.* **2013**, *250–251*, 272–291. [CrossRef] [PubMed]
2. Zachara, J.M.; Ainsworth, C.; Brown, G.E., Jr.; Catalano, J.G.; McKinley, J.P.; Qafoku, O.; Smith, S.C.; Szecsody, J.E.; Traina, S.J.; Warner, J.A. Chromium speciation and mobility in a high level nuclear waste vadose zone plume. *Geochim. Cosmochim. Acta* **2004**, *68*, 13–30. [CrossRef]
3. Zhou, X.; Korenaga, T.; Takahashi, T.; Moriwake, T.; Shinoda, S. A process monitoring/controlling system for the treatment of wastewater containing Chromium (VI). *Water Res.* **1993**, *27*, 1049–1054. [CrossRef]
4. Tiravanti, G.; Petruzzelli, D.; Passino, R. Pretreatment of Industrial Wastewaters II, Selected Proceedings of the Second IAWQ International Conference on Pretreatment of Industrial Wastewaters. *Water Sci. Technol.* **1997**, *36*, 197–207. [CrossRef]
5. Pagilla, K.R.; Canter, L.W. Biosorption of Hexavalent Chromium from Aqueous Medium with Opuntia BiomassAthens, Greece. *Environ. J. Eng.* **1999**, *125*, 243–248. [CrossRef]
6. Bruce Manning, A.; Goldberg, S. Modeling Competitive Adsorption of Arsenate with Phosphate and Molybdate on Oxide Minerals. *Soil. Sci. Soc. Am. J.* **1996**, *60*, 121–131. [CrossRef]
7. Kongsricharoern, N.; Polprasert, C. Removal of Heavy Metals from Wastewater by Adsorption and Membrane Processes: A Comparative Study. *Water Sci. Technol.* **1996**, *34*, 109–116. [CrossRef]
8. Hu, J.; Chen, G.; Lo, I.M.C. Removal and recovery of Cr (VI) from wastewater by maghemite nanoparticles. *Water Res.* **2005**, *39*, 4528–4536. [CrossRef] [PubMed]
9. Dahbi, S.; Azzi, M.; de la Guardia, M. Removal of hexavalent chromium from wastewaters by bone charcoal. *Fresenius J. Anal. Chem.* **1999**, *363*, 404–407. [CrossRef]
10. Dahbi, S.; Azzi, M.; Saib, N.; de la Guardia, M.; Faure, R.; Durand, R. Removal of trivalent chromium from tannery waste waters using bone charcoal. *Anal. Bioanal. Chem.* **2002**, *374*, 540–546. [PubMed]
11. Hyder, A.H.M.G.; Shamim, A.B.; Nosa, O.E. Adsorption isotherm and kinetic studies of hexavalent chromium removal from aqueous solution onto bone char. *J. Environ. Chem. Eng.* **2015**, *3*, 1329–1336. [CrossRef]
12. Yoon, J. Phosohate-Induced Lead Immobilization in Contaminated Soil. Master's Thesis, University of Florida, Gainesville, FL, USA, May 2005.

13. Chowdhury, S.R.; Ernest Yanful, K. Arsenic and chromium removal by mixed magnetite-maghemite nanoparticles and the effect of phosphate on removal. *J. Environ. Manag.* **2010**, *91*, 2238–2247. [CrossRef] [PubMed]

14. Lee, J.Y.; Elzinga, E.J.; Reeder, R. Lead and zinc removal with storage period in porous asphalt pavement. *J. Environ. J. Sci. Technol.* **2005**, *39*, 4042–4048. [CrossRef]

15. Qureshi, M.; Varshney, K.G. *Inorganic Ion Exchangers*; Chemical Analysis CRC Press: Boca Raton, FL, USA, 2002; p. 282.

16. Correa, F.G.; Martínez, J.B.; Gómez, J.S. Synthesis and characterization of calcium phosphate and its relation to Cr (VI) adsorption properties. *Rev. Int. Contam. Ambient.* **2010**, *26*, 129–134.

17. Conca, J.L.; Lu, N.; Parker, G.; Moore, B.; Adams, A.; Wright, J.V.; Heller, P. Remediation of Chlorinated and Recalcitrant Compounds. *J. Environ. Qual.* **2000**, *7*, 319–326.

18. Saxena, S.; D'Souza, S.F. Heavy metal pollution abatement using rock phosphate mineral. *Environ. Int.* **2006**, *32*, 199–202. [CrossRef] [PubMed]

19. Aklil, A.; Mouflih, M.; Sebti, S. Removal of Cadmium from Water Using Natural Phosphate as Adsorbent. *Hazard J. Mater. A* **2004**, *112*, 183–190. [CrossRef] [PubMed]

20. Corami, A.; D'Acapito, F.; Mignardia, S.; Ferrini, V. Removal of Cu from Aqueous Solutions by Synthetic Hydroxyapatite: EXAFS Investigation. *Mater. Sci. Eng. B* **2008**, *149*, 209–213. [CrossRef]

21. Zhao, X.-Y.; Zhu, Y.-J.; Zhao, J.; Lu, B.-Q.; Chen, F.; Qi, C.; Wu, J. Hydroxyapatite nanosheet-assembled microspheres: Hemoglobin-templated synthesis and adsorption for heavy metal ions. *J. Colloid Interface Sci.* **2014**, *416*, 11–18. [CrossRef] [PubMed]

22. Serrano-Gómez, J.; Ramírez-Sandoval, J.; Bonifacio-Martínez, J.; Granados-Correa, F.; Badillo-Almaraz, V.E. Uptake of CrO42-Ions by Fe-Treated Tri-Calcium Phosphate. *J. Mex. Chem. Soc.* **2010**, *54*, 34–39.

23. Razzouki, B.; El Hajjaji, S.; ElYahyaoui, A.; Lamhamdi, A.; Jaafar, A.; Azzaoui, K.; Boussaoud, A.; Zarrouk, A. Kinetic investigation on arsenic (III) adsorption onto iron hydroxide (III). *Pharm. Lett.* **2015**, *7*, 53–59.

24. Soundarrajan, M.; Gomathi, T.; Sudha, P.N. Understanding the Adsorption Efficiency of Chitosan Coated Carbon on Heavy Metal Removal. *Int. J. Sci. Res. Publ.* **2013**, *3*, 2250–3153.

25. Sakai, S.; Anada, T.; Tsuchiya, K.; Ymazaki, H.; Margouli, H.C.; Suzuki, O. Comparative study on the resorbability and dissolution behavior of octacalcium phosphate, β-tricalcium phosphate, and hydroxyapatite under physiological conditions. *Dent. Mater. J.* **2016**, *35*, 216–224. [CrossRef] [PubMed]

26. Osamu, S. Octacalcium phosphate (OCP)-based bone substitute materials. *Jpn. Dent. Sci. Rev.* **2013**, *49*, 58–71.

27. Samake, D. Treatment of Tannery Wastewater Using Clay Materials. Ph.D. Thesis, Joseph Fourier-Grenoble University, Bamako, Mali, December 2008.

28. Zachara, J.M.; Cowan, C.E.; Schmidt, R.L.; Ainsworth, C.C. Chromate Adsorption by Kaolinite. *Clays Clay Miner.* **1988**, *36*, 317–326. [CrossRef]

29. Grossl, V.; Eick, M.; Sparks, D.; Goldberg, S.; Ainsworth, C.C. Arsenate and Chromate Retention Mechanisms on Goethite. 2. Kinetic Evaluation Using a Pressure-Jump Relaxation Technique. *Environ. Sci. Technol.* **1997**, *31*, 321–326. [CrossRef]

30. Lorphensri, O.; Intravijit, J.; Sabatini, D.A.; Kibbey, T.C.G.; Osathaphan, K.; Saiwan, C. Characterization of Commercial Ceramic Adsorbents and its Application on Naphthenic Acids Removal of Petroleum Distillates. *Water Res.* **2006**, *40*, 1481–1491. [CrossRef] [PubMed]

31. Harding, I.S.; Rashid, N.; Hing, K.A. urface charge and the effect of excess calcium ions on the hydroxyapatite surface. *Biomaterials* **2005**, *26*, 6818–6826. [CrossRef] [PubMed]

32. Zachara, J.M.; Girvin, D.C.; Scmidt, R.L.; Resch, C.T. Chromate adsorption on amorphous iron oxyhydroxide in presence of major ground water ion. *Environ. Sci. Technol.* **1987**, *21*, 589–594. [CrossRef] [PubMed]

33. Demetriou, A.; Pashalidis, I. Spectrophotometric studies on the competitive adsorption of boric acid (B(iii)) and chromate (Cr(Vi)) onto iron (oxy) hydroxide (Fe(O)OH). *Glob. Nest J.* **2012**, *14*, 32–39.

34. Hsia, T.H.; Lo, S.L.; Lin, C.F.; Lee, D.Y. Chemica and spectroscopie evidence for specifie adsorption of chromate on hydrous iron oxide. *Chemosphere* **1993**, *26*, 1897–1904. [CrossRef]

35. Aoyama, M.; Tsuda, M. Removal of Cr(VI) from aqueous solutions by larch bark. *Wood Sci. Technol.* **2001**, *35*, 425–434. [CrossRef]

36. Odeh, L.; Odeh, I.; Khamis, M.; Khatib, M.; Qurie, M.; Shakhsher, Z.; Qutob, M. Hexavalent Chromium Removal and Reduction to Cr (III) by Polystyrene Tris(2-aminoethyl)amine. *Sci. Res. Acad. Publ.* **2016**, *6*, 26–37. [CrossRef]

37. Chang, L. Metal Ions Removal from polluted Waters by Sorption onto Exhausted Coffee Waste. Application to Metal Finishing Industries Wastewater Treatment. Ph.D. Thesis, University of Girona, Girona, Spain, September 2014.

38. Stankiewicz, A.I.; Moulijn, J.A. Process Intensification: Transforming Chemical Engineering. *Chem. Eng. Prog.* **2000**, *96*, 22–34. [CrossRef]

39. Oldham, D.J.; Mohsen, E.A. A technique for predicting the performance of self-protecting buildings with respect to traffic noise. *Noise Control Eng. J.* **1980**, *15*, 11–19. [CrossRef]

40. Bhattacharjee, S.; Swain, S.K.; Sengupta, D.K.; Singh, B.P. Effect of heat treatment of hydroxyapatite on its dispersibility in aqueous medium. *J. Colloids Surf.* **2006**, *277*, 164–170. [CrossRef]

41. Wu, X.W.; Ma, H.W.; Zhang, Y.R. Adsorption of chromium (VI) from aqueous solution by a mesoporous aluminosilicate synthesized from microcline. *Appl. Clay Sci.* **2010**, *48*, 538–541. [CrossRef]

42. Pandey, P.K.; Sharma, S.K.; Sambi, S.S. Kinetics and equilibrium study of chromium adsorption on zeoliteNaX. *Int. J. Environ. Sci. Technol.* **2010**, *7*, 395–404. [CrossRef]

43. Schmuhl, R.; Krieg, H.M.; Keizer, K. Adsorption of Cu(II) and Cr(VI) ions by chitosan: Kinetics and equilibrium studies. *Water SA* **2001**, *27*, 1–8. [CrossRef]

44. Asgari, G.; Rahmani, A.R. Preparattion of an Adsorbent from Pumice Stone and Its Adsorption Potential for Removal of Toxic Recalcitrant Contaminants. *J. Res. Health Sci.* **2013**, *13*, 53–57. [PubMed]

45. Dzombak, D.A.; Morel, F.M.M. *Adsorption of Inorganic Contaminants in Ponded Effluents from Coal-Fired Power Plants*; Massachusetts Institute of Technology Cambridge: Cambridge, MA, USA, 1985.

46. Farley, K.J.; Dzombak, D.A.; Morel, F.M.M. Distinguishing Adsorption from Surface Precipitation. *J. Colloid Surf. Sci.* **1985**, *106*, 226–242. [CrossRef]

47. Correa, F.G.; Gómez, J.S.; Martínez, J.B. Synthesis and Characterization of Inorganic Materials to Be Used as Adsorbents of Toxic Metals and of Nuclear Interest. In *Contributions of the National Institute of Nuclear Research to the Advance of Science and Technology in Mexico. Commemorative Edition 2010*; Instituto Nacional de Investigaciones Nucleares: Ocoyoacac, Mexico, 2010; pp. 195–210.

48. Müller, A.; Duffek, A. Similar Adsorption Parameters for Trace Metals with Different Aquatic Particles. *J. Aquat. Geochem.* **2001**, *7*, 107–126. [CrossRef]

49. Fendorf, S.; Eick, M.J.; Grossl, P.; Sparks, D. Arsenate and Chromate Retention Mechanisms on Goethite. 1. Surface Structure. *Environ. Sci. Technol.* **1997**, *31*, 315–320. [CrossRef]

50. Waychunas, G.A.; Fuller, C.C.; Davis, J.A. Surface complexation and precipitate geometry for aqueous Zn (II) sorption on ferrihydrite. *Geochim. Cosmochim. Acta* **2002**, *66*, 1119–1137. [CrossRef]

51. Mario, V.; Maya, A.T.; James, O.L. Surface Complexation Modeling of Carbonate Effects on the Adsorption of Cr(VI), Pb(II), and U(VI) on Goethite. *Environ. Sci. Technol.* **2001**, *35*, 3849–3856.

52. Raven, K.P.; Jain, A.; Loeppert, R.H. rsenite and Arsenate Adsorption on Ferrihydrite: Kinetics, Equilibrium, and Adsorption Envelopes. *Environ. Sci. Technol.* **1998**, *32*, 344–349. [CrossRef]

53. Perrone, J.; Fourest, B.; Giffaut, E. Surface and Physicochemical Characterization of Phosphates Vivianite, $Fe_2(PO_4)_3$ and Hydroxyapatite, $Ca_5(PO_4)_3OH$. *J. Colloid Interface Sci.* **2002**, *249*, 441–452. [CrossRef] [PubMed]

54. Smičiklas, I.; Dimović, S.; Plećaš, I.; Mitrić, M. Adsorption and removal of strontium in aqueous solution by synthetic hydroxyapatite. *Water Res.* **2006**, *40*, 2267–2274. [CrossRef] [PubMed]

55. Antelo, J.; Fiol, S.; Gondar, D.; López, R.; Arce, F. Comparison of arsenate, chromate and molybdate binding on schwertmannite: Surface adsorption vs. anion-exchange. *J. Colloid Interface Sci.* **2012**, *386*, 338–343. [CrossRef] [PubMed]

56. Gunnarsson, M. *Surface Complexation at the Iron Oxide/Water Interface Experimental Investigations and Theoretical Developments*; Institution för kemi Göteborgs University: Göteborgs, Germany, 2002.

57. Jain, A.; Raven, K.P.; Loeppert, R.H. Understanding Arsenate Reaction Kinetics with Ferric Hydroxides. *Environ. Sci. Technol.* **1999**, *33*, 1179–1184. [CrossRef]

58. Álvarez-Ayusço, E.; García-Sánchez, A.; Querol, X. Adsorption of Heavy Metals from Aqueous Solutions on Synthetic Zeolite. *Hazard J. Mater.* **2007**, *142*, 191–198.

59. Ellouzi, K.; Elyahyaoui, A.; Bouhlassa, S. Octacalcium phosphate: Microwave-assisted hydrothermal synthesisand potentiometric determination of the Point of Zero Charge (PZC) and isoelectric point (IEP). *Pharm. Lett.* **2015**, *7*, 152–159.
60. Jin-Wook, K. *The Modeling of Arsenic Removal from Contaminated Water Using Coagulation and Sorption*; Texas A&M University: College Station, TX, USA, 2005.
61. Singh, S.P.N.; Mattigod, S.V. Modelling boron adsorption on kaolinite. *Clays Clay Miner.* **1992**, *40*, 192–205. [CrossRef]
62. Eick, M.J.; Peak, J.D.; Brady, W.D. The effect of oxyanions on the oxalatepromoted dissolution of goethite. *Soil. Sci. Soc. Am. J.* **1999**, *63*, 1133–1141. [CrossRef]
63. Tanuma, Y.; Anada, T.; Honda, Y.; Kawai, T.; Kamakura, S.; Echigo, S.; Suzuki, O. Granule Size–Dependent Bone Regenerative Capacity of Octacalcium Phosphate in Collagen Matrix. *Tissue Eng. A* **2011**, *18*, 546–557. [CrossRef] [PubMed]

applied
sciences

MDPI

Article

Suppressing Salt Transport through Composite Pervaporation Membranes for Brine Desalination

Lin Li [1], Jingwei Hou [1,*], Yun Ye [1], Jaleh Mansouri [1,2], Yatao Zhang [3] and Vicki Chen [1]

[1] The United Nations Educational, Scientific and Cultural Organization (UNESCO) Centre for Membrane Science and Technology, School of Chemical Engineering, University of New South Wales, Sydney 2052, Australia; lin.li@student.unsw.edu.au (L.L.); yun.ye@unsw.edu.au (Y.Y.); j.mansouri@unsw.edu.au (J.M.); v.chen@unsw.edu.au (V.C.)
[2] Cooperative Research Centre for Polymers, Notting Hill 3168, Australia
[3] School of Chemical Engineering and Energy, Zhengzhou University, Zhengzhou 450001, China; zhangyatao@zzu.edu.cn
* Correspondence: Jingwei.hou@unsw.edu.au

Received: 30 June 2017; Accepted: 16 August 2017; Published: 19 August 2017

Featured Application: The pervaporation membranes fabricated in this study can be potentially used for brine treatment.

Abstract: Pervaporation membranes have gained renewed interest in challenging feedwaters desalination, such as reverse osmosis (RO) concentrated brine wastewater. In this study, composite polyvinyl alcohol (PVA)/polyvinylidene fluoride (PVDF) pervaporation membranes were prepared for brine treatment. The composite membrane was firstly studied by adjusting the cross-linking density of PVA by glutaraldehyde: the membrane with higher cross-linking density exhibited much higher salt rejection efficiency for long-term operation. A trace of salt on the permeate side was found to diffuse through the membrane in the form of hydrated ions, following solution-diffusion mechanism. To further suppress the salt transport and achieve long-term stable operation, graphene oxide (GO) was incorporated into the PVA layer: the addition of GO had minor effects on water permeation but significantly suppressed the salt passage, compared to the pure PVA/PVDF membranes. In terms of brine wastewater containing organic/inorganic foulant, improved anti-fouling performance was also observed with GO-containing membranes. Furthermore, the highest flux of 28 L/m^{2}h was obtained for the membrane with 0.1 wt. % of GO using 100 g/L NaCl as the feed at 65 °C by optimising the pervaporation rig, with permeate conductivity below 1.2 μS/cm over 24 h (equivalent to a salt rejection of >99.99%).

Keywords: brine wastewater treatment; pervaporation; composite PVA/PVDF membrane; graphene oxide; anti-fouling properties

1. Introduction

Desalination has been widely utilized to relieve the shortage of fresh water in many parts of the world. However, a large amount of brine wastewater is also produced as a by-product of desalination. For example, seawater reverse osmosis (RO) retentate may contain concentrated salt, humic acid and other dissolved solids. The disposal of brine wastewater into the ocean or inland could lead to environmental and ecological problems [1]. Due to high osmotic pressures, RO cannot be utilized to treat such wastewater. On the other hand, the treatment efficiency of thermally driven processes like membrane distillation (MD) is less dependent on the feed solution concentration, and can be regarded as a promising candidate in brine wastewater treatment [2,3]. However, the major problems for the industrial application of MD are membrane fouling/scaling and pore wetting.

More recently, to address the pore wetting, a dense hydrophilic pervaporation membrane has been explored for brine treatment. Pervaporation is also a thermally driven process, and the main water transport mechanism is solution-diffusion [4]. It has been extensively studied for organic dehydration [5,6]. The purpose of both desalination and organic solvents dehydration is to separate water from the bulk feed solution; however, feed properties vary significantly. For organic solvent dehydration, water usually only accounts for less than 10% of the bulk solution [7], while water is the major component in feed for desalination (more than 90%). Thus, the major problem of applying many conventional pervaporation membranes for desalination is that the membranes are not stable in the solution with both high water fractions and elevated temperatures.

Polyvinyl alcohol (PVA) pervaporation membranes have high water vapor permeability, satisfactory membrane-forming capability and excellent anti-fouling property. Recently, PVA membranes have been applied for pervaporation desalination, but a certain amount of salt can still diffuse through the polymeric layers in the form of hydrated salt ions [8,9]. This unfavorable effect can be partially mitigated by polymer cross-linking. However, a higher salt rejection would be always compromised by a reduced permeability [10,11]. Aside from the organic cross-linker, the addition of inorganic nanofillers can also enhance the stability of the composite pervaporation membranes. For example, the PVA/maleic acid/silica freestanding membranes exhibited a reduced swelling degree with higher silica concentration [12,13]. Furthermore, graphene oxide (GO), a unique 2D inorganic material, has good interfacial compatibility with PVA due to the formation of hydrogen bonds [14], showing good potential to enhance PVA stability. GO has been extensively investigated for gas and liquid separation membranes [15,16]. For porous filtration membranes, the incorporation of GO can improve the permeation flux, salt rejection and anti-fouling properties due to improved hydrophilicity [17–23]. For a porous hydrophobic membrane for MD desalination applications, the immobilization of GO by PVDF as the binder material on the surface of polytetrafluoroethylene (PTFE) membranes showed enhanced flux due to selective sorption, nanocapillary effect, reduced temperature polarization and polar functional groups in GO [24]. It can also improve the thermal, mechanical and electrical properties of the composite membranes [25,26]. Considering its 2D structure, the incorporation of GO can potentially suppress the passage of hydrated salt ions within the PVA polymer. In addition, stacked GO laminates can also form highly efficient molecular sieving channels within polymeric matrix [27,28]. However, these aspects have yet to be explored for pervaporation membranes.

Furthermore, most studies on pervaporation desalination so far only use single monovalent salt solutions as feed, and the systems are only evaluated for a relatively short period. The long-term performance of composite pervaporation membranes is crucial for evaluating their feasibility for practical application, especially for complex brine treatment. Thus, in this study, a series of composite PVA/PVDF membranes were fabricated by coating a thin layer of PVA onto commercial hydrophobic PVDF membranes. Different cross-linking density was investigated to understand their effect on the membrane performance. In addition, GO nanosheets were blended into the PVA layer to further suppress salt transport through the membrane. To better understand the properties of the GO/PVA composite membrane, freestanding GO/PVA membranes were fabricated to investigate the interactions between GO and PVA, as well as the water/salt diffusion process within the composite layer. Furthermore, the composite membrane's anti-fouling performance was explored using highly concentrated brine solutions containing humic and calcium salts. Lastly, the pervaporation rig optimization was carried out to promote the operational flux for the composite membranes.

2. Materials and Methods

2.1. Materials

Commercial hydrophobic microfiltration PVDF flat sheet membrane (Millipore, GVHP 0.22 μm, and thickness 125 μm, Billerica, MA, USA) was used in this study. Polyvinyl alcohol (PVA, Mw 89k–98k, 99+ % hydrolysed) was purchased from Sigma-Aldrich (Sigma-Aldrich Corp., St. Louis, MO, USA).

Sodium chloride (NaCl) and potassium hydroxide (KOH) pellets were supplied by Ajax Finechem (Cheltenham, VIC, Australia). Dextran (Mw 9k–11k) and glutaraldehyde (25% in water) were obtained from Alfa Aesar (Thermo Fisher Scientific, Heysham, UK). Hydrochloric acid (HCl, 32% aqueous solution) was obtained from RCl Labscan (RCl Labscan Limited, Bangkok, Thailand). Graphene oxide nanosheets were fabricated from graphene with a modified Hummer's method. The detailed process can be found in our previous publication [28].

2.2. Composite Membrane Preparation

The commercial supporting membrane was firstly pre-treated with 1 M KOH solution at 65 °C water bath for different length (0.5, 1, 2, 6 and 8 h) [29,30]. The KOH treatment led to reduction in water contact angle due to the hydrophilization effect. In this work, 2 h pre-treatment time was selected as longer treatment time did not lead to the increase of surface hydrophilicity (Figure S1). Subsequently, the membrane was rinsed with Milli-Q water and dried at room temperature. For the casting solution, aqueous PVA solution (10 wt. %) was prepared by dissolving PVA powder in Milli-Q water at 80 °C water bath with mechanical stirring for at least 6 h. Then the PVA solution was cooled down to room temperature. A certain amount of glutaraldehyde, and quencher methanol were added to the aqueous PVA solution and the mixture was magnetically stirred for another 30 min, which was followed by the addition of catalyst HCl and stirring for another 5 min. The molar ratio of glutaraldehyde/PVA repeat unit (denoted as MR value) was varied, while the ratio of methanol/HCl/PVA was maintained invariant, i.e., 1 g PVA was mixed with 2 mL of 10% methanol and 0.4 mL of 1 M HCl. To prepare the composite membrane with MR value of 0.2, for 1 g PVA prepared for the casting solution, the glutaraldehyde used (25% in water) was 1.72 mL. The PVA concentration in the final casting solution was maintained at 5 wt. % by adjusting the amount of Milli-Q water in the casting solution. The casting machine (Sheen 1133N automatic film applicator, Sheen Instruments, Surry, UK) was used to cast the PVA membrane on PVDF supports with a casting speed of 50 mm/s under 50% humidity. Then the coated membrane was dried at room temperature overnight, followed by oven drying at 80 °C for 30 min.

For the GO-containing PVA composite membrane fabrication, a certain amount of GO was evenly dispersed into the PVA casting solution containing glutaraldehyde and methanol. To initiate the cross-linking, HCl was then added to the above casting solution and stirred for 5 min. The membrane casting and drying procedures are identical to the previous pure PVA composite membranes. The weight ratio of GO to PVA (0.1, 0.2, 0.3 wt. %) was used to denote different samples, e.g., PVA0.1GO/PVDF denoted composite membrane with 0.1 wt. % of GO in PVA matrix. For the PVA-GO composite membranes, the molar ratio of glutaraldehyde to PVA repeat unit was kept invariant of 0.2.

In order to understand the interfacial interactions between GO and PVA, the freestanding PVA-GO membranes were prepared by pouring the above-mentioned casting solution in Petri dishes, followed by the same drying and post-heat treatment procedure as the cast coated membrane.

2.3. Pervaporation Desalination

Pervaporation desalination experiments were conducted by using a laboratory scale pervaporation unit, shown in Figure 1 A membrane with an effective surface area of 40 cm^2 was placed in the middle of the module, where four thermocouples were installed to monitor the inlet and outlet temperature of feed and permeate, in order to have an accurate measurement of the temperature difference on both feed and permeate sides across the membrane. During the pervaporation test, the feed solution was preheated in a water bath and pumped to the membrane module with a cross-flow velocity (CFV) of 0.625 m/s, while maintaining feed inlet temperature to the module at 65 ± 1 °C. The permeate water vapor was withdrawn by a pump on the permeate side with vacuum pressure around 24 kPa (unless otherwise stated), and collected in the beaker on the balance after condensing (cold water maintained at ~10 °C). Both the feed pump and permeate pump used in this experiment were Masterflex L/S variable speed digital peristaltic pump (Cole-Parmer, Vernon Hills, USA). Spacers (commercial RO module spacer, with filament thickness of 0.66 mm, mesh length of 3.1 mm and spacer thickness of 1.18 mm, Synderfiltration, Vacaville, USA) were used in both of the membrane feed and permeate sides to support the membrane. During the operation, the feed tank was topped up using the solution with the same concentration as the original feed.

The salt rejection (R) was calculated by the following equation:

$$R = \frac{C_f - C_p}{C_f} \times 100\% = \frac{\kappa_f - \kappa_p}{\kappa_f} \times 100\% \tag{1}$$

where C_f and C_p referred to the feed and permeate salt concentration, κ_f and κ_p referred to the feed and permeate conductivity.

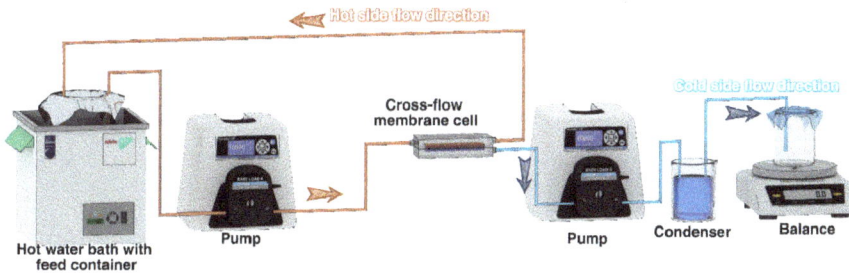

Figure 1. Schematic diagram of cross-flow pervaporation setup with flat-sheet membrane module.

2.4. Membrane Characterization

2.4.1. Scanning Electron Microscopy (SEM)

The surface and cross-section of the membranes were characterized by field emission scanning electron microscopy (FE-SEM, FEI Nova NanoSEM, Hillsboro, OR, USA). The cross-sectional SEM images of the composite membrane were obtained by snapping the membrane in liquid nitrogen. Samples for FE-SEM were prepared by sputter coating a thin layer of chromium under vacuum to generate conductivity. The qualitative surface chemistry of the membrane after desalination was investigated by energy dispersive X-ray spectroscopy (EDX) (FEI Nova NanoSEM and Hitachi S3400, Hitachi Ltd, Tokyo, Japan) to detect the presence of salt.

2.4.2. Membrane Hydrophobicity/Hydrophilicity Characterization

Static contact angles were measured using a contact angle goniometer (KSV CAM 200, Biolin Scientific, Gothenburg, Sweden) by the sessile drop method. Reported values were the average of at least 5 measurements.

2.4.3. Equilibrium Water Content (EWC)

EWC was measured to assess the cross-linked PVA's swelling property, as the ratio of the weight of water in the hydrogel to the weight of the hydrogel at equilibrium hydration. A freestanding hydrogel film was weighted and then immersed in 50 mL Milli-Q water for at least 24 h at room temperature, in order to achieve the equilibrium hydration state. The surface of the film was then blotted with an absorbent tissue paper (Kimwipe, Holcomb Bridge Road Roswell, GA, Canada) to remove excess water present on the surface and weighed again. The EWC value was calculated as follows:

$$EWC = \frac{m_w - m_d}{m_w} \tag{2}$$

where m_w and m_d represented the weight of cross-linked PVA at equilibrium hydration and dry state.

2.4.4. Salt Desorption Test

Salt transport in hydrogel film was measured using kinetic desorption experiment at room temperature. Dense freestanding membrane was prepared following the same procedure as for the EWC test. A hydrogel film with known thickness was immersed in 50 mL of 100 g/L NaCl for 24 h, which was long enough for the film to absorb an equilibrium amount of NaCl from the solution. Then the surface of the film was dried by Kimwipe and placed in a beaker with 50 mL Milli-Q water and stirred by magnetic stirring at around 500 rpm speed. Solution conductivity increased as the salt diffused out of the film into the Milli-Q water and the solution conductivity was recorded. Calculation of the NaCl diffusivity in the hydrogels was done using a Fickian analyses of NaCl desorption from a planar film.

2.4.5. Pore size Characterization

Liquid entry pressure (LEP) and mean pore size of the composite membrane were measured using capillary flow porometer from Porous Materials Inc. (PMI, Ithaca, NY, USA), based on wet/dry flow method [31]. A membrane sample with a diameter of 13 mm was placed in the chamber between two O-rings. In addition, then the membrane surface was covered by the wetting agent (Galwick®, PMI, Ithaca, NY, USA) with defined surface tension (15.9 dynes/cm). The gas flow rate through the membrane was measured as a function of the differential pressure. All porometry tests were performed at room temperature.

2.4.6. Pressurized Dead-End Filtration Test

For the membrane pore size smaller than the porometer instrument capability, a pressurized filtration test was used to determine the membrane pore size with a dead-end filtration membrane cell. Membrane resistance and permeability were also investigated using the same technique. Before the test, membrane sample was soaked in Milli-Q water overnight. Permeate flux over time was monitored with different feed solutions, i.e., Milli-Q water, 30 g/L NaCl, and 100 mg/L dextran. In addition, the dextran content of both feed and permeate was measured using the total organic carbon analyser (TOC, Shimadzu V-CSH, Kyoto, Japan). The entire filtration test was performed at room temperature.

2.4.7. X-ray Diffraction (XRD)

To determine whether GO sheets were dispersed as separated sheets in PVA matrix, XRD measurements (Empyrean X-ray diffraction system, PANalytical, 7602 EA Almelo, the Netherlands) was carried out by using Cu K_α radiation.

2.4.8. Fourier Transform Infrared Spectroscopy (FT-IR)

FT-IR (Spotlight 400, PerkinElmer, Waltham, MA, USA) was used to characterize the surface properties of composite membranes.

2.4.9. Differential Scanning Calorimetry (DSC)

Thermal analysis of the freestanding membrane was conducted using a DSC Q20 (TA Instruments, Inc., New Castle, DE, USA). The samples were heated from $-30\,°C$ to $300\,°C$ at a rate of $10\,°C/min$ in a nitrogen atmosphere. Approximately 6–8 mg of sample is used for each DSC measurement.

3. Results

3.1. Investigation on Cross-linking of PVA by Glutaraldehyde

3.1.1. Pervaporation Performance Using Single Salt Brine

PVA/PVDF membranes fabricated with different molar ratio of glutaraldehyde to PVA repeat unit (MR values of 0.025, 0.1, 0.2) were investigated regarding their pervaporation desalination performance using 100 g/L NaCl as the feed (Figure 2). In this work, each pervaporation test was carried out with at least two membrane samples, and the difference in permeation flux and salt rejection was less than 10% throughout the whole testing process.

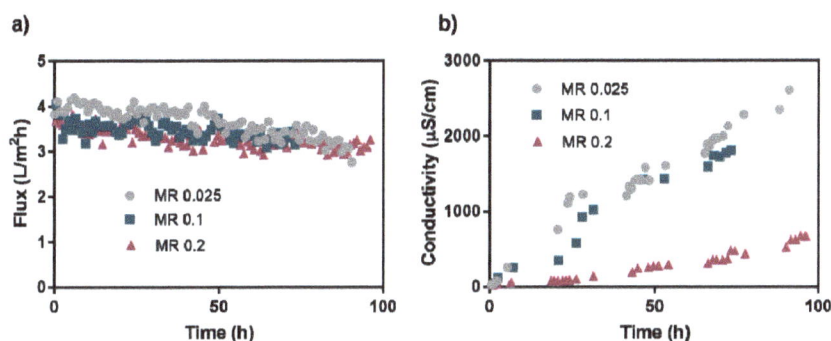

Figure 2. (a) Flux and (b) permeate conductivity profile of composite PVA/PVDF membrane with different molar ratio of glutaraldehyde to PVA repeat unit (MR 0.025, 0.1, 0.2) using a feed solution of 100 g/L NaCl at 65 °C.

As shown in Figure 2a, all membrane samples had comparable water flux under the same operating condition although a slightly lower flux was observed for membranes with higher MR value. Besides, all the membranes experienced continuous flux declines during the whole testing process due to the gradual increase of the feed concentration. In terms of the permeate conductivity, all membrane samples showed an increased conductivity with a longer operation time, and the membrane with the highest MR value (0.2) showed the most stable salt rejection after 96 h operation. This trend suggested that the cross-linking of PVA can benefit the operational stability of the membrane, which can be attributed to more acetal/ether linkages with higher MR value, leading to the improved structural stability with reduced free volume and swelling tendency [32,33]. Comparatively, the virgin supporting membrane, hydrophobic PVDF membrane with nominal pore size of 0.22 μm, showed a permeate conductivity of 2 mS/cm after 10 min of operation under the same operating condition. This indicated that the salt rejection efficiency of the pervaporation process was not attributed to the supporting membrane used in the composite membrane.

The water flux results in this work were comparable to the values reported in previous literatures of pervaporation membranes, when taking the operating temperature and feed solution properties into consideration. For example, composite membranes containing 100–1000 nm PVA surface coating on polysulfone membrane yielded fluxes of 4.6–7.4 L/m^{2}h with 30 g/L NaCl solution at 70 °C [34]. In another line of research, a slightly higher water flux (5.57 L/m^{2}h) was obtained with a PVA/polyacrylonitrile (PAN)/polyethylenimine (PEI) triple-layered membrane when using 50 g/L NaCl solution as feed under room temperature [35]. In terms of the salt rejection, the PVA membrane in this work exhibited higher salt rejections even with more concentrated feed solution compared with the literatures [34,35].We also explored the membrane stability by recycling the membrane and testing the membrane using different feed concentrations and salt types, as shown in Figures S2 and S3.

However, considering the permeate flux was relatively invariant, the gradual increase of permeate conductivity for all the PVA/PVDF membranes indicated a more rapid transport of salt ions as the pervaporation test proceeded. The following characterizations and discussion were aimed to better understand the transport mechanism of salt through the composite PVA/PVDF membrane, to provide guidance for the further modification of such membranes.

3.1.2. Transport Mechanism of Salt

The salt transport through the membrane was further confirmed by the EDX results (Figure 3): the presence of salt crystals was observed on the membrane feed side as well as within the porous substrate after 96 h pervaporation test. The transport of salt ions through the PVA layer could be originated via the free volume between PVA polymeric chains. In this work, the gradual increase of the conductivity suggested a change of PVA chain structure overtime, possibly due to the swelling effect by hot water feed during the pervaporation process.

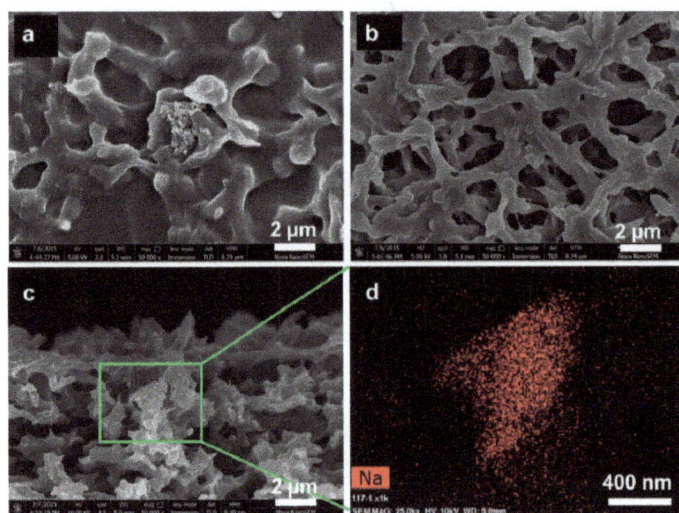

Figure 3. SEM and EDX autopsy of the composite membrane after 96 h pervaporation test (membrane with MR of 0.2). (**a**) Feed side, (**b**) permeate side and (**c**) cross-section of the composite membrane; and (**d**) EDX mapping the membrane cross-section beneath the PVA coating layer.

To further explore the PVA membrane pore structure, a porometer test was carried out with the composite membrane (both fresh membranes and membranes after 96 h permeation test) using Galwick as a wetting agent at room temperature. However, no direct liquid penetration through the membrane was observed for all the membranes with different MR values under maximum feed of 10 bar. To further explore the pore size (free volume) range of the composite membrane, we conducted the dead-end filtration test with 15 bar feed pressure using fresh composite membranes at room temperature. As suggested in Table 1, the resistance for MR 0.2 membrane was in the range of RO membranes, with a pure water flux of 1.4 ± 0.4 L/m^{2}h. The rejection of dextran (1–2 nm) can remain ~83.4% over 10 h operation. In comparison, the rejection of NaCl (diameter of hydrated Na$^+$ 0.716 nm, hydrated Cl$^-$ 0.664 nm) was much lower at around 33%, which was lower than the reported PVA composite nanofiltration (NF) membrane [36,37]. Thus, it was believed that the pore size (free volume) of the composite membrane (MR of 0.2) was in the range of NF membrane (below 1 nm). Deng et al. also reported that the diameter of casted/cross-linked PVA membrane's free volume was around 0.426 nm tested by bulk positron annihilation lifetime spectroscopy (PALS) [38].

Table 1. Pressurized filtration test results of composite PVA/PVDF membrane using feed solutions of Milli-Q water, 30 g/L NaCl or 100 mg/L dextran, under TMP of 15 bar at room temperature. The flux was recorded after the stabilization of 30 min. The membrane resistance (R_m) was calculated based on $R_m = \frac{\Delta P}{\mu J}$, where ΔP, μ and J indicated the trans-membrane pressure, viscosity of Milli-Q water and corresponded pure water flux, respectively.

Pressurized Filtration Test Results	Unit	Commercial PVDF	MR0.025	MR0.1	MR0.2
Pure water flux	L/m²h	>15,000	240 ± 56	5.5 ± 0.27	1.4 ± 0.4
Permeability	L/m² h·bar	>1000	16 ± 3.5	0.37 ± 0.018	0.093 ± 0.027
Average membrane resistance	m⁻¹	<3.9 × 10¹¹	~2.5 × 10¹³	~1.1 × 10¹⁵	4.3 × 10¹⁵
Flux using 30 g/L NaCl as the feed	L/m²h	/	129 ± 16.5	3.74 ± 0.23	0.54 ± 0.1
Salt rejection using 30 g/L NaCl as the feed	%	/	~0	~9	~33
Flux using 100 mg/L Dextran as the feed	L/m²h	/	142 ± 21	4.69 ± 0.42	0.99 ± 0.18
Rejection using 100 mg/L Dextran as the feed	%	/	~0	53.9%	83.4

It also indicated that the change of MR value had a significant effect on the PVA layer pore (free volume) structure. With the increase of MR value, the membrane resistance gradually increased, accompanied with the increase of rejections for salt ions and dextran. In this work, the lowest resistance was observed for the membrane with MR 0.025: it fell into the range of UF or even MF membranes.

Figure 4 showed the SEM and atomic force microscope (AFM) analysis of membrane with different MRs. With the increase of MR value, the PVA layer thickness became more homogeneous, accompanied with the loss of surface roughness. The surface roughness of PVDF membrane after KOH pre-treatment was ~200 nm, which was comparable to the composite membrane with MR of 0.025. This indicated that the coating of PVA with low MR value (0.025) did not significantly change the membrane's surface roughness. Furthermore, the change of MR value did not only alter the membrane pore (free volume) structure; it can also affect the chemical component within the selective layer. Theoretically, the cross-linking of PVA with glutaraldehyde can consume the hydroxyl groups for the polymer. It can also increase the polymer chain structure rigidity. Both aspects can lead to the reduced EWC value for the membrane as shown in Table 2, which well aligns other studies [32,33].

Table 2. Characterization results of composite PVA/PVDF membrane (Diffusivity of NaCl in pure water is 14.7×10^{-6} cm²/s). MR: Molar Ratio. EWC: Equilibrium Water Content.

Composite Membrane	Surface Roughness (nm)	EWC (%) in Water	EWC (%) in 100 g/L NaCl	NaCl Diffusivity (10^{-6} cm²/s)
MR0.025	281.5 ± 7.78	184.5 ± 35.3	129.7 ± 16.8	2.02 ± 0.97
MR0.1	140 ± 31.1	60.4 ± 5.2	56.6 ± 2.5	1.42 ± 0.36
MR0.2	106.4 ± 13.58	49.6 ± 7.0	38.1 ± 6.3	0.64 ± 0.13

This could well explain the observation of slightly lower permeate flux but improved salt rejection for the membrane with higher MR values (Figure 2): the transport rate of bulkier penetrant (hydrated salt ions) was more sensitive to the changes in free volume than those of smaller penetrant (water vapor) [39,40]. Therefore, the decreased diffusivity of salt through the membrane with higher MR (Table 2) can be attributed to its lower free volume.

During the pervaporation process, salt ions could diffuse through the PVA layer in the form of hydrated ions and condensed together with water due to the temperature drop in the permeate side. The increase of cross-linking density can retard but not completely block the passage of salt ions. At the same time, the gradual up-take of water can lead to the swelling of PVA due to the preferable adsorption of water molecules by the hydroxyl groups on PVA chains. In the swollen PVA especially the one with large free volume, more water was adsorbed by the hydroxyl groups of PVA chains, where a small fraction of hydrated salt ions could fit into such enlarged free volume, allowing their gradual passage. As a result, the increased free volume allows fast dissolution and diffusion of bulkier hydrated salt ions through the membrane [34,41,42], which explained the gradual loss of the salt rejection during the extended pervaporation test (Figure 5). Thus, in the following session, GO nanosheets, impermeable hydrophilic nanosheets, were incorporated into the PVA matrix to limit the free volume increase with time in PVA layer and further suppress salt transport through the membrane. MR value of 0.2 was applied to fabricate the membranes with GO because of its highest salt rejection efficiency compared to membranes with lower MR values.

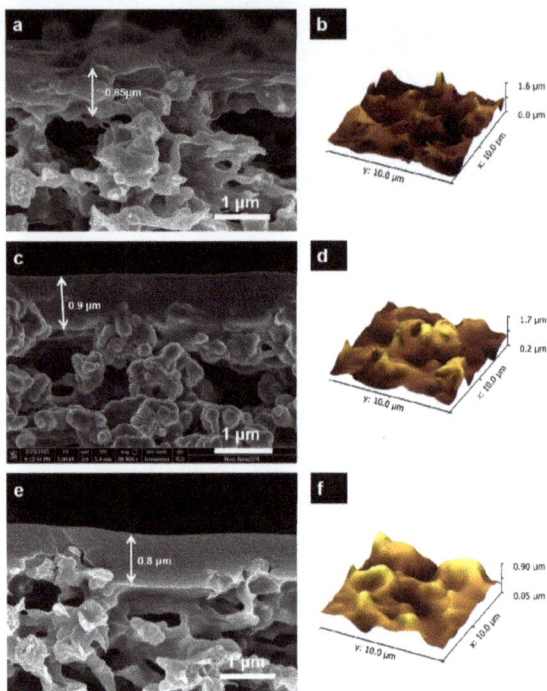

Figure 4. Cross-sectional SEM and AFM surface height images of PVA/PVDF composite membrane with MR of (**a**) and (**b**) 0.025; (**c**) and (**d**) 0.1; (**e**) and (**f**) 0.2.

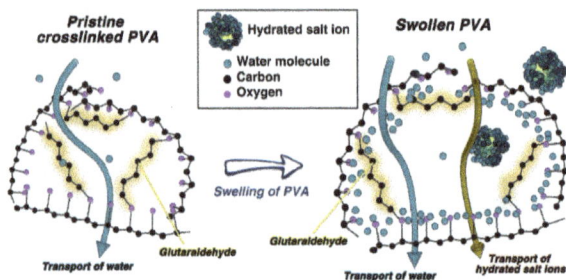

Figure 5. Schematic diagram of salt passage through the swollen PVA membrane (not to scale).

3.2. Incorporation of GO into PVA Matrix

3.2.1. Characterizations

The addition of a small amount of GO into PVA formed small protrusions on the membrane surface, and this became more significant with the increase of GO loading within the coating layer

(Figure 6). Similar surface morphology was observed after incorporating GO into polyether block amide (Pebax) matrix using cast coating method [43]. In terms of the cross-sectional images (Figure S4), they showed good agreement with the surface SEM images. The pure PVA coating was even with a thickness of around 0.8 μm. The incorporation of GO reduced the smoothness for the PVA-GO coating layer, especially when the GO loading was relatively high. During the formation of PVA-GO surface coating, the gradual evaporation of the solvent (water) allowed the cross-linking of the PVA chains, forming the final coating layer of 0.5–1 μm. In terms of the GO, the evaporation-induced capillary flow during the drying process, together with the viscous PVA solution, disrupted the entropy-driven phase transition of GO, leading to the formation of surface protrusions [44].

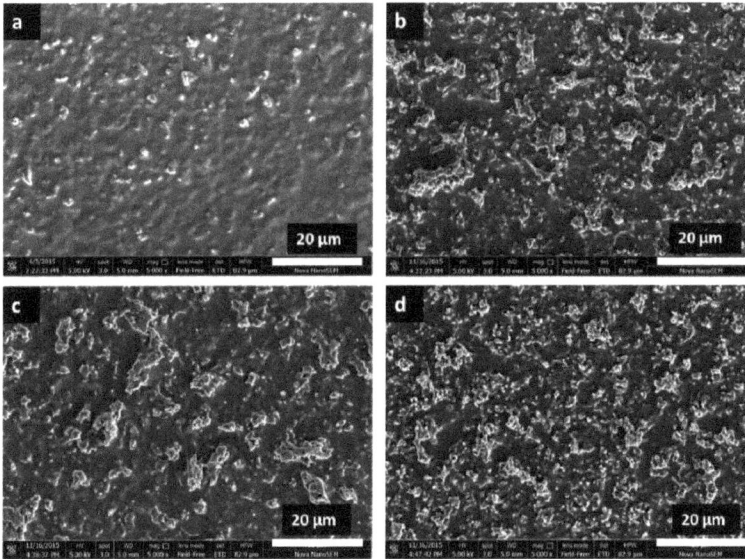

Figure 6. Surface SEM images of composite membranes fabricated with cast coating method: (**a**) pure PVA coating; and PVA-GO hybrid coating with (**b**) 0.1 wt. %, (**c**) 0.2 wt. %, (**d**) 0.3 wt. % GO loading.

The XRD pattern can reveal the GO nanosheets arrangement within the hybrid coating layer. As shown in Figure 7, for the pure GO coated membrane, it had a clear peak at $10.6°$, suggesting the GO nanosheets were piled together with an average interlayer distance of 0.83 nm. After incorporating the GO into PVA matrix, no clear GO peak can be observed. This observation was in good agreement with the previous researches [14,45]. This can be attributed to the good interfacial compatibility between GO and PVA: the presence of carboxyl groups and hydroxyl groups on GO can form hydrogen bonding with hydroxyl groups on PVA. As a result, the GO nanosheets can be well dispersed within the PVA matrix at the molecular level [26].

The successful incorporation of GO was further evidenced by the FT-IR patterns, as shown in Figure 8 The characteristic peaks of GO powder at around $3280 \ cm^{-1}$ and $1725 \ cm^{-1}$ corresponded to O–H and C=O stretch, which indicated the existence of hydroxyl groups and carbonyl groups on GO. Besides, the peak at around $2950 \ cm^{-1}$ confirmed a large amount of C–H stretch on GO. Compared to pure PVA coating, the emerging peaks for the PVA-GO composite coating layer at the region of $3100–3010 \ cm^{-1}$ can be ascribed to alkenyl C–H stretch due to the presence of GO. In addition, the right shift of the O–H stretching peaks were detected with the addition of GO in PVA matrix (from $3375 \ cm^{-1}$ for pure PVA to $3345 \ cm^{-1}$ for PVA with GO of 0.3 wt. %). This observation confirmed the formation of hydrogen bonding between hydroxyl/carboxylate groups on GO and hydroxyl groups on PVA chains, which was consistent with those of other studies [14,45,46].

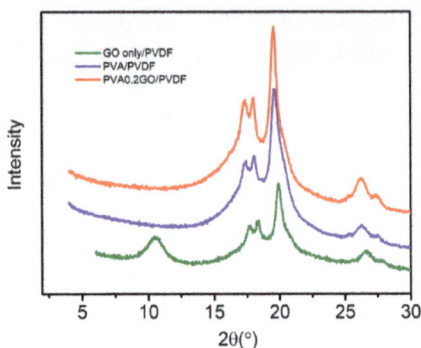

Figure 7. XRD curves of composite membranes.

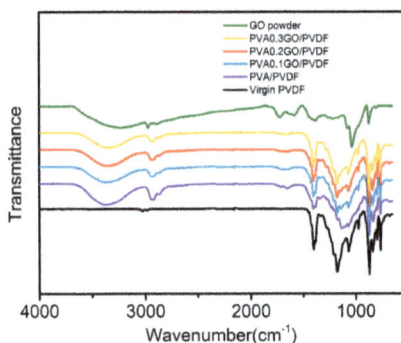

Figure 8. FT-IR curves of PVA-GO/PVDF membranes with GO loading of 0.1, 0.2 and 0.3 wt. %.

Regarding the contact angle shown in Table S1, the addition of GO into PVA matrix slightly increased the contact angle of the membrane surface. For example, the pure PVA surface coating has a contact angle of $30.2 \pm 3.5°$, compared with the $37.1 \pm 2.3°$ for the composite membrane containing 0.2 wt. % GO. This can be attributed to the presence of a large amount non-polar benzene rings on GO surface, as well as the formation of hierarchical surface morphology after the addition of GO [47].

To better understand the effect of GO incorporation on the coating layer properties, freestanding PVA-GO films were also prepared and characterized to explore their surface morphology, thermal and swelling properties. The thickness of the freestanding films was around 100 μm. From the digital photo (Figure S5), with higher GO loading, the freestanding film color changed from almost transparent to yellowish brown. Similar to the composite membrane fabricated via cast coating, the addition of GO introduced wrinkles and protrusions onto the membrane surface, as shown in Figure S6. As the freestanding films were fabricated via the solvent evaporation method, without the shear force induced alignment during the casting process, the GO nanosheets randomly oriented within the original membrane casting solution, and eventually led to the wrinkled paper-like structure on the surface due to the capillary flow during the solvent evaporation process. Based on the SEM images, GO was homogeneously dispersed within the PVA matrix. Same as the composite membranes, no GO peak was observed in freestanding films after the incorporation of GO in PVA matrix from the

XRD patterns (Figure S7), indicating a good interfacial compatibility between the GO nanosheets and the PVA matrix.

As shown in Table 3, the DSC results exhibited an increased Tg for the freestanding films after the incorporation of GO nanosheets (from around 48 °C to around 60 °C). This further confirmed the formation of hydrogen bonding between GO nanosheets and PVA matrix, which reduced the polymer chain mobility. This could be also lead to the rigidification of polymer chains near the GO sheets [15]. Similar trends were also observed by incorporating a small amount of GO into PVA matrix [45,48].

Table 3. Thermal and swelling property of the freestanding films (PVA0.1GO denoted membrane with 0.1 wt. % of GO in PVA matrix). PVA: polyvinyl alcohol. GO: graphene oxide.

Freestanding Films	Tg (°C)	EWC in Milli-Q Water (%)	EWC in 100 g/L NaCl (%)	NaCl Diffusivity $(10^{-6}$ cm^2/s)
PVA only	48	49.6 ± 7	38.1 ± 6.3	0.64 ± 0.13
PVA0.1GO	59	25.35 ± 1.41	28.06 ± 1.07	0.712 ± 0.001
PVA0.2GO	60	38.15 ± 2.05	34.98 ± 1.66	0.549 ± 0.066
PVA0.3GO	60	35.35 ± 4.37	30.68 ± 0.54	0.526 ± 0.005

The freestanding films' swelling properties were also studied as it is closely related to the membrane's cross-linking density and free volume [12]. The equilibrium water content (EWC) values of freestanding films (Table 3) in both Mill-Q water and 100 g/L NaCl aqueous solution were investigated. Among different membranes, the pure PVA freestanding film exhibited the highest EWC values for both Milli-Q water and salt water. The initial addition of GO (0.1 wt. %) significantly reduced the EWC value. This can be attributed to the impermeable feature of GO and formation of hydrogen bonding between GO and PVA, which reduced the solubility of water in the PVA polymer chains. However, with the increase of GO loading (0.2 and 0.3 wt. %), the EWC values increased again due to the water adsorption on the nanosheet surface, but they were still lower than the pure PVA films. Another potential reason could be the aggregation of GO sheets in PVA matrix created new pathways for water adsorption and diffusion. Smaller free volume fraction was also obtained by adding GO into PVA matrix from other studies [45].

By comparing the EWC value in Milli-Q water and salt water, it was found that for pure PVA films the EWC value in salt water was significantly lower than that in Milli-Q water. While for the films with GO incorporated, the difference is less significant. This indicated that the swelling ability of PVA with GO was less affected by the salt content, compared to pure PVA. Furthermore, with the addition of GO into the PVA matrix, the salt diffusivity tended to decrease. This was caused by much longer and more torturous transport pathways for salt to pass through the film, since GO is an impermeable nanosheet to salt ions. It is worth mentioning that in the salt diffusivity test, with the addition of GO, the amount of salt absorbed by the membrane decreases significantly. After immersing the hydrated films into Milli-Q water, the equilibrium solution conductivity dropped from around 120 µS/cm for pure PVA films down to 20 µS/cm for PVA with 0.1 wt. % GO. It took similar time (~200 s) for both films to achieve equilibrium during the immersion process. This clearly indicates that less salt is adsorbed by the membrane with GO incorporated: i.e., the GO incorporation considerably decreased the membrane's salt solubility in the PVA films.

From the above characterizations on both composite membranes and freestanding films with GO incorporated into PVA matrix, a good dispersion of GO in PVA was observed by SEM and XRD results; and the hydrogen bonding formed between PVA and GO is detected from FT-IR and DSC analysis. The EWC and NaCl diffusivity results showed that the incorporation of impermeable GO decreased the size of the free volume in PVA matrix and also the water and salt solubility in the matrix.

3.2.2. Pervaporation Performance Using Single Salt Brine

The composite membranes with different GO loadings in PVA matrix (0.1, 0.2, and 0.3 wt. %) were tested in the pervaporation setup, which was then compared with the pure PVA/PVDF membrane (MR value of 0.2). For each membrane, Milli-Q water permeation tests were firstly conducted for at least 4 h. Compared with pure PVA/PVDF membrane, the incorporation of 0.1 wt. % GO showed the negligible effect on the Milli-Q water flux (4.38 L/m^{2}h). However, further increase of the GO loading can lead to a marginal loss of the flux: 94.4 and 85.5% of the original PVA/PVDF flux can be obtained for composite membranes containing 0.2 and 0.3 wt. % GO (results not shown). In highly cross-linked PVA polymer chains, the free volume was organized in a tortuous manner, which led to the random and non-directional water passage pathway [49]. After the addition of GO, the presence of a large amount of hydrophobic aromatic rings on its surface can facilitate the rapid "sliding" of water molecules along its surface [50]. This aspect would facilitate the water transport through the membrane. On the other hand, the incorporation of impermeable GO sheets also increased the transport pathway length of water vapor. Meanwhile, lower free volume, measured as the EWC value, was obtained as a result of hydrogen bonding between GO and PVA chains. As a result, the initial addition of GO (0.1 wt. %) had a negligible effect on the Milli-Q water flux and a further increase of the GO loading (0.2 and 0.3 wt. %) slightly reduced the permeation flux.

In terms of the brine desalination tests with 100 g/L NaCl solution, similar flux profiles (Figure 9) were observed as to the Milli-Q results: the initial addition of 0.1 wt. % GO had a negligible effect on the brine solution flux, and a further increase of the GO loading can lead to more obvious loss of the permeation flux. In terms of permeate conductivity, it constantly increased to over 650 µS/cm (salt rejection of over 99.6%) after 96 h operation for the pure PVA/PVDF membrane, while the composite membrane containing 0.1 wt. % GO had a much more stable performance: the permeate conductivity gradually increased to less than 150 µS/cm (salt rejection of over 99.9%) for 120 h operation while maintaining lower than 50 µS/cm in the first 50 h. As discussed above, the presence of GO nanosheets within the PVA matrix can facilitate the selective water molecule transport along with its "slippery" surface, which suppressed the transport of salt ions through the composite membrane. The reduced free volume of PVA matrix had a more significant blocking effect on the larger hydrated salt ions compared with smaller water molecules. As a result, the addition of GO can improve the salt rejection for the composite membrane.

When 0.2 wt. % GO was added into the PVA matrix, the permeate conductivity was more stable (less than 30 µS/cm in 82 h operation). However, in this case, the permeate flux was much lower than the PVA0.1GO/PVDF membrane, and a significant flux decline was observed after 10 h operation. Due to the reduced free volume after GO addition, the blockage effect by the hydrated salt ions to the smaller water molecules can be more significant. This can also explain the more significant flux loss for membrane containing 0.3 wt. % GO. In addition, for the membrane containing 0.3 wt. % GO, the permeate conductivity increased more rapidly compared with a membrane containing 0.1 and 0.2 wt. % GO. The possible reason was that some defects or cracks may exist due to strong filler interactions and agglomerations. Similar results were also observed with a composite membrane containing silica nanoparticles and ordered mesoporous carbon (OMC), less water uptake and permeable flux were detected with higher loading of nanoparticles in the thin-film layer, and the overdose of nanofillers can also lead to the loss of membrane selectivity [51–53].

Upon the completion of the long-term brine desalination test, the membranes were re-tested with Milli-Q water to investigate their permeation flux recovery. The results suggested even though the initial flux was much lower, it can gradually recover to the original Milli-Q water flux prior to the brine desalination test in less than 2 h, shown in Figure S8. This observation confirmed flux decline during the brine desalination test was mainly due to the blockage of water transport pathway by the bulkier hydrated salt ions. In all, with the increase of GO loading in PVA matrix, the water flux tended to decrease for both Milli-Q water and brine, while the salt rejection firstly increased then decreased. Similar membrane behavior has been reported for the PVA-GO-based composite pervaporation membrane for organic solution dehydration [54].

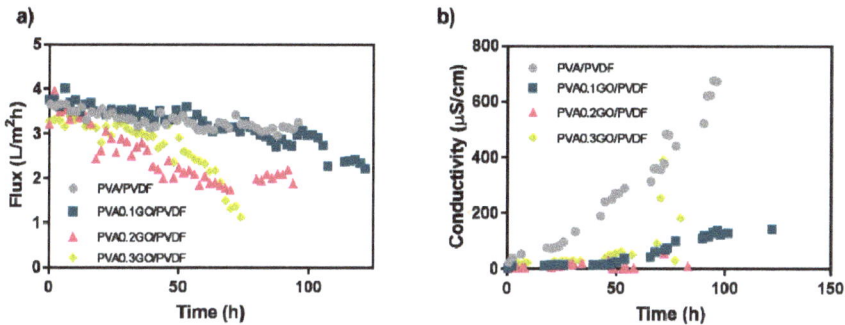

Figure 9. (a) Flux and (b) permeate conductivity profiles of the composite membrane using a feed solution of 100 g/L NaCl at 65 °C.

3.3. Anti-Fouling Performance

GO nanosheets are known to improve the anti-fouling property for the mixed matrix filtration membranes [18,20–23]. This is ascribed to the improved membrane's hydrophilicity due to the hydrophilic groups on GO. However, whether GO can promote the anti-fouling performance of pervaporation membranes has not been explored yet. The fouling study on pure PVA/PVDF composite membranes (Figure 10) showed that the presence of 10 mg/L humic acid in 100 g/L NaCl aqueous solution had a negligible effect on the flux, but increased the permeate conductivity significantly (from 370 µS/cm to 2.8 mS/cm at the end of 72 h operation). This may indicate that the attachment of humic acid on the surface of PVA layer and the binding between humic acid and salt ions accelerated the salt transport through the membrane [55]. With GO incorporated in the PVA coating layer, increased permeate conductivity was also observed (from 140 to 355 µS/cm at the end of 120 h operation), which well aligned with the membrane without GO. However, the increase was less significant. This indicated that the incorporation of GO in PVA matrix stabilized the composite membrane's performance when processing brine containing organic foulant (humic acid).

We further investigated the effect of $CaCl_2$ on the membrane performance by adding 1.26 g/L $CaCl_2$ in the mixture of 100 g/L NaCl and 10 mg/L humic acid: the presence of the inorganic salt ions had a negligible effect on the permeation flux. It is interesting that the presence of calcium slightly reduced the permeation conductivity for both pure PVA/PVDF and PVA-GO/PVDF composite membranes, which can be attributed to the formation of the bulkier calcium-humic complex [55]. The results showed that a much lower permeate conductivity was obtained for the composite membrane containing GO than the pure PVA/PVDF membrane, which clearly demonstrated the improved anti-fouling property for the composite membrane after the addition of GO. This observation also suggested better operational stability for the pervaporation desalination process compared with the MD process: the presence of $CaCl_2$ can lead to severe fouling and flux decline for virgin hydrophobic PVDF (flux dropping to 0 with permeate conductivity rising to 1 mS/cm at the end of 20 h' operation) and modified superhydrophobic membrane (flux dropping to 0 with permeate conductivity rising to 3 mS/cm at the end of 40 h' operation) during the vacuum MD process, as suggested by our previous work [56].

The permeate solution composition at the different operating time was analyzed by ICP when using feed solution containing 100 g/L NaCl, 10 mg/L humic acid and 1.26 g/L $CaCl_2$ with the composite membrane containing 0.1 wt. % GO (Table 4). With the progress of the pervaporation operation, the permeate conductivity increased due to the gradual diffusion of the sodium ions and calcium ions through the membrane, and the diffusion rate of the smaller Na ions was higher than the bulkier Ca ions. In all, the incorporation of small amount of GO in PVA matrix efficiently slowed down the transport of salt ions through the membrane during the pervaporation process, although this was not completely prevented.

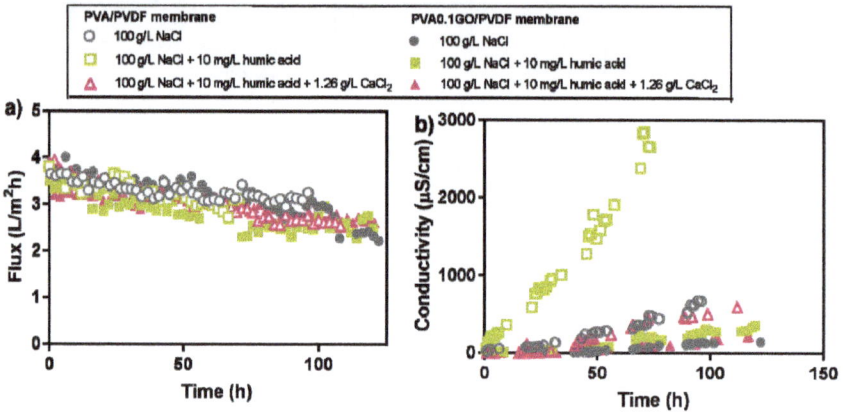

Figure 10. (**a**) Flux and (**b**) permeate conductivity profiles of the composite membrane PVA/PVDF, and PVA0.1GO/PVDF prepared by cast coating method using different feed solution.

Table 4. Permeate property of membrane PVA0.1GO/PVDF at different operation time by using feed solution of 100 g/L NaCl, 10 mg/L humic acid and 1.26 g/L $CaCl_2$, where initial feed conductivity is around 155 mS/cm, with Na^+ 39.4 g/L and Ca^{2+} 0.164 g/L.

Operation Time (h)	Conductivity (µS/cm)	Na^+ (mg/L)	Ca^{2+} (mg/L)
19.2	3.7	0.55	0.00
66.5	78.8	13.6	0.03
116.4	205	37.1	0.18
164.25	493	127	1.07

3.4. Effect of Permeate Pressure on Pervaporation Performance

As the pervaporation process is a thermal-driven process, the vapor pressure difference across the membrane plays a major role in the membrane water vapor permeability. While the main goal of this study was to suppress the salt transport through the composite membrane by modifications on PVA layer, there is still a need to suggest how the permeate pressure influenced on the membranes pervaporation performance, especially the membrane productivity. Conducting a thorough study could be an independent endeavor, however, in this study a limited number of tests were conducted to explore the effect of permeate pressure on membrane performance. A secondary condensation unit was applied next to the membrane cell in the permeate side, in order to maintain the permeate pressure at 3 kPa (instead of the initial pressure of 24 kPa). The membrane PVA/PVDF and PVA0.1GO/PVDF (both with 0.2 MR value and effective membrane area of 14.7 cm²) were further tested, with both Milli-Q water and 100 g/L NaCl as the feeds. For the Milli-Q water feed, PVA/PVDF membrane and PVA0.1GO/PVDF membrane had very comparable flux (average flux 39.7 and 39.2 L/m²h for 4 h operation after stabilization, results not shown). The pervaporation performance using highly concentrated NaCl as the feed was shown in Figure 11 an initial volume of 2 L feed solution with a concentration of 100 g/L was used without top-up during a 24 h operation. Both membranes showed an initial flux of 27–28 L/m²h, which was followed by obvious flux decline (around 25%) as a result of continuously increased feed concentration. After the 24 h pervaporation test using salt water, a flux of around 39 L/m²h was restored using Milli-Q water as the feed for both membranes (results not shown). In terms of permeate conductivity, the membrane PVA0.1GO/PVDF obtained a constant value of 1–1.2 μS/cm, while the membrane PVA/PVDF showed a value of 15 μS/cm at the end of 24 h testing. This also confirmed that the addition of 0.1 wt. % GO effectively suppressed salt diffusion through the membrane. More importantly, the flux value using the same feed concentration was 9 times to that of using permeate pressure of 24 kPa. This suggested the further direction for optimization for such pervaporation wastewater treatment processes.

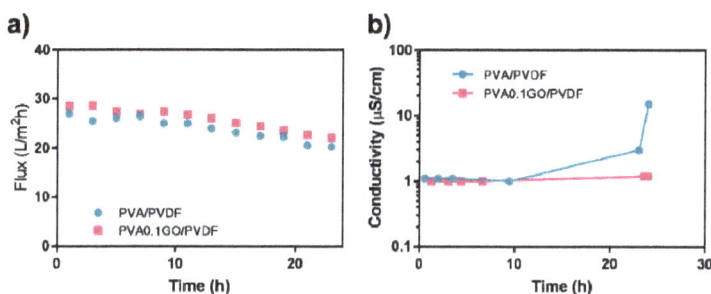

Figure 11. (**a**) Flux and (**b**) permeate conductivity profiles of composite PVA/PVDF membrane (MR value of 0.2) and PVA0.1GO/PVDF membrane using 2 L of 100 g/L NaCl as the feed without top-up at operating temperature of 65 °C.

4. Conclusions

Initially, in this work, composite membrane desalination performance was optimized by varying the cross-linking density of PVA. The results suggested the transport of salt ions through the membrane followed the solution-diffusion mechanism in the form of hydrated ions. Then, by incorporating a small amount of GO (0.1 wt. %) into PVA matrix, a high salt rejection can be maintained during the extended desalination operation, while the flux was relatively unaffected. This can be attributed to the impermeable properties of GO to salt ions and fast water molecule transport along its surface. In addition, the presence of polar groups on GO can also form hydrogen bonds with PVA, which rigidified the PVA chains and suppressed the transport of bulkier salt ions. The GO/PVA composite membrane also exhibited satisfactory anti-fouling properties during the treatment of calcium and humic-containing brine wastewater. Lastly, the study on permeate pressure indicated its crucial role in affecting the membrane productivity for such PVA-based pervaporation membranes.

Supplementary Materials: The following are available online at www.mdpi.com/2076-3417/7/8/856/s1. Figure S1: Contact angle after membrane pre-treatment, Figure S2: Flux and conductivity profile of composite PVDF membrane with KOH/MR 0.2/5 wt. % by one layer casting, where the membrane was cleaned with in-situ cleaning method: (top) "96 h 100 g/L NaCl + 24 h Milli-Q water cycle; (bottom) "23 h 100 g/L NaCl + 1 h Milli-Q water cycle". The feed inlet temperature is 65 ± 1 °C, Figure S3: Flux and conductivity profile of composite PVA/PVDF membrane with KOH/MR 0.2/5 wt. % by one layer casting using a feed solution of 30 g/L, 100 g/L NaCl and PAC solution with the same conductivity as 30 g/L NaCl (~55 mS/cm). The feed inlet temperature was 65 ± 1 °C, Figure S4: Cross-sectional SEM images of the composite membrane fabricated with the cast coating: (a,b) PVA/PVDF, (c,d) PVA0.1GO/PVDF, Figure S5: Digital photo of freestanding PVA samples with different GO loadings: (a) 0, (b) 0.1 wt. %, (c) 0.2 wt. %, and (d) 0.3 wt. %, Figure S6: SEM images of freestanding PVA membrane with GO loading of (a) 0, (b) 0.1 wt. %, (c) 0.2 wt. %, (d) magnified 0.1 wt. %, Figure S7: XRD curves of GO only/PVDF, freestanding PVA, and freestanding PVA with GO loading of 0.1 wt. %, Figure S8: Flux profile of composite membrane PVA0.1GO/PVDF using Milli-Q water, 100 g/L NaCl and Milli-Q water as the feed, sequentially, Table S1: Contact angles of the composite membrane samples with different GO loadings.

Acknowledgments: This research was supported under Australian Research Council's Discovery Projects funding scheme (DP130104048).

Appl. Sci. **2017**, *7*, 856

Author Contributions: Lin Li, Jingwei Hou and Yatao Zhang conceived the idea of suppressing salt transport by GO incorporation. The experimental work was carried out by Lin Li. Yun Ye and Jingwei Hou built the pervaporation setup. Lin Li, Jingwei Hou, Yun Ye, Jaleh Mansouri, Yatao Zhang and Vicki Chen contributed to the manuscript writing and editing.

Conflicts of Interest: The authors declare no conflict of interest.

References

1. Lattemann, S.; Höpner, T. Environmental impact and impact assessment of seawater desalination. *Desalination* **2008**, *220*, 1–15. [CrossRef]

2. Mariah, L.; Buckley, C.A.; Brouckaert, C.J.; Curcio, E.; Drioli, E.; Jaganyi, D.; Ramjugernath, D. Membrane distillation of concentrated brines—Role of water activities in the evaluation of driving force. *J. Membr. Sci.* **2006**, *280*, 937–947. [CrossRef]

3. Mericq, J.-P.; Laborie, S.; Cabassud, C. Vacuum membrane distillation of seawater reverse osmosis brines. *Water Res.* **2010**, *44*, 5260–5273. [CrossRef] [PubMed]

4. Khayet, M.; Matsuura, T. Pervaporation and vacuum membrane distillation processes: Modeling and experiments. *AIChE J.* **2004**, *50*, 1697–1712. [CrossRef]

5. Feng, X.; Huang, R.Y. Pervaporation with chitosan membranes. I. Separation of water from ethylene glycol by a chitosan/polysulfone composite membrane. *J. Membr. Sci.* **1996**, *116*, 67–76. [CrossRef]

6. Chapman, P.D.; Oliveira, T.; Livingston, A.G.; Li, K. Membranes for the dehydration of solvents by pervaporation. *J. Membr. Sci.* **2008**, *318*, 5–37. [CrossRef]

7. Shah, D.; Kissick, K.; Ghorpade, A.; Hannah, R.; Bhattacharyya, D. Pervaporation of alcohol–water and dimethylformamide–water mixtures using hydrophilic zeolite NaA membranes: Mechanisms and experimental results. *J. Membr. Sci.* **2000**, *179*, 185–205. [CrossRef]

8. Quiñones-Bolaños, E.; Zhou, H.; Soundararajan, R.; Otten, L. Water and solute transport in pervaporation hydrophilic membranes to reclaim contaminated water for micro-irrigation. *J. Membr. Sci.* **2005**, *252*, 19–28. [CrossRef]

9. Wang, Q.; Li, N.; Bolto, B.; Hoang, M.; Xie, Z. Desalination by pervaporation: A review. *Desalination* **2016**, *387*, 46–60. [CrossRef]

10. Bolto, B.; Tran, T.; Hoang, M.; Xie, Z. Crosslinked poly (vinyl alcohol) membranes. *Prog. Polym. Sci.* **2009**, *34*, 969–981. [CrossRef]

11. Praptowidodo, V.S. Influence of swelling on water transport through pva-based membrane. *J. Mol. Struct.* **2005**, *739*, 207–212. [CrossRef]

12. Xie, Z.; Hoang, M.; Duong, T.; Ng, D.; Dao, B.; Gray, S. Sol–gel derived poly (vinyl alcohol)/maleic acid/silica hybrid membrane for desalination by pervaporation. *J. Membr. Sci.* **2011**, *383*, 96–103. [CrossRef]

13. Xie, Z.; Ng, D.; Hoang, M.; Duong, T.; Gray, S. Separation of aqueous salt solution by pervaporation through hybrid organic–inorganic membrane: Effect of operating conditions. *Desalination* **2011**, *273*, 220–225. [CrossRef]

14. Huang, H.-D.; Ren, P.-G.; Chen, J.; Zhang, W.-Q.; Ji, X.; Li, Z.-M. High barrier graphene oxide nanosheet/poly (vinyl alcohol) nanocomposite films. *J. Membr. Sci.* **2012**, *409*, 156–163. [CrossRef]

15. Tan, B.; Thomas, N.L. A review of the water barrier properties of polymer/clay and polymer/graphene nanocomposites. *J. Membr. Sci.* **2016**, *514*, 595–612. [CrossRef]

16. Cui, Y.; Kundalwal, S.I.; Kumar, S. Gas barrier performance of graphene/polymer nanocomposites. *Carbon* **2016**, *98*, 313–333. [CrossRef]

17. Ganesh, B.; Isloor, A.M.; Ismail, A.F. Enhanced hydrophilicity and salt rejection study of graphene oxide-polysulfone mixed matrix membrane. *Desalination* **2013**, *313*, 199–207. [CrossRef]

18. Zinadini, S.; Zinatizadeh, A.A.; Rahimi, M.; Vatanpour, V.; Zangeneh, H. Preparation of a novel antifouling mixed matrix pes membrane by embedding graphene oxide nanoplates. *J. Membr. Sci.* **2014**, *453*, 292–301. [CrossRef]

19. Yin, J.; Zhu, G.; Deng, B. Graphene oxide (GO) enhanced polyamide (PA) thin-film nanocomposite (TFN) membrane for water purification. *Desalination* **2016**, *379*, 93–101. [CrossRef]

20. Bano, S.; Mahmood, A.; Kim, S.-J.; Lee, K.-H. Graphene oxide modified polyamide nanofiltration membrane with improved flux and antifouling properties. *J. Mater. Chem. A* **2015**, *3*, 2065–2071. [CrossRef]

21. Zhao, H.; Wu, L.; Zhou, Z.; Zhang, L.; Chen, H. Improving the antifouling property of polysulfone ultrafiltration membrane by incorporation of isocyanate-treated graphene oxide. *Phys. Chem. Chem. Phys.* **2013**, *15*, 9084–9092. [CrossRef] [PubMed]

22. Wu, H.; Tang, B.; Wu, P. Development of novel SiO_2–GO nanohybrid/polysulfone membrane with enhanced performance. *J. Membr. Sci.* **2014**, *451*, 94–102. [CrossRef]

23. Zhao, C.; Xu, X.; Chen, J.; Yang, F. Effect of graphene oxide concentration on the morphologies and antifouling properties of pvdf ultrafiltration membranes. *J. Environ. Chem. Eng.* **2013**, *1*, 349–354. [CrossRef]

24. Bhadra, M.; Roy, S.; Mitra, S. Desalination across a graphene oxide membrane via direct contact membrane distillation. *Desalination* **2016**, *378*, 37–43. [CrossRef]

25. Yin, J.; Deng, B. Polymer-matrix nanocomposite membranes for water treatment. *J. Membr. Sci.* **2015**, *479*, 256–275. [CrossRef]

26. Yang, H.-C.; Hou, J.; Chen, V.; Xu, Z.-K. Surface and interface engineering for organic–inorganic composite membranes. *J. Mater. Chem. A* **2016**, *4*, 9716–9729. [CrossRef]

27. Shen, J.; Liu, G.; Huang, K.; Jin, W.; Lee, K.-R.; Xu, N. Membranes with fast and selective gas-transport channels of laminar graphene oxide for efficient CO_2 capture. *Angew. Chem. Int. Ed.* **2015**, *54*, 578–582.

28. Zhang, Y.; Shen, Q.; Hou, J.; Sutrisna, P.D.; Chen, V. Shear-aligned graphene oxide laminate/pebax ultrathin composite hollow fiber membranes using a facile dip-coating approach. *J. Mater. Chem. A* **2017**, *5*, 7732–7737. [CrossRef]

29. Liang, B.; Zhan, W.; Qi, G.; Nan, Q.; Liu, Y.; Lin, S.; Cao, B.; Pan, K. High performance graphene oxide/polyacrylonitrile composite pervaporation membranes for desalination applications. *J. Mater. Chem. A* **2015**. [CrossRef]

30. Li, X.; Chen, Y.; Hu, X.; Zhang, Y.; Hu, L. Desalination of dye solution utilizing PVA/PVDF hollow fiber composite membrane modified with TiO_2 nanoparticles. *J. Membr. Sci.* **2014**, *471*, 118–129. [CrossRef]

31. Li, L.; Hashaikeh, R.; Arafat, H.A. Development of eco-efficient micro-porous membranes via electrospinning and annealing of poly (lactic acid). *J. Membr. Sci.* **2013**, *436*, 57–67. [CrossRef]

32. Kim, K.-J.; Lee, S.-B.; Han, N.W. Effects of the degree of crosslinking on properties of poly (vinyl alcohol) membranes. *Polym. J.* **1993**, *25*, 1295–1302. [CrossRef]

33. Zhang, L.; Yu, P.; Luo, Y. Dehydration of caprolactam–water mixtures through cross-linked pva composite pervaporation membranes. *J. Membr. Sci.* **2007**, *306*, 93–102. [CrossRef]

34. Chaudhri, S.G.; Rajai, B.H.; Singh, P.S. Preparation of ultra-thin poly (vinyl alcohol) membranes supported on polysulfone hollow fiber and their application for production of pure water from seawater. *Desalination* **2015**, *367*, 272–284. [CrossRef]

35. Liang, B.; Pan, K.; Li, L.; Giannelis, E.P.; Cao, B. High performance hydrophilic pervaporation composite membranes for water desalination. *Desalination* **2014**, *347*, 199–206. [CrossRef]

36. Peng, F.; Huang, X.; Jawor, A.; Hoek, E.M. Transport, structural, and interfacial properties of poly (vinyl alcohol)–polysulfone composite nanofiltration membranes. *J. Membr. Sci.* **2010**, *353*, 169–176. [CrossRef]

37. Baroña, G.N.B.; Choi, M.; Jung, B. High permeate flux of PVA/PSf thin film composite nanofiltration membrane with aluminosilicate single-walled nanotubes. *J. Colloid Interface Sci.* **2012**, *386*, 189–197. [CrossRef] [PubMed]

38. Deng, Y.H.; Chen, J.T.; Chang, C.H.; Liao, K.S.; Tung, K.L.; Price, W.E.; Yamauchi, Y.; Wu, K.C.W. A drying-free, water-based process for fabricating mixed-matrix membranes with outstanding pervaporation performance. *Angew. Chem. Int. Ed.* **2016**, *55*, 12793–12796. [CrossRef] [PubMed]

39. Ju, H.; Sagle, A.C.; Freeman, B.D.; Mardel, J.I.; Hill, A.J. Characterization of sodium chloride and water transport in crosslinked poly (ethylene oxide) hydrogels. *J. Membr. Sci.* **2010**, *358*, 131–141. [CrossRef]

40. Xie, Z. *Hybrid Organic-Inorganic Pervaporation Membranes for Desalination*; Victoria University: Toronto, ON, Canada, 2012.

41. Kusumocahyo, S.P.; Sano, K.; Sudoh, M.; Kensaka, M. Water permselectivity in the pervaporation of acetic acid–water mixture using crosslinked poly (vinyl alcohol) membranes. *Sep. Purif. Technol.* **2000**, *18*, 141–150. [CrossRef]

42. Xu, R.; Lin, P.; Zhang, Q.; Zhong, J.; Tsuru, T. Development of ethenylene-bridged organosilica membranes for desalination applications. *Ind. Eng. Chem. Res.* **2016**, *55*, 2183–2190. [CrossRef]

43. Zhao, D.; Ren, J.; Qiu, Y.; Li, H.; Hua, K.; Li, X.; Deng, M. Effect of graphene oxide on the behavior of poly (amide-6-b-ethylene oxide)/graphene oxide mixed-matrix membranes in the permeation process. *J. Appl. Polym. Sci.* **2015**, *132*. [CrossRef]

44. Qin, L.; Zhao, Y.; Liu, J.; Hou, J.; Zhang, Y.; Wang, J.; Zhu, J.; Zhang, B.; Lvov, Y.; Van der Bruggen, B. Oriented clay nanotube membrane assembled on microporous polymeric substrates. *ACS Appl. Mater. Interfaces* **2016**, *8*, 34914–34923. [CrossRef] [PubMed]

45. Sharma, S.; Prakash, J.; Pujari, P. Effects of the molecular level dispersion of graphene oxide on the free volume characteristics of poly (vinyl alcohol) and its impact on the thermal and mechanical properties of their nanocomposites. *Phys. Chem. Chem. Phys.* **2015**, *17*, 29201–29209. [CrossRef] [PubMed]

46. Loryuenyong, V.; Saewong, C.; Aranchaiya, C.; Buasri, A. The improvement in mechanical and barrier properties of poly(vinyl alcohol)/graphene oxide packaging films. *Packag. Technol. Sci.* **2015**, *28*, 939–947. [CrossRef]

47. Hou, J.; Dong, G.; Ye, Y.; Chen, V. Enzymatic degradation of bisphenol-a with immobilized laccase on tio 2 sol–gel coated pvdf membrane. *J. Membr. Sci.* **2014**, *469*, 19–30. [CrossRef]

48. Liang, J.; Huang, Y.; Zhang, L.; Wang, Y.; Ma, Y.; Guo, T.; Chen, Y. Molecular-level dispersion of graphene into poly (vinyl alcohol) and effective reinforcement of their nanocomposites. *Adv. Funct. Mater.* **2009**, *19*, 2297–2302. [CrossRef]

49. Ma, H.; Burger, C.; Hsiao, B.S.; Chu, B. Highly permeable polymer membranes containing directed channels for water purification. *ACS Macro Lett.* **2012**, *1*, 723–726. [CrossRef]

50. Joshi, R.; Carbone, P.; Wang, F.-C.; Kravets, V.G.; Su, Y.; Grigorieva, I.V.; Wu, H.; Geim, A.K.; Nair, R.R. Precise and ultrafast molecular sieving through graphene oxide membranes. *Science* **2014**, *343*, 752–754. [CrossRef] [PubMed]

51. Yin, J.; Kim, E.-S.; Yang, J.; Deng, B. Fabrication of a novel thin-film nanocomposite (TFN) membrane containing MCM-41 silica nanoparticles (NPs) for water purification. *J. Membr. Sci.* **2012**, *423*, 238–246. [CrossRef]

52. Kim, E.-S.; Deng, B. Fabrication of polyamide thin-film nano-composite (PA-TFN) membrane with hydrophilized ordered mesoporous carbon (H-OMC) for water purifications. *J. Membr. Sci.* **2011**, *375*, 46–54. [CrossRef]

53. Wang, X.; Chen, X.; Yoon, K.; Fang, D.; Hsiao, B.S.; Chu, B. High flux filtration medium based on nanofibrous substrate with hydrophilic nanocomposite coating. *Environ. Sci. Technol.* **2005**, *39*, 7684–7691. [CrossRef] [PubMed]

54. Wang, N.; Ji, S.; Li, J.; Zhang, R.; Zhang, G. Poly (vinyl alcohol)–graphene oxide nanohybrid "pore-filling" membrane for pervaporation of toluene/n-heptane mixtures. *J. Membr. Sci.* **2014**, *455*, 113–120. [CrossRef]

55. Meng, S.; Ye, Y.; Mansouri, J.; Chen, V. Fouling and crystallisation behaviour of superhydrophobic nano-composite pvdf membranes in direct contact membrane distillation. *J. Membr. Sci.* **2014**, *463*, 102–112. [CrossRef]

56. Meng, S.; Ye, Y.; Mansouri, J.; Chen, V. Crystallization behavior of salts during membrane distillation with hydrophobic and superhydrophobic capillary membranes. *J. Membr. Sci.* **2015**, *473*, 165–176. [CrossRef]

applied
sciences

MDPI

Article

Functionalization of a Hydrophilic Commercial Membrane Using Inorganic-Organic Polymers Coatings for Membrane Distillation

Lies Eykens [1,2,*], Klaus Rose [3], Marjorie Dubreuil [1], Kristien De Sitter [1], Chris Dotremont [1], Luc Pinoy [4] and Bart Van der Bruggen [2,5]

[1] VITO-Flemish Institute for Technological Research, Boeretang 200, 2400 Mol, Belgium; marjorie.dubreuil@vito.be (M.D.); kristien.desitter@vito.be (K.D.S.); chris.dotremont@vito.be (C.D.)
[2] Department of Chemical Engineering, KU Leuven, Celestijnenlaan 200F, B-3001 Leuven, Belgium; bart.vanderbruggen@kuleuven.be
[3] Fraunhofer Institute for Silicate Research ISC, Neunerplatz 2, 97082 Würzburg, Germany; Klaus.rose@isc.fraunhofer.de
[4] Department of Chemical Engineering, Cluster Sustainable Chemical Process Technology, KU Leuven, Gebroeders Desmetstraat 1, B-9000 Ghent, Belgium; luc.pinoy@kuleuven.be
[5] Faculty of Engineering and the Built Environment, Tshwane University of Technology, Private Bag X680, Pretoria 0001, South Africa
* Correspondence: lies.eykens@vito.be; Tel.: +32-14-335-663

Academic Editor: Faisal Hai
Received: 3 May 2017; Accepted: 16 June 2017; Published: 20 June 2017

Abstract: Membrane distillation is a thermal separation technique using a microporous hydrophobic membrane. One of the concerns with respect to the industrialization of the technique is the development of novel membranes. In this paper, a commercially available hydrophilic polyethersulfone membrane with a suitable structure for membrane distillation was modified using available hydrophobic coatings using ORMOCER® technology to obtain a hydrophobic membrane that can be applied in membrane distillation. The surface modification was performed using a selection of different components, concentrations, and application methods. The resulting membranes can have two hydrophobic surfaces or a hydrophobic and hydrophilic surface depending on the application method. An extensive characterization procedure confirmed the suitability of the coating technique and the obtained membranes for membrane distillation. The surface contact angle of water could be increased from 27° up to 110°, and fluxes comparable to membranes commonly used for membrane distillation were achieved under similar process conditions. A 100 h test demonstrated the stability of the coating and the importance of using sufficiently stable base membranes.

Keywords: hydrophobic coatings; direct contact membrane distillation (DCMD); polyethersulfone; ORMOCER®; wetting

---

1. Introduction

Membrane distillation (MD) is a thermal separation technique using a hydrophobic microporous membrane as a contactor between two liquid phases. The membrane allows vapors (e.g., water vapor) to permeate, whereas the liquid phase including the dissolved components (e.g., salts) is retained by the membrane. A temperature difference induces the driving force and allows vapors to permeate from the hot feed side to the cold permeate side. The technique was initially proposed as an alternative technology for reverse osmosis in seawater desalination. However, due to the benefits of very high retentions and less dependence on salinity, it is recently also proposed for applications beyond the scope of reverse osmosis. The applications can include but are not limited to desalination and brine treatment [1], waste water treatment [2,3], and resource recovery [4], where the dissolved components can be salts, proteins [5], acids [2,6], and minerals [4].

Currently, hydrophobic microfiltration membranes are used in membrane distillation, although these membranes are not optimized for the MD process [7,8]. The specific requirements for membrane distillation membranes are described in literature [9–11]. Most importantly, the membrane must consist of at least one layer that is not wetted by the liquid stream under the operational pressures used in the module. The minimum pressure required to wet a hydrophobic membrane is the liquid entry pressure (*LEP*), which depends on the membrane characteristics as well as on the feed composition and is defined by the following equation:

$$LEP = \frac{-2B\gamma_l \cos(\theta)}{r_{max}} \tag{1}$$

where γ_l is the surface tension $(N \cdot m^{-1})$ of the liquid, θ the contact angle (°), r_{max} the maximum pore size (µm), and B is a geometric factor. To ensure proper operation under fluctuating pressures and temperatures, an *LEP* of 2.5 bar is required [12]. To achieve a sufficient *LEP*, membranes with maximum pore diameter between 0.1 and 1 µm with a water contact angle above 90° are recommended for membrane distillation [9,13,14]. Moreover, it is generally agreed that a high membrane porosity is one of the most important membrane parameters in membrane distillation for both flux and energy efficiency, regardless of the MD configuration [15–19]. Additionally, membranes with a thickness between 30 up to 60 µm are recommended; however, it was shown recently that this optimal value depends on salinity, and at high salinity, thicker membranes are preferred [20]. Currently, most commercial systems use membranes not specifically developed for membrane distillation (i.e., hydrophobic polyethylene (PE), polyvinylidene fluoride (PVDF) and polytetrafluoroethylene (PTFE) microfiltration membranes). However, in the literature, many efforts are described to improve the membrane performance. These efforts include the optimization of the phase inversion process, mainly using the hydrophobic polyvinylidene fluoride (PVDF) or poly (vinylidene fluoride-co-hexafluoropropylene) (PVDF-HFP) as polymer [21–24] and the use of surface modifying macromolecules [25–27] and electrospinning [28,29]. In addition to the optimization of the membrane structure, research is also oriented toward the enhancement of the surface properties for membrane distillation, including but not limited to plasma treatment [30,31], fluorination of a TiO$_2$ coating [32], or the use of fluoroalkylsilane coatings on Tunisian Clay membranes [33]. Currently, these coatings are only applied in lab scale experiments and are not yet commercially available. In this publication, the use of hydrophobic sol-gel coatings forming an organic-inorganic network on hydrophilic polyethersulfone (PES) membranes is presented. These types of coatings are already used on a commercial scale, showing excellent stability in other applications, including scratch- and abrasion-resistant coatings for plastics [34], functional coatings on glass [35], and gas-sensitive layers [36]. Because of its easy scalable production method, excellent stability, and ability to functionalize the surface properties, this coating material was selected to be applied on a commercially available hydrophilic membrane with the required structure for membrane distillation [14]. The inorganic network is formed by Si-O-Si bonds, whereas the organic network is formed by reactive and polymerizable organic functional groups. The choice of hydrophobic fluorosilanes results in a surface with a hydrophobic character, whereas the unique formation of an organic-inorganic network results in a scratch- and leach-resistant coating. For the first time, these readily available coatings are applied for tuning the hydrophobicity of a cheap hydrophilic membrane to enable application in membrane distillation.

2. Materials and Methods

2.1. Membranes

The commercial hydrophilic microfiltration membrane used as base material in this study is the MicroPES® 2F (3M, Wuppertal, Germany). PVDF GVHP (Merck Chemicals N.V., Overijse, Belgium) is a hydrophobic membrane commonly used in the membrane distillation literature as a reference for comparison of the performance of newly synthesized membrane distillation membranes [37–39]. PE (Solupor, Lydall, Manchester, CT, USA) is also added as a reference membrane, because it is used in commercially available membrane distillation modules.

2.2. Coatings

The commercially available and patented ORMOCER® technology was used to apply a hydrophobic coating on the MicroPES membrane. Three different combinations of silanes were explored. The three different fluorosilanes investigated in this study have a different structure, different chain length, and different hydrophobizing properties. The monomers exhibit a bifunctional character and are able to form a stable combined inorganic-organic network. The synthesis procedure is presented in Figure 1.

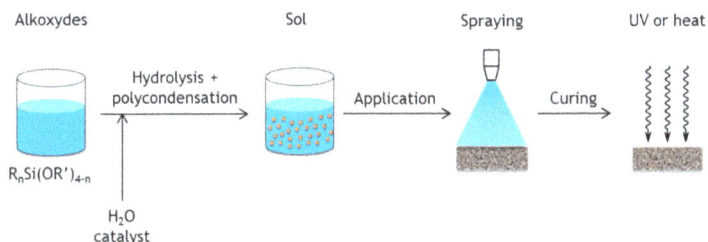

Figure 1. Coating procedure.

The process starts from an alcoholic solution of the R'Si(OR)$_3$ monomers, where R' is a functional non-reactive hydrophobic group or a polymerizable group, e.g., acryl or vinyl. R represents simple aliphatic groups, e.g., methyl or ethyl. By the addition of water and catalyst, both hydrolysis and polycondensation reactions can occur, resulting in the formation of Si-O-Si covalent bonds forming the inorganic network (Figure 2). The hydrolysis reaction results in the cleavage of a chemical bond by the addition of water. During the polycondensation reactions, two molecules combine to form a larger molecule by splitting a small molecule. Two of the OH groups formed after hydrolysis on the silica components can form H$_2$O (polycondensation 1), or the unreacted OR group can react with an OH group to form ROH (polycondensation 2). After multiple polycondensation steps, an inorganic polymer network with a Si-O backbone is formed, resulting in a disperse solution, called the sol.

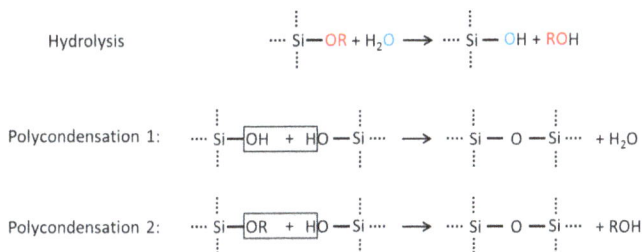

Figure 2. Mechanism of the hydrolysis and polycondensation reactions.

In a second step, this sol is applied on the membrane surface using a bar coater, a roll-to-roll system, or spray coating. Finally, the coating is cured using a thermal or photochemicalcuring step using ultraviolet (UV) light, in which the polymerizable groups present in the solution will form the organic network. The nature of the chemicals used (acryl, vinyl, etc.) determines the organic network type. More details on the system can be found in the literature [34,40].

Figure 3 shows the structure of the perfluorodecyl (PFD) silane. After the sol-gel processing this single component only forms the silica network in an alcoholic solution. No polymerizable group is present in this solution and hence, no organic network is formed in this system. The PFD system was only thermally dried at 80 °C for 30 min. i.e., thermal based evaporation of the solvents (water/alcohol), which led to a solid film.

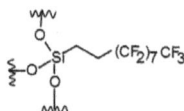

Figure 3. Single component system with perfluorodecyl (PFD), ∿∿ represents the inorganic network.

The second system is a four-component system where a mono-acrylic (Ak) component forms the reactive site for acrylic polymerization. The methyl (T) and dimethyl (D) silanes as well as the highly fluorinated silane (BTFO2N) are participating as hydrophobizing component in the organic network (Figure 4). A composition of Ak/T/D/BTFO2N of 20/44/34/2 wt. % was used.

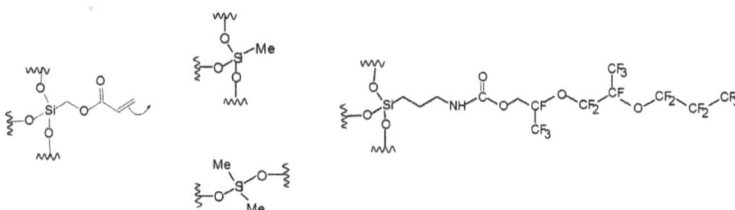

Figure 4. The second system (Ak/T/D/BTFO2N system), ∿∿ represents the inorganic network, ⌣ represents the organic network formation.

The third is a three component system including a vinyl silane (V), a 3-mercaptopropyl silane (Mc) and a perfluoro-octyl silane (F13) with composition 49/49/2 wt. % V/Mc/F13 (Figure 5). The inorganic silica network is formed by the three components, while the organic network is formed by an addition reaction of the thiol-group to the vinyl group [41]. The hydrophobicity is provided by the perfluoro-octyl group. The UV-curing for the systems Ak/T/D/BTFO2N and V/Mc/F13 was performed using a mercury UV lamp, running with 1200 Watt power, a UV dose of 5000 mJ/m^2, and a UV curing duration 20 s. The UV-curing temperature is ca. 60–80 °C, which evolves from the UV-lamps.

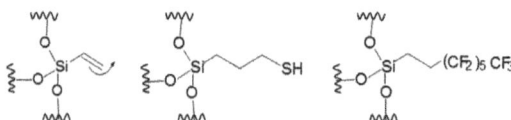

Figure 5. The third system (V/Mc/F13 system), ∿∿ represents the inorganic network, ⌣ represents the organic network formation.

Table 1 shows the composition, the application method and the final network of the different coatings applied in this study. Coatings 1, 2, and 4 were applied using a roll-to-roll system, coatings 3 and 5 were applied using a bar coater system. Coatings 6 and 7 use the same components as for coating 5, but are applied using spray coating (6 as single side coating, 7 as double side coating). Coatings 1 and 2 have no reactive organic group and only differ in the mass fraction of the fluorinated alkylsiloxane. After thermal curing, these solutions only form an inorganic network. Coatings 3 to 7 are multiple component systems and contain an organic-inorganic network. For membrane 4, a higher mass fraction of the components was used.

Table 1. Composition of the coatings used in this study (perfluorodecyl: PFD; the second system: Ak/T/D/BTFO2N; the third system: V/Mc/F13).

Coating System	Coating Components	Mass Fraction in Coating Solution	Application Method	Network Formation
1	PFD solution	5 wt. %	Roll-to-roll system	No organic crosslinking
2	PFD solution	10 wt. %	Roll-to-roll system	No organic crosslinking
3	Ak/T/D/BTFO2N	5 wt. %	Bar coater	Acrylic polymerization
4	Ak/T/D/BTFO2N	30 wt. %	Roll-to-roll system	Acrylic polymerization
5	V/Mc/F13	5 wt. %	Bar coater	Vinyl + SH addition
6	V/Mc/F13	5 wt. %	Spray coater	Vinyl + SH addition
7	V/Mc/F13	5 wt. %	Spray coater	Vinyl + SH addition

2.3. Characterization Methods

The contact angle of the membranes was measured with the OCA 15EC Contact Angle System (DataPhysics Instruments GmbH, Filderstadt, Germany) using the static sessile drop method. The liquid entry pressure was determined as described by Khayet et al. [42]. The pressure was increased slowly by 0.1 bar each 30 s, until a flow was detected. A Porolux^{TM} 1000 device (Porometer N.V., Eke, Belgium) using the wet/dry capillary flow porometry method measured the pore size distribution as described by Francis et al. [43]. Porefil with a liquid surface tension of 16 mN·m^{-1} was used as wetting liquid and the shape factor was assumed to be 1. The porosity of the membranes was calculated using the following equation suggested by Smolders and Franken [44]:

$$\epsilon = 1 - \frac{\rho_m}{\rho_{pol}} \tag{1}$$

with ρ_m and ρ_{pol} representing the density of the membrane and the polymer, respectively, in g·cm^{-3}. The density of the membrane was obtained by measuring the mass of a circular membrane cut with a circular mold with diameter of 5 cm. The density of the polymer was measured using gas pycnometry with a He-pycnometer (Micromeretics, Norcross, GA, USA) [20]. A cold field emission scanning electron microscope (SEM) type JSM6340F (JEOL, Tokyo, Japan) was used to study membrane cross-sections at an acceleration voltage of 5 keV. Cross-sections were obtained by a cross-section polisher type SM-09010 (JEOL, Tokyo, Japan) using an argon ion beam. All samples were coated with a thin Pt/Pd layer (~1.5 nm) using a Cressington HR208 high-resolution sputter-coater (Cressington Scientific Instruments, Watford, UK) to avoid charging by the e-beam. The images of the cross-sections were analyzed in ImageJ [20].

2.4. Membrane Distillation Setup

The membrane distillation performance was evaluated with a lab-scale DCMD setup (Figure 6). The flat-sheet module had a feed and permeate channel with dimensions of 6 cm width and 18 cm length. The channel height and spacer thickness was 2 mm. On the permeate side, purified water with electrical conductivity below 20 μS·cm^{-1} was used. The feed and distillate were circulated counter-currently on their respective sides of the membrane with a flow velocity of 0.13 m/s using peristaltic pumps (Watson-Marlow, 520DuN/R2, Zwijnaarde, Belgium). $T_{f,in}$ and $T_{p,out}$ was kept constant at 60 °C and 45 °C, respectively, for all experiments. The temperatures were kept constant using two heating baths (Huber, Ministat 230w-cc-NR, Offenburg, Germany) and monitored using four thermocouples (Thermo Electric Company, PT100 TF, Balen, Belgium). The flux was measured by evaluating the weight variations in the feed and distillate tank, using an analytical balance (Sartorius GmbH, ED8801-CW, Goettingen, Germany). The average of at least two experiments is reported. The electrical conductivity at the feed and permeate side were monitored by portable conductivity meters (WTW GmbH, pH/Cond 340i, Weilheim, Germany).

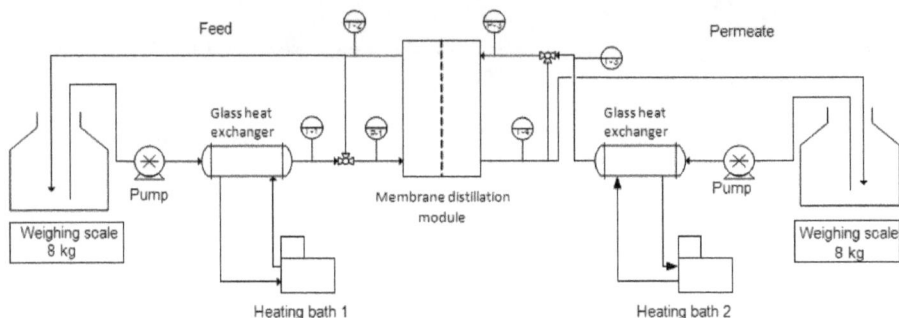

Figure 6. Schematic of the membrane distillation setup.

The energy efficiency (*EE*) of the process is defined in Equation (3). The total heat transfer through the membrane Q_m is considered to be equal to the heat transfer in the feed channel, as described by Khayet et al. [45].

$$EE \ (\%) \ = \ \frac{N \cdot \Delta H \cdot A}{F \cdot C_p \cdot (T_{in} \ - \ T_{out})} \tag{3}$$

N (kg·m^{-2}·h^{-1}) is the water flux and ΔH (J·kg^{-1}) the enthalpy of evaporation. F is the mass flow rate in the channels expressed in kg·s^{-1}, A (m^2) is the effective membrane surface area, C_p is the specific heat capacity of the solution (J·kg^{-1}·°C^{-1}), T_{in} and T_{out} are the bulk temperatures at the channel inlet and outlet of the module expressed in °C, respectively. The calculations were carried out for the feed and permeate channel and the average and standard deviation are reported.

Long term stability tests were performed using 35 g·L^{-1} NaCl as feed concentration. The experimental conditions were chosen differently from the screening tests, because these experiments run overnight. The flux should be limited, to prevent spilling over of the permeate vessel. Therefore, lower temperatures have been chosen for these experiments: T_f and T_p were 45 °C and 40 °C respectively with a cross flow velocity of the feed of 0.1 m·s^{-1} and a salinity of 35 g·L^{-1}. The goal of this experiment was mainly to investigate the stability of the coating under constant shear of the feed liquid, not the thermal stability. In general, the thermal stability of the hybrid coatings is 150 °C and higher, up to 300 °C. This thermal stability has been measured for another application elsewhere [4]. The temperature stability of the coating material is much higher compared to the temperatures generally used in MD up to 90 °C and the temperature stability of this coating is not an issue for these coatings.

3. Results

3.1. Characterization of the Membranes

The measured properties of the MicroPES 2F membrane used as base membrane for the coatings are shown in Table 2, together with the properties of the commercial hydrophobic membranes commonly used in membrane distillation.

SEM images reveal a difference in the surface porosity on both sides of the PES-membrane (Table 3). While the pore size is larger on the surface side 2, the pore density and porosity are the highest on surface side 1. The cross-section shows an hourglass-shaped pore structure [46]. The densest zone of the membrane is located in the region of 10 to 50 μm distance from surface 1. All coatings are applied on the surface 1 because this side has the lowest pore size, which is preferred to increase the wetting resistance.

Table 2. Characteristics of the hydrophilic support membrane (polyethersulfone (PES)) and the membranes commonly used for membrane distillation (polyethylene (PE) and polyvinylidene fluoride (PVDF)).

Membrane	Θ (°)	d_{max} (µm)	d_{av} (µm)	Δ (µm)	E (%)	LEP (bar)
PES (Membrana, MicroPES)	26 ± 4	0.61 ± 0.04	0.56 ± 0.03	115 ± 1	75.2 ± 0.3	0
PE (Lydall, Solupor)	120 ± 1	0.32 ± 0.02	0.43 ± 0.02	99 ± 1	75.6 ± 0.6	3.9 ± 0.1
PVDF (Millipore, GVHP)	120 ± 3	0.44 ± 0.01	0.60 ± 0.01	119 ± 1	65.7 ± 0.9	2.3 ± 0.1

Table 3. Surface and cross-section images of the commercial polyethersulfone (PES) membrane and properties obtained using image analysis.

SEM	Surface 1	Surface 2	Cross-Section
Images			
Properties	$\varepsilon_s = 22 \pm 4\%$ $d_{min,s} = 0.03 \pm 0.1$ µm $d_{av,s} = 0.47 \pm 0.2$ µm $d_{max,s} = 1.90 \pm 0.5$ µm Pore density = 4.1 ± 0.7	$\varepsilon_s = 8.8 \pm 0.5\%$ $d_{min,s} = 0.24 \pm 0.1$ µm $d_{av,s} = 0.91 \pm 0.7$ µm $d_{max,s} = 2.2 \pm 0.3$ µm Pore density = 1.2 ± 0.2	$\delta = 113 \pm 1$ µm

The SEM images in Table 4 show a difference in membrane structure for the stretched PE membrane, which is more porous compared to the PVDF membrane. Both membranes have a symmetric cross-section. The PVDF membrane has a more open surface structure compared to the PES-membrane, while it has lower bulk porosity.

Table 4. SEM (Cold field emission scanning electron microscope) images of PE (polyethylene) and PVDF (polyvinylidene fluoride) commercial hydrophobic membrane.

PE1 Stretched	PVDF Phase Inversion

3.2. Wetting Resistance of the Coating

Table 5 shows the water contact angle of the membranes coated in this study. The uncoated membrane has a hydrophilic water contact angle of 27°. After the application of the coatings, surface 1 is hydrophobic and has a water contact angle of at least 100°, confirming the hydrophobic character of the coatings. Without organic crosslinking agent (PFD solution only), the highest water contact angle of 117° for 5 wt. % solution (membrane 1) and 118° for a 10 wt. % solution (membrane 2) are achieved. This difference in water contact angle is statistically insignificant and therefore, the contact angle is considered to be independent of the concentration in the range from 5 and 10 wt. % for this system. When adding components with polymerizable functionalities to form the polymeric network, the fluorinated fraction decreases (membranes 3–5). This is visualized in a slightly reduced hydrophobicity on the coated surface compared to membrane 1 and 2, with a water contact angle ranging from 100° to 110°. For the Ak/T/D/BTFO2N system (3 and 4), the water contact angle is increased from 100° to 110° using a 5 and 30 wt. % solution respectively. The V/Mc/F13 system (5) achieves a water contact angle of 109° with a concentration of only 5 wt. % and including a polymeric network. The water contact angle of untreated surface 2 equals 28° after 0.5 s, whereas after the coating procedure the water contact angle after 0.5 s varies from 83° to 100° for membrane 1–4. For these membranes, the water contact angle continuously decreases over time. For membrane 1–3 the droplet disperses in the membrane only after 2 min. For membrane 4, the droplet sinks into the membrane within 2 min of contact time. For membrane 5, the uncoated side even shows a stable contact angle of 110° similar to the contact angle on the coated side. These observations show that the surface hydrophobicity on the uncoated side is also affected by the coating process, indicating that a part of the coating is able to pass through the entire membrane cross-section and is also applied (partially) on the uncoated side of the membrane. However, despite the increase in water contact angle, the membranes 1–4 are still wetted in membrane distillation and are considered as hydrophobic/hydrophilic membranes. Membrane 5 is considered as an entirely hydrophobic membrane. The measured liquid entry pressure for the different membranes varies from 1.6 to 3.5 bar, with a large variation for multiple measurements of the same membrane. This variation in liquid entry pressure is attributed to the inhomogeneity of the coatings, meaning that the coating is not applied equally on the entire membrane surface on some parts of the surface are not sufficiently hydrophobic. This becomes visible after submerging the membranes in water. While most regions are not wetted, some areas of the membrane surface are wetted, showing that the bar coater and the roll-to-roll system are not the preferred application methods when applying the coatings on porous membranes. The contact angles reported in Table 5 only consider the area of the membrane that was not wetted after submersion of the membrane in water. However, the inhomogeneous application of the coating with barcoater or the roll-to-roll system was also visible in the large spreading of the contact angle measurements of 20° when measuring at random spots on the membrane.

Table 5. Contact angle.

Membrane	Coating Components	Mass Fraction in Coating Solution	θ (°)	
			Surface 1 (Coated Side)	Surface 2 (Uncoated Side)
Uncoated	-	-	27 ± 6 [1]	28 ± 4 [1]
1	PFD solution	5 wt. %	117 ± 1	$90 \pm 5 \rightarrow 62 \pm 5$ [2]
2	PFD solution	10 wt. %	118 ± 1	$96 \pm 1 \rightarrow 73 \pm 6$ [2]
3	Ak/T/D/BTFO2N	5 wt. %	100 ± 3	$100 \pm 1 \rightarrow 85 \pm 5$ [2]
4	Ak/T/D/BTFO2N	30 wt. %	110 ± 1	83 ± 2 [1]
5	V/Mc/F13	5 wt. %	109 ± 1	110 ± 1

Legend: [1] Droplet wets the membrane within 2 min; [2] Contact angle after 0.5 s $\rightarrow$ Contact angle after 2 min.

To improve the coating homogeneity, coating system 5 (V/Mc/F13, 5 wt. %) was selected for spray coating application. The single component systems (1 and 2) are not sufficiently stable (see Section 3.4), whereas the Ak/T/D/BTFO2N system (3 and 4) requires a higher amount of silica components to achieve the same contact angle (Table 5). This spray coating process ensures an improved homogeneity compared to the bar coater system. The coating was applied in two ways: (a) only on surface 1, resulting in a membrane with a hydrophobic and a hydrophilic side and (b) on both sides of the membrane, producing a membrane with two hydrophobic sides (Table 6). Membrane 6, coated on surface 1 only shows a slightly lower liquid entry pressure compared to membrane 7, coated on both sides.

Table 6. Contact angle and LEP of the coatings applied by the spray system.

Membrane	Coated Side	θ (°)		LEP (bar)
		Surface 1	Surface 2	
6	Surface side 1 only	97 ± 1	41 ± 3 *	1.8 ± 0.2
7	Both sides	102 ± 1	107 ± 1	2.2 ± 0.1

* Droplet wets the membrane within 2 min.

3.3. Structure of the Coating

Apart from the contact angle measurements, another important issue is the possibility that the coating will block the pores. This reduces the porosity and pore size, and in the ultimate case, a dense membrane is obtained, obstructing the mass transport. Pore blockage can easily be seen by porometry because it reduces both the gas flow through the membrane and the pore size compared to the untreated membrane. For membranes 1–3 and 5–7, the pore sizes and the gas flows of the uncoated and coated membranes are equal. As an example, the pore size distribution obtained using porometry is presented for an untreated membrane and membrane 3 and 4 are given in Figure 7. For membrane 4, coated with a 30 wt. % solution, no pores are detected using porometry, indicating that the pores are completely blocked by the coating in this case. These measurements reveal that a 30% solution is not suitable for membrane modification since it significantly affects the porosity.

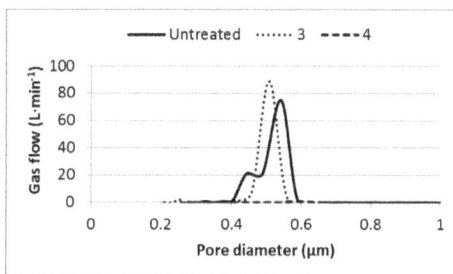

Figure 7. Pore size distribution untreated PES-membrane, membrane 3, and membrane 4 using porometry.

The position of the coating is investigated using energy-dispersive X-ray spectroscopy (EDX). As an example, Figure 8 shows the EDX spectrum of membrane 3. The peaks of the oxygen and sulfur atoms in the spectrum correlate with the presence of the PES membrane material or pores at the measured position. In the first 20 μm at surface 1, a first increase of the silicon and fluorine atoms is observed, which correlates with the position of the coating. A second increase is observed between 30 and 50 μm, which is also the densest zone of the membrane (Table 3) and therefore, more surface is available to deposit the coating. In the part between 50 and 120 μm, the silicon and fluorine content is lower. The permeation of the coating might be inhibited by the denser structure of the membrane. Further in the membrane cross-section, the silicon and fluorine content decreases, but are not equal to zero. The increase of the oxygen and silicon at 120 to 140 μm is caused by the silicon glue used to fix the sample. Based on the EDX on the chemical composition, the structure is hydrophobic until a depth of at least 50 μm. Unfortunately, it is impossible to indicate the exact hydrophobicity or hydrophilicity in terms of a water contact angle at a certain point of the membrane cross-section based on the elemental composition obtained with EDX. Therefore, the exact hydrophobic thickness cannot be derived from these EDX figures.

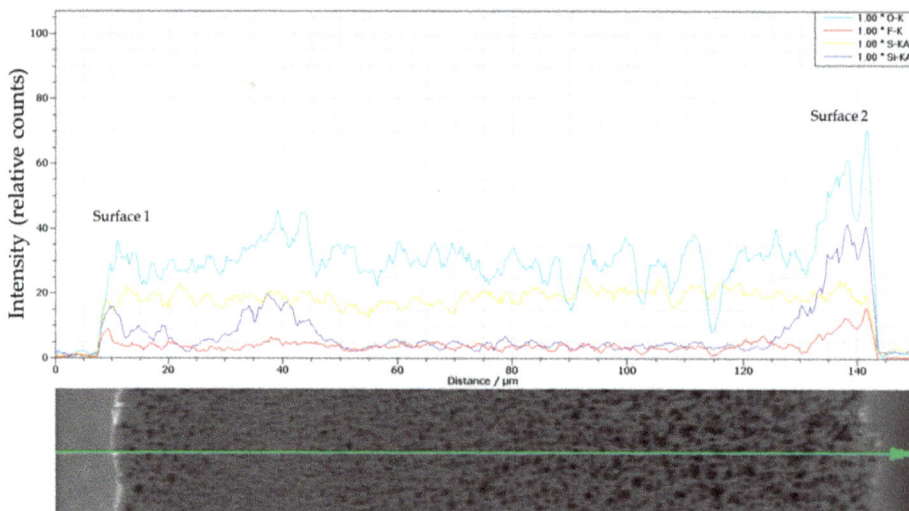

Figure 8. EDX (energy-dispersive X-ray spectroscopy) spectra of the cross section for membrane 3.

In summary, these results indicate that no difference in membrane structure in terms of thickness, pore size, and porosity was found for the coatings applied using less than 10 wt. % silanes in the coating solution. Only coating 4, with very high load of silanes, the MD-flux and N2 flux during porometry decreased to zero, showed a strong difference, which is a strong indication for pore blocking.

3.4. Membrane Distillation Performance

The flux and energy efficiency of the membranes produced in this study are compared to the PVDF GVHP membrane from Millipore commonly used in the literature and to a PE membrane currently used in pilot scale membrane distillation modules (Table 2).

All coated membranes are coated on the same base membrane structure, except for membrane 4, where pore blocking occurs. The other membranes have equal porosity, pore size, and total thickness. However, the position of the coating and hence the thickness of the hydrophobic layer can be different. Since this thickness can affect the flux and energy efficiency of the membrane for desalination applications [20], a difference in MD performance is expected. Flux and energy efficiencies of the different membranes are summarized in Table 7.

Membranes 1 and 2 are single component systems without organic network formation and show immediate breakthrough of the salts. The membrane wetting is also visually observed after demounting the module. This indicates that the inorganic network formed by these coatings is not sufficient to provide stable coatings for the process conditions used in membrane distillation.

Membranes 3 and 4 use the same components, but with a difference in mass fraction of 5 wt. % and 30 wt. % respectively (Table 1). Membrane 3 shows a much higher flux of 16.2 kg·m^{-2}·h^{-1} compared to membrane 4 with a flux of only 0.4 kg·m^{-2}·h^{-1}. This low flux of membrane 4 is caused by the pore blocking shown in Figure 7, which hinders the transport of water vapor through the membrane. This shows that a mass fraction of 5 wt. % silanes in the coating solutions is balancing between a sufficient hydrophobicity and avoiding obstruction of the pores, which occurred at concentrations of 30 wt. %.

The hydrophobic/hydrophilic membranes 3 and 6 show higher fluxes of about 16 kg·m^{-2}·h^{-1} compared to membranes 5 and 7 with a flux of about 14 kg·m^{-2}·h^{-1}, which have two hydrophobic sides (Table 7). The effect of partial pore wetting on flux and heat transfer is comprehensively described in reference [47], where it was shown that an increase of the depth of pore wetting results in an increase of the flux. This difference in flux is explained by a different in hydrophobic non-wetted membrane thickness, which imposes the mass transport resistance for vapor transport. As discussed in the literature, the optimal hydrophobic thickness ranges from 10 to 60 μm using 35 g·L^{-1} NaCl [20,48–50]. While the membranes with two hydrophobic surfaces (5 and 7) are probably fully hydrophobic or at least they do not contain a wetted part, the cross-section of the hydrophobic/hydrophilic membrane structures (3 and 6) is partially wetted by the permeate liquid on surface side 2. Therefore, the hydrophobic layer thickness is much closer to the optimal values of 10–60 μm for the hydrophobic/hydrophilic membranes. The fluxes achieved are higher compared to the commercial PVDF membrane, whereas the commercial PE membrane shows a flux slightly higher compared to the membranes with two hydrophobic sides, but lower compared to the membranes with a hydrophobic/hydrophilic structure.

The energy efficiency varies from 43% to 55% and is lower compared to the commercial membranes and appears mainly to depend on the membrane base structure and porosity (Section 3.1). PE has the highest porosity and surface porosity and shows the highest energy efficiency as well. The energy efficiency of the PVDF membrane in negatively affected by the lower bulk porosity [14]. The coated PES membrane has relatively high bulk porosity, but as shown in Table 3, the membrane is not symmetric and has a more dense structure at the surface and in the first 100 μm. This causes less mass transport and more heat transport through the membrane in the first 60 μm, reducing the energy efficiency. Membranes applied with the same application systems (membrane 3 and 5 and membrane 6 and 7) have equal energy efficiency, regardless of the fact that the resulting membrane is a hydrophobic/hydrophilic membrane or a membrane with 2 hydrophobic surfaces. This can be explained by the independence of the energy efficiency as function of membrane thickness, which is shown by different authors [20,48]. This independency occurs because both heat transfer due to flux and heat transfer due to conduction are approximately inversely proportional to the membrane thickness. A high salt retention above 99.9% was measured for all membranes.

Table 7. Flux, energy efficiency and salt retention. Process conditions: T_f = 60 °C, T_p = 45 °C, v = 0.13 m·s^{-1}, NaCl concentration = 35 g·L^{-1}.

Membrane	Structure	Flux (kg·m^{-2}·h^{-1})	Energy Efficiency (%)	Salt Retention (%)
1	hydrophobic/hydrophilic		Wetted	
2	hydrophobic/hydrophilic			
3	hydrophobic/hydrophilic	16.2 ± 0.5	55 ± 2	99.98 ± 0.01
4	hydrophobic/hydrophilic	0.4 ± 0.3	-	99.99 ± 0.01
5	2 hydrophobic surfaces	14.5 ± 0.5	50 ± 2	99.99 ± 0.01
6	hydrophobic/hydrophilic	16.1 ± 0.1	44 ± 3	99.92 ± 0.06
7	2 hydrophobic surfaces	13.9 ± 0.1	43 ± 5	99.99 ± 0.01
PVDF	hydrophobic	12.0 ± 0.1	52 ± 2	99.99 ± 0.01
PE	hydrophobic	15.3 ± 0.4	67 ± 4	99.99 ± 0.01

3.5. Medium Term Performance

The coating material applied on membranes 5–7 is advantageous for membrane distillation based on its superior wetting resistance and homogeneity. Membranes 5 and 6 were tested for a longer period. The flux and salt retention for membrane 6 are shown as an example in Figure 9, but similar results are obtained for membrane 5. The steady decrease in flux is caused by the increasing salt concentration, while jump in flux is explained by the addition of water to the feed solution after a certain amount of time to maintain the concentration at 35 g·L^{-1}. During a period of 80 h, salt retention was always above 99.9% and flux remained constant. To confirm the stability of the coating on the active membrane surface, the LEP of the used membrane was measured. The initial LEP was 1.8, while the LEP of the membrane after 80 h of operation equals 1.4 bar. This reduction is caused by the reduced surface tension after long term exposure to salts, which is also reported in the literature [51]. However, this reduction is not severe enough to indicate that the coating is washed off during operation. In that case, 0.1 bar would already result in the penetration of liquid through the membrane.

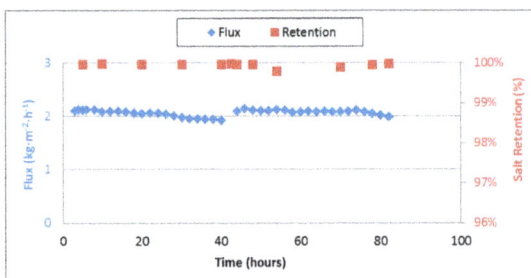

Figure 9. Long term experiment of membrane 6, T_f = 45 °C, T_p = 40 °C, v = 0.13 m·s^{-1}, NaCl concentration = 35 g·L^{-1}.

4. Conclusions

Sol-gel coatings have proven their stability and excellent performance in many other applications and were successfully applied for the first time on a hydrophilic PES membrane with a suitable pore size, porosity, and thickness for application in membrane distillation. The sol-gel coatings provide sufficient hydrophobicity and resistance against membrane wetting, and the surface contact angle can be increased from 27° up to 110°. Based on this study, the V/Mc/F13 system was recommended because of its higher hydrophobicity at 5 wt. % loading of siloxanes. This allows for keeping the pores open for vapor transport. Moreover, it is possible to produce a membrane with two hydrophobic sides using the spraying technique on both sides of the membrane, whereas a membrane with a hydrophobic/hydrophilic structure is obtained when spraying the coating on only one side of the membrane. The membranes with a hydrophobic/hydrophilic structure are recommended for seawater desalination because the hydrophobic layer thickness is closer to the optimal thickness for flux. While the coated membranes achieve comparable fluxes, the energy efficiency is relatively low compared to the commercial membranes in the same conditions. The energy efficiency was found to be independent of the coating procedure, but is dependent on the base membrane structure. Therefore, further optimization of the base membrane structure is required to further improve the membrane performance of these types of membranes in membrane distillation.

Acknowledgments: L. Eykens thankfully acknowledge a Ph.D. scholarship provided by VITO.

Author Contributions: L.E. performed the membrane characterization and M.D. testing and wrote the paper. K.R. prepared the inorganic-organic coatings on the PES-substrate and corrected the manuscript. M.D., K.D.S., C.D., L.P. and B.V.D.B. guided the experiments, analysis and the writing process.

Conflicts of Interest: The authors declare no conflict of interest.

References

1. Zhang, P.; Knötig, P.; Gray, S.; Duke, M. Scale reduction and cleaning techniques during direct contact membrane distillation of seawater reverse osmosis brine. *Desalination* **2015**, *374*, 20–30. [CrossRef]
2. Kesieme, U.K. Mine Waste Water Treatment and Acid Recovery Using Membrane Distillation and Solvent Extraction. Ph.D. Thesis, Victoria University, Footscray, Australia, 2015.
3. Calabrò, V.; Drioli, E.; Matera, F. Membrane distillation in the textile wastewater treatment. *Desalination* **1991**, *83*, 209–224. [CrossRef]
4. Hickenbottom, K.L.; Cath, T.Y. Sustainable operation of membrane distillation for enhancement of mineral recovery from hypersaline solutions. *J. Membr. Sci.* **2014**, *454*, 426–435. [CrossRef]
5. Christensen, K.; Andresen, R.; Tandskov, I.; Norddahl, B.; du Preez, J.H. Using direct contact membrane distillation for whey protein concentration. *Desalination* **2006**, *200*, 523–525. [CrossRef]
6. Tomaszewska, M.; Gryta, M.; Morawski, A.W. Study on the concentration of acids by membrane distillation. *J. Membr. Sci.* **1995**, *102*, 113–122. [CrossRef]
7. Teoh, M.M.; Chung, T.S. Membrane distillation with hydrophobic macrovoid-free PVDF-PTFE hollow fiber membranes. *Sep. Purif. Technol.* **2009**, *66*, 229–236. [CrossRef]
8. Bonyadi, S.; Chung, T.S. Flux enhancement in membrane distillation by fabrication of dual layer hydrophilic-hydrophobic hollow fiber membranes. *J. Membr. Sci.* **2007**, *306*, 134–146. [CrossRef]
9. El-Bourawi, M.S.; Ding, Z.; Ma, R.; Khayet, M. A framework for better understanding membrane distillation separation process. *J. Membr. Sci.* **2006**, *285*, 4–29. [CrossRef]
10. Alkhudhiri, A.; Darwish, N.; Hilal, N. Membrane distillation: A comprehensive review. *Desalination* **2012**, *287*, 2–18. [CrossRef]
11. Lawson, K.W.; Lloyd, D.R. Membrane distillation. *J. Membr. Sci.* **1997**, *124*, 1–25. [CrossRef]
12. Schneider, K.; Hölz, W.; Wollbeck, R.; Ripperger, S. Membranes and modules for transmembrane distillation. *J. Membr. Sci.* **1988**, *39*, 25–42. [CrossRef]
13. Khayet, M.; Matsuura, T. *Membrane Distillation: Principles and Applications*; Elsevier B.V.: Amsterdam, The Netherland, 2011.
14. Eykens, L.; De Sitter, K.; Dotremont, C.; Pinoy, L.; van der Bruggen, B. How to Optimize the Membrane Properties for Membrane Distillation: A Review. *Ind. Eng. Chem. Res.* **2016**, *55*, 9333–9343. [CrossRef]
15. Alklaibi, A.M.; Lior, N. Transport analysis of air-gap membrane distillation. *J. Membr. Sci.* **2005**, *255*, 239–253. [CrossRef]
16. Abu Al-Rub, F.A.; Banat, F.; Beni-Melhim, K. Parametric Sensitivity Analysis of Direct Contact Membrane Distillation. *Sep. Sci. Technol.* **2002**, *37*, 3245–3271. [CrossRef]
17. Jönsson, A.S.; Wimmerstedt, R.; Harrysson, A.C. Membrane distillation—A theoretical study of evaporation through microporous membranes. *Desalination* **1985**, *56*, 237–249. [CrossRef]
18. Ali, M.I.; Summers, E.K.; Arafat, H.A.; Lienhard, J.H.V. Effects of membrane properties on water production cost in small scale membrane distillation systems. *Desalination* **2012**, *306*, 60–71. [CrossRef]
19. Bahmanyar, A.; Asghari, M.; Khoobi, N. Numerical simulation and theoretical study on simultaneously effects of operating parameters in direct contact membrane distillation. *Chem. Eng. Process. Process Intensif.* **2012**, *61*, 42–50. [CrossRef]
20. Eykens, L.; Hitsov, I.; de Sitter, K.; Dotremont, C.; Pinoy, L.; Nopens, I.; van der Bruggen, B. Influence of membrane thickness and process conditions on direct contact membrane distillation at different salinities. *J. Membr. Sci.* **2016**, *498*, 353–364. [CrossRef]
21. Thomas, R.; Guillen-Burrieza, E.; Arafat, H.A. Pore structure control of PVDF membranes using a 2-stage coagulation bath phase inversion process for application in membrane distillation (MD). *J. Membr. Sci.* **2014**, *452*, 470–480. [CrossRef]
22. Song, Z.W.; Jiang, L.Y. Optimization of morphology and perf ormance of PVDF hollow fiber for direct contact membrane distillation using experimental design. *Chem. Eng. Sci.* **2013**, *101*, 130–143. [CrossRef]
23. Tang, Y.; Li, N.; Liu, A.; Ding, S.; Yi, C.; Liu, H. Effect of spinning conditions on the structure and performance of hydrophobic PVDF hollow fiber membranes for membrane distillation. *Desalination* **2012**, *287*, 326–339. [CrossRef]

24. García-Payo, M.C.; Essalhi, M.; Khayet, M. Effects of PVDF-HFP concentration on membrane distillation performance and structural morphology of hollow fiber membranes. *J. Membr. Sci.* **2010**, *347*, 209–219. [CrossRef]

25. Kim, Y.; Rana, D.; Matsuura, T.; Chung, W.J. Influence of surface modifying macromolecules on the surface properties of poly(ether sulfone) ultra-filtration membranes. *J. Membr. Sci.* **2009**, *338*, 84–91. [CrossRef]

26. Essalhi, M.; Khayet, M. Surface segregation of fluorinated modifying macromolecule for hydrophobic/hydrophilic membrane preparation and application in air gap and direct contact membrane distillation. *J. Membr. Sci.* **2012**, *417–418*, 163–173. [CrossRef]

27. Khayet, M.; Matsuura, T. Application of surface modifying macromolecules for the preparation of membranes for membrane distillation. *Desalination* **2003**, *158*, 51–56. [CrossRef]

28. Tijing, L.D.; Choi, J.S.; Lee, S.; Kim, S.H.; Shon, H.K. Recent progress of membrane distillation using electrospun nanofibrous membrane. *J. Membr. Sci.* **2014**, *453*, 435–462. [CrossRef]

29. Essalhi, M.; Khayet, M. Self-sustained webs of polyvinylidene fluoride electrospun nanofibers at different electrospinning times: 1. Desalination by direct contact membrane distillation. *J. Membr. Sci.* **2013**, *433*, 167–179. [CrossRef]

30. Tian, M.; Yin, Y.; Yang, C.; Zhao, B.; Song, J.; Liu, J.; Li, X.M.; He, T. CF_4 plasma modified highly interconnective porous polysulfone membranes for direct contact membrane distillation (DCMD). *Desalination* **2015**, *369*, 105–114. [CrossRef]

31. Wu, Y.; Kong, Y.; Lin, X.; Liu, W.; Xu, J. Surface-modified hydrophilic membranes in membrane distillation. *J. Membr. Sci.* **1992**, *72*, 189–196. [CrossRef]

32. Razmjou, A.; Arifin, E.; Dong, G.; Mansouri, J.; Chen, V. Superhydrophobic modification of TiO_2 nanocomposite PVDF membranes for applications in membrane distillation. *J. Membr. Sci.* **2012**, *415–416*, 850–863. [CrossRef]

33. Khemakhem, S.; Amar, R.B. Modification of Tunisian clay membrane surface by silane grafting: Application for desalination with Air Gap Membrane Distillation process. *Colloids Surf. A Physicochem. Eng. Asp.* **2011**, *387*, 79–85. [CrossRef]

34. Haas, K.-H.; Amberg-Schwab, S.; Rose, K. Functionalized coating materials based on inorganic-organic polymers. *Thin Solid Films* **1999**, *351*, 198–203. [CrossRef]

35. Kron, J.; Amberg-schwab, S.; Schottner, G. Functional coatings on glass using ORMOCER®-systems. Code: BP20. *J. Sol-Gel Sci. Technol.* **1994**, *2*, 189–192. [CrossRef]

36. Matějec, V.; Rose, K.; Hayer, M.; Pospíšllová, M.; Chomát, M. Development of organically modified polysiloxanes for coating optical fibers and their sensitivity to gases and solvents. *Sens. Actuators B Chem.* **1997**, *39*, 438–442. [CrossRef]

37. Liao, Y.; Wang, R.; Tian, M.; Qiu, C.; Fane, A.G. Fabrication of polyvinylidene fluoride (PVDF) nanofiber membranes by electro-spinning for direct contact membrane distillation. *J. Membr. Sci.* **2013**, *425–426*, 30–39. [CrossRef]

38. Khayet, M.; Khulbe, K.C.; Matsuura, T. Characterization of membranes for membrane distillation by atomic force microscopy and estimation of their water vapor transfer coefficients in vacuum membrane distillation process. *J. Membr. Sci.* **2004**, *238*, 199–211. [CrossRef]

39. Hou, D.; Dai, G.; Wang, J.; Fan, H.; Zhang, L.; Luan, Z. Preparation and characterization of PVDF/nonwoven fabric flat-sheet composite membranes for desalination through direct contact membrane distillation. *Sep. Purif. Technol.* **2012**, *101*, 1–10. [CrossRef]

40. Haas, K.H.; Amberg-Schwab, S.; Rose, K.; Schottner, G. Functionalized coatings based on inorganic-organic polymers (ORMOCER S) and their combination with vapor deposited inorganic thin films. *Surf. Coat. Technol.* **1999**, *111*, 72–79. [CrossRef]

41. Schottner, G.; Rose, K.; Posset, U. Scratch and Abrasion Resistant Coatings on Plastic Lenses—State of the Art, Current Developments and Perspectives. *J. Sol-Gel Sci. Technol.* **2003**, *27*, 71–79. [CrossRef]

42. Khayet, M.; Matsuura, T. Preparation and Characterization of Polyvinylidene Fluoride Membranes for Membrane Distillation. *Ind. Eng. Chem. Res.* **2001**, *40*, 5710–5718. [CrossRef]

43. Francis, L.; Ghaffour, N.; Alsaadi, A.S.; Nunes, S.P.; Amy, G.L. Performance evaluation of the DCMD desalination process under bench scale and large scale module operating conditions. *J. Membr. Sci.* **2014**, *455*, 103–112. [CrossRef]

44. Smolders, K.; Franken, A.C.M. Terminology for Membrane Distillation. *Desalination* **1989**, *72*, 249–262. [CrossRef]
45. Khayet, M. Solar desalination by membrane distillation: Dispersion in energy consumption analysis and water production costs (a review). *Desalination* **2013**, *308*, 89–101. [CrossRef]
46. Khare, V.P.; Greenberg, A.R.; Krantz, W.B. Vapor-induced phase separation—Effect of the humid air exposure step on membrane morphology: Part I. Insights from mathematical modeling. *J. Membr. Sci.* **2005**, *258*, 140–156. [CrossRef]
47. Gilron, J.; Ladizansky, Y.; Korin, E. Silica Fouling in Direct Contact Membrane Distillation. *Ind. Eng. Chem. Res.* **2013**, *52*, 10521–10529. [CrossRef]
48. Martínez, L.; Rodríguez-Maroto, J.M. Membrane thickness reduction effects on direct contact membrane distillation performance. *J. Membr. Sci.* **2008**, *312*, 143–156. [CrossRef]
49. Laganà, F.; Barbieri, G.; Drioli, E. Direct contact membrane distillation: Modelling and concentration experiments. *J. Membr. Sci.* **2000**, *166*, 1–11. [CrossRef]
50. Wu, H.Y.; Wang, R.; Field, R.W. Direct contact membrane distillation: An experimental and analytical investigation of the effect of membrane thickness upon transmembrane flux. *J. Membr. Sci.* **2014**, *470*, 2257–2265. [CrossRef]
51. Boubakri, A.; Bouguecha, S.A.-T.; Dhaouadi, I.; Hafiane, A. Effect of operating parameters on boron removal from seawater using membrane distillation process. *Desalination* **2015**, *373*, 86–93. [CrossRef]

![applied sciences logo] *applied sciences*

MDPI

Review

The Performance and Fouling Control of Submerged Hollow Fiber (HF) Systems: A Review

Ebrahim Akhondi [1,2], Farhad Zamani [1,3], Keng Han Tng [4,5], Gregory Leslie [4,5], William B. Krantz [1,6], Anthony G. Fane [1] and Jia Wei Chew [1,3,*]

[1] Singapore Membrane Technology Center, Nanyang Environment and Water Research Institute, Nanyang Technological University, Singapore 639798, Singapore; EBRA0001@e.ntu.edu.sg (E.A.); FARH0004@e.ntu.edu.sg (F.Z.); krantz@colorado.edu (W.B.K.); a.fane@unsw.edu.au (A.G.F.)
[2] Young Researchers and Elite Club, Central Tehran Branch, Islamic Azad University, Tehran 1469669191, Iran
[3] School of Chemical and Biomedical Engineering, Nanyang Technological University, Singapore 637459, Singapore
[4] UNESCO Centre for Membrane Science and Technology, Sydney, NSW 2052, Australia; k.h.tng@unsw.edu.au (K.H.T.); g.leslie@unsw.edu.au (G.L.)
[5] School of Chemical Engineering, The University of New South Wales, Sydney, NSW 2052, Australia
[6] Department of Chemical & Biological Engineering, University of Colorado, Boulder, CO 80309-0424, USA
[*] Correspondence: JChew@ntu.edu.sg; Tel.: +65-6316-8916

Received: 27 June 2017; Accepted: 24 July 2017; Published: 28 July 2017

Abstract: The submerged membrane filtration concept is well-established for low-pressure microfiltration (MF) and ultrafiltration (UF) applications in the water industry, and has become a mainstream technology for surface-water treatment, pretreatment prior to reverse osmosis (RO), and membrane bioreactors (MBRs). Compared to submerged flat sheet (FS) membranes, submerged hollow fiber (HF) membranes are more common due to their advantages of higher packing density, the ability to induce movement by mechanisms such as bubbling, and the feasibility of backwashing. In view of the importance of submerged HF processes, this review aims to provide a comprehensive landscape of the current state-of-the-art systems, to serve as a guide for further improvements in submerged HF membranes and their applications. The topics covered include recent developments in submerged hollow fiber membrane systems, the challenges and developments in fouling-control methods, and treatment protocols for membrane permeability recovery. The highlighted research opportunities include optimizing the various means to manipulate the hydrodynamics for fouling mitigation, developing online monitoring devices, and extending the submerged HF concept beyond filtration.

Keywords: submerged hollow fiber membranes; water treatment; fouling mitigation; critical flux; module design

1. Introduction

Low-pressure membrane processes, such as microfiltration (MF) and ultrafiltration (UF), are popular technologies in the water industry due to their proven efficiency in removing particles, colloids, and high molecular weight organics [1,2]. MF and UF membranes either are contained in a closed pressurized module or incorporated in an uncontained module that is submerged (immersed) in a tank. In submerged systems, the feed enters the tank at atmospheric pressure and the permeate is removed by applying suction on the permeate side of the membrane, which limits the transmembrane pressure (TMP) to <1 atmosphere and more typically to <0.5 atmosphere. The submerged concept is now well-established in the water industry, with applications in surface-water treatment, pretreatment prior to reverse osmosis (RO) in desalination and water reclamation, and membrane bioreactors

(MBRs) [3]. Submerged membranes are either in a flat sheet (FS) format (vertically aligned) or hollow fiber (HF) form either horizontally aligned or more typically vertically aligned, often with suction applied at both ends. Submerged FS membranes are used only in MBRs, whereas submerged HFs cover a range of applications and consequently are much more common. The submerged HF concept has the advantages of a higher packing density, the ability to induce movement by mechanisms such as bubbling, and the feasibility of backwashing [4–8]. The focus of this review is the submerged HF concept.

Submerged membrane systems have to deal with fouling, which represents a major drawback that restricts the application of membrane processes in the water industry [8]. In general, fouling is a widespread and costly problem that affects membrane performance and is a complex function of the feed characteristics, membrane properties, and operating conditions [9–11]. Fouling mechanisms include physical and chemical adsorption, precipitation of sparingly soluble salts, the growth of biofilms, and the deposition of suspended matter onto or into the membrane [12]. Inadequate pretreatment, poor fluid management (process hydrodynamics), extreme operating conditions, and improper membrane selection are factors that exacerbate fouling [13,14].

The key parameters influencing fouling deposition in submerged HF membranes are the membrane characteristics (e.g., membrane material, the structure of membranes/fibers, fiber diameter, length, and tautness), feed properties (e.g., foulant characteristics, concentration, viscosity), operating conditions (e.g., temperature, flux), and hydrodynamic conditions (e.g., surface shear, air flowrate) [15,16]; these parameters are discussed in detail in this review. For feeds with a high solids concentration, such as membrane bioreactors (MBRs), cross-flow operation is required for the constant application of surface shear to mitigate concentration polarization and fouling deposition [14,17]. A common practice for submerged membranes in MBRs is two-phase bubbly flow [18–20]; other approaches could include mechanical vibrations [21–23] and particle fluidization [24], all of which are discussed in this review. An important development that coincided with the introduction of submerged HFs was the realization that dead-end filtration was attractive for dilute feeds (surface waters, RO pretreatment). In this case the filtration is operated in cycles, with the dead-end forward flux interrupted by backwashing and periodic surface flushing [25] or relaxation [26]. The submerged HF module makes backwashing feasible for polymeric membranes.

As HF membrane performance continues to improve, submerged HF systems are increasingly becoming more attractive for water treatment, particularly in membrane bioreactor (MBR) applications [27,28]. In spite of the importance of submerged HF processes and the extensive research literature on advancing such systems, a comprehensive review summarizing the current state-of-the-art systems with respect to performance and fouling control remains a gap in the literature. Therefore, this review focuses on recent developments in submerged hollow fiber (HF) membrane systems, the challenges and developments in fouling control methods, and treatment protocols for membrane permeability recovery.

2. Submerged Membrane-Filtration Applications and Benefits

Compared to conventional water-treatment techniques, the most popular of which involves an integrated system consisting of coagulation, flocculation, sedimentation, and disinfection, the production of drinking water via membrane technology is acknowledged to be attractive, especially in terms of higher quality water and ease of implementation [29–34]. The use of submerged hollow fiber membranes can be classified into three main application areas, namely, surface-water treatment for drinking purposes (Section 2.1), pretreatment for RO desalination and reclamation (Section 2.2), and membrane bioreactors (MBRs) (Section 2.3). The former two usually are operated in the dead-end filtration mode with intermittent backwashing, while the third is usually operated as a continuous filtration process with bubbling for inducing tangential shear to mitigate fouling. Table 1 summarizes the submerged membrane-filtration applications and benefits.

Table 1. Submerged membrane-filtration applications and benefits.

Application	Operation Mode	Intermittent Fouling Control	Is Bubbling Implemented?	Advantages
Surface-water treatment	Dead-end with intermittent foulant removal	Backwashing, relaxation, chemical cleaning	With or without bubbling during foulant removal	Less chemical requirements; consistent quality of the filtrate [35–37]
Pretreatment of RO (reverse osmosis)	Dead-end with intermittent foulant removal	Backwashing, relaxation, chemical cleaning	With or without bubbling during foulant removal	Improved water quality; smaller footprint; less chemical requirements; consistent quality of the filtrate; lowered energy cost for RO plants [38–41]
Membrane bioreactors (MBRs)	Cross-flow with tangential shear	Continuous bubbling, sometimes backwash and relaxation	Continuous bubbling	Small footprint; complete solid-liquid separation; high volumetric organic removal rate; higher effluent quality [19,20,42]

2.1. Surface-Water Treatment

The submerged membrane process is employed to remove microparticles and macromolecules, which generally includes inorganic particles, microorganisms, and dissolved organic matter (DOM) [34,43,44]. The microparticles and macromolecules present in the feed tend to affect the membrane pores adversely through pore blocking (i.e., sealing off the membrane pore entrance), pore constriction (i.e., narrowing the membrane pore channels), and/or cake-layer formation, all of which result in a decrease in the membrane permeability [35,45,46]. DOM 'particles', whose size approximates that of the membrane pores, can cause pore blocking, while microparticles and macromolecules larger than the size of the membrane pores result in a fouling layer on the membrane surface. Reversible fouling can be removed by hydraulic flushing/backwashing with air bubbles as scouring agents, whereas irreversible fouling binds more stubbornly to the membrane, thereby necessitating chemical cleaning [35–37,47].

2.2. Pretreatment for RO Desalination and Reclamation

Adequate pretreatment of the feed to reverse osmosis (RO) systems is essential to ensure optimal performance. Low-pressure membrane pretreatment is increasingly implemented prior to the RO unit operation in seawater reverse osmosis (SWRO) plants, and also RO plants for the treatment of surface water and treated municipal effluent [7,18,44,48–51]. Similar to surface-water treatment, the low solids content in these applications allows the low-pressure membranes to be operated in the dead-end filtration mode with intermittent backwashing. The main advantages of membrane filtration compared to conventional pretreatments such as coagulation, flocculation, and sand filtration are improved water quality, smaller footprint, less chemical requirements, and consistent quality of the filtrate [38,39,52,53]. Higher energy demand and membrane fouling are the main disadvantages of having a membrane-filtration system for RO pretreatment. RO membranes are very sensitive to foulants, so enhanced pretreatment via low-pressure membranes can significantly improve the performance and reduce the energy cost of RO plants [51]. In particular, RO systems with submerged membrane pretreatment have been proven to exhibit a consistently lower silt-density index (SDI) relative to conventional pretreatment [7,40,41,54,55]. Gravity-driven membrane filtration, which was initially developed as a low-energy process for surface water and diluted wastewater treatment, has also shown potential for seawater pretreatment that requires less energy and no chemical cleaning [56,57].

2.3. Membrane Bioreactors (MBRs)

Membrane bioreactor (MBR) technology, which combines conventional activated sludge treatment with low-pressure membrane filtration, is widely used for the treatment of wastewater [18,19,58]. The considerable growth of MBR is driven by the high quality of the water produced, increased water scarcity, and decreasing specific energy requirements [28,59]. The anaerobic membrane bioreactor

(AnMBR), a combination of an anaerobic bioreactor and membrane filtration, also is a promising option for anaerobic treatment of wastewater [60–63].

A small footprint, complete solid-liquid separation, high volumetric organic removal rate, and higher effluent quality are some of the key advantages of the MBR and AnMBR [20,64–66]. In the submerged hollow fiber MBR, the membranes are directly immersed in the aeration tank. The results of bench-scale experiments, as well as many industrial and municipal operations, demonstrate high treatment efficiencies for chemical oxygen demand (COD), total suspended solids (TSS), and turbidity [19,42,63,67,68]. Although the MBR technology has been applied in many full-scale plants worldwide for treating municipal and industrial wastewater, membrane fouling and correspondingly increased energy consumption remain chief obstacles, as highlighted in a recent review [28]. Specifically, because membrane fouling diminishes productivity, fouling mitigation measures such as air scouring and frequent cleaning of the membrane are needed to restore the membrane permeability, which increases the energy requirement; furthermore, frequent cleaning shortens the membrane lifespan and results in higher membrane replacement costs. Aeration, bubbling, or gas sparging are the most common methods for mitigating membrane fouling; the important features of the interaction of bubbles with submerged hollow fibers (HF) are discussed in Section 5.2.

3. Fouling and Concentration Polarization in Submerged HF Systems

As noted earlier, submerged HFs can be operated in either a dead-end or cross-flow mode. In dead-end filtration, tangential shear is absent, while in cross-flow, there is shear on the membrane due to bubbling, vibration, or particle scouring. Figure 1 shows a schematic of dead-end and cross-flow filtration in a submerged system.

The operation of submerged HF membrane systems under either dead-end or cross-flow (usually induced by bubbles) conditions involves very different dynamics. Ideally the process is at steady-state with a fixed flux and TMP for cross-flow, while the process is cyclic with repeatable and regular changes in the TMP for dead-end filtration. However, irrespective of the mode of operation, some degree of fouling inevitably occurs, although membrane fouling is relatively less extensive in the cross-flow mode due to the continuous tangential shear on the membrane.

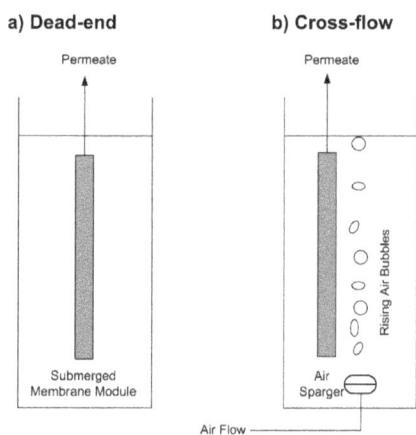

Figure 1. A schematic of dead-end (**a**) and cross-flow; (**b**) submerged filtration.

Accumulation of retained species on the membrane surface is unavoidable in membrane-based separation technologies for liquid feeds. In submerged HF membrane processes, depending on the membrane pore size, the retained species are particulates and macromolecules. The localized accumulation of particles or dissolved species on the membrane leads to concentration polarization

(CP), which is the primary reason for flux decline or TMP rise during the initial period of a membrane-based separation in low-pressure processes [3,69,70]. CP is considered to be reversible and can be controlled in a membrane module by means of optimizing the process hydrodynamics. Unfortunately, membrane fouling resulting from CP can lead to an irreversible loss of membrane permeability [69,71]. Membrane fouling, the process by which foulants, namely colloidal (e.g., clays, flocs), biological (e.g., bacteria, fungi), organic (e.g., oils, polyelectrolytes, humic substances), and scaling (e.g., mineral precipitates in RO systems) foulants, deposit onto the membrane surface or in the membrane pores [72,73], may take different forms, the main mechanisms of which are adsorption (physical and/or chemical), pore blocking, deposition of a cake layer, and gel formation [74–79]. The extent of fouling, which stems from the nature of foulant-membrane interaction, is a complex function of the feed characteristics (e.g., foulant type, foulant concentration, and physicochemical properties of the foulants such as the functional groups, charge, size, and conformation [72,80–82]), operating conditions (e.g., inadequate pretreatment, inadequate control of the hydrodynamics of the system, excessive flux, and low cross-flow velocity (in cross-flow systems) [72,82–84]), and membrane properties (e.g., pore-size distribution, surface roughness, charge properties, and hydrophobicity [70,85–87]).

3.1. Fouling in Submerged Dead-End Filtration

In dead-end filtration, tangential shear is absent and particles are convected by the permeate flow to deposit on the membrane surface, thereby forming a growing cake layer with time. Physical cleaning approaches are typically implemented periodically for the effective removal of the fouling layer in order to prolong the filtration process and membrane lifespan in submerged membrane systems. Such approaches include relaxation (i.e., intermittent cessation of permeation), backwashing (i.e., reversal of permeate flow through the pores), and air backwashing with or without air scouring [25,88–91]. Filtration duration, backwash and relaxation durations, and backwashing flowrate are important parameters in the fouling mitigation of submerged HF membranes [26,90,92,93]. However, a major challenge in the application of these techniques is that the imposed permeate fluxes have to be elevated to maintain a given water production, which in turn could result in a higher fouling rate [94]. More details on backwashing and relaxation can be found in Section 7.

3.2. Fouling in Submerged Cross-Flow Filtration

For filtration with cross-flow, particle back-transport can be caused by the mechanisms of Brownian motion, shear-induced diffusion and/or inertial lift depending on the foulant size and the tangential shear rate [95]. For a submerged HF module, the major hydrodynamic technique to mitigate particle deposition on the fibers is bubbling, which induces unsteady-state shear at the membrane surface through turbulent eddies, fiber oscillations, particle scouring, and the recirculation of the bioreactor liquid [14,15,22,96,97]. The critical flux phenomenon (the flux below which negligible fouling occurs) also is observed in bubbled submerged HFs [98–100]. Bubbling intensity can increase the critical flux [100–103] and eventually reach a plateau beyond which it has little effect. Judd [104] gave typical values of the bubbling intensity (specific aeration demand) in submerged MBRs, defined either as the air flowrate per unit membrane area ($m^3/h\,m^2$) or airflow per unit permeate (m^3/m^3). For complex feeds, such as those encountered in an MBR, the imposed flux is usually 'subcritical' for the biofloc, but could be above critical for the supernatant colloids and macrosolutes. Under these conditions the submerged HF MBR typically initially shows a slow TMP rise that eventually becomes more rapid or displays a TMP 'jump' [105,106]. The TMP jump is not specific to submerged MBRs and can have a number of possible causes [106]. Importantly, earlier TMP jumps occur with higher fluxes and/or inadequate bubbling in the submerged HF MBR. More details on the role of bubbles and the attendant hydrodynamics are in Section 5.2.

4. Blocking and Blocking Mitigation in Submerged HF Modules

Blocking, clogging, or sludging in submerged HF modules obstructs local flows in the fiber bundle and results in an uneven flow distribution; this is detrimental to the performance and can promote membrane fouling [58,107]. It should be noted that this section is targeted at the obstruction in the module rather than membrane fouling. The root cause has been attributed to the accumulation of solids, and the growth and merger of cake layers formed on the individual fibers in the module. In such cases it has been shown that the overall performance of a module packed with HFs is worse than that of a single HF [108]. This aforementioned blocking is distinct from two other forms of blocking that can occur in submerged HF modules: (i) the term 'blocking' can also refer to that of the aerators especially in MBR systems [104]; (ii) it also can refer to the blocking or closure of the pores of the HFs. Both of the latter forms of blocking are distinct from the blocking that occurs due to solids accumulation within the spaces between the fibers in the membrane module that is the focus here.

The hydrodynamics within the module represents the key factor influencing the blocking phenomenon [107,109]. Stagnant regions caused by poor local flow lead to lower shear on the fibers, which in turn results in cake buildup and eventually local blocking [110]. The packing of the fibers within a module has been observed to be very different axially [111], which causes some non-uniformity in the flow. Due to the complexity of the hydrodynamics in the module [111–116], the buildup of cake deposits is not likely to be uniform either among the fibers or along the fiber surface. The misdistribution of fouling deposits, which is acknowledged to be a direct function of the non-uniformity of the flow within the HF module [109,115,117–122], can result in large variations in the performance of fibers at the same position in the module and in poorer performance of the fibers in the middle of the module [108].

The adherence of fibers to one another is traced to the buildup and eventual merging of cake layers, which in turn causes the fibers to foul more rapidly due to hindered local flow; hence, in time it results in blocking. It has been observed that some fibers tend to be held tightly together by the dense cake layers, while other fibers remain freely suspended [108]. The principal difference between constant-pressure and constant-flux operation is that the former is self-limiting whereas the latter is self-accelerating when fouling or blocking occurs [110]. The self-acceleration in the commonly used constant-flux operation is because (i) incipient blocking will reduce the local flow that thereby enhances the deposition [108], and (ii) the flux decline in some fibers has to be matched by a flux increase in other fibers to maintain a constant net flux that thereby accelerates blocking. Either local or global non-uniformities caused by the operating conditions or design parameters [104,109], including high packing densities, low cross-flow velocities, high feed concentrations, high TMP, or lack of means to promote unsteady-state shear [123] (e.g., bubbling [14]), contribute to a greater tendency for blocking.

The control of fouling and blocking in practice is primarily through employing some or all of five strategies [109]: (i) pretreatment of the feed, (ii) physical or chemical cleaning protocols, (iii) flux reduction, (iv) aeration enhancement, and (v) chemical or biochemical modification of the feed. For (i), it is widely acknowledged that upgrading the pretreatment, in particular the screening, is pivotal to the successful retrofitting of an ASP (activated sludge process) or SBR (sequencing batch reactor) with an MBR [109]. Hair, rags, and other debris tend to aggregate at the top of the submerged HF module and become entwined with the filaments, thereby preventing their effective removal by backwashing [124]. Therefore, adequate removal of the solids before the submerged HF module is key to mitigating blocking. Methanogenic [125] and chlorinated [126] pretreatments have also been explored, as well as hybrid processes, both of which add an additional unit operation before the membrane module [127–130]. With respect to (ii), physical cleaning tends to remove reversible fouling while chemical cleaning can remove some irreversible fouling. The primary physical cleaning parameters include duration, frequency, and backwash flux [88,107,109,131,132], while the type and concentration of reagents are the important parameters for effective chemical cleaning [124,131,133–136]. Regarding (iii), flux reduction is not as cost-effective as might be expected, but the attendant extension of membrane lifespan and smoother operation may counterbalance some of the increased cost. Flux reduction is commonly implemented as a sustainable permeability operation (i.e., at a lower flux to maintain stable operation) or as an intermittent operation (i.e., at a higher operating flux with intermittent remedial measures such as relaxation and/or backwashing). As for (iv), aeration is acknowledged to be the main mixing mechanism [137,138] that generates a large amount of momentum that in turn reduces dead-zones and short-circuiting [139]. More details on aeration are given in Section 6.2. In connection with (v), the nature of the biomass in MBRs can be partially controlled through the addition of coagulants or flocculants [124,140–144] and adsorbent agents (e.g., activated carbon [145–151]). In addition, the bioprocess parameters, such as the sludge retention time (SRT), influence the biomass and supernatant as well as the fouling [152].

A suitable mitigation means for blocking has to be determined for each MBR plant. For example, a Mitsubishi Rayon unit failed to resolve the blocking problem via measures such as overnight relaxation and intensive or regular chemical cleaning, but regular backwashing for 30 s during each permeate production cycle achieved a stable low permeation flux [104]. To limit blocking, the PURON® system is designed with free movement of the filaments at the top of the module. This allows for larger solids, such as hair and agglomerated cellulose fibers, to escape without causing clogging in this region. The fibers are reinforced by an inner braid, to withstand the lateral movement of the filaments that subjects them to mechanical stress [153].

A major difficulty in the mitigation of module blocking in submerged HFs is the lack of simple techniques for in situ monitoring of the phenomenon. One approach that has been developed [122] is based on the assumption that blocking within the module is initially localized, which would result in a localized drop in flux relative to the overall average flux. Using a simple array of flow detectors mounted in the permeate header, it is possible to measure local fluxes and identify fiber regions that become less productive. It was shown that shifts in the standard deviation of the local fluxes was much more sensitive to local blocking than shifts in the overall system TMP. In the aforementioned study, the flow detectors were based on constant temperature anemometry strips, but other low cost detectors could be used.

5. Parameters Affecting the Performance of a Submerged Hollow Fiber System

Several factors, including the feed characteristics, membrane and module properties (e.g., fiber length, diameter, and looseness), and hydrodynamic properties (e.g., flowrates), collectively affect the performance of submerged HF systems.

5.1. Membrane Properties and Module Configurations

5.1.1. Membrane Materials and Surface Morphology

Membrane characteristics such as material, surface charge, hydrophilicity, pore size, and pore morphology significantly impact membrane performance and the fouling potential. Membrane surfaces can be modified to combat and mitigate adhesive fouling [25]. The membrane material largely influences the initial rates of deposition of foulants due to the tendency of some materials to adsorb certain solutes or particulates more readily, as quantifiable by the Gibbs free energy of foulant-membrane interaction [154–157]. When the adsorption becomes such that the effective pore size of the membrane is reduced, the flux is adversely affected [158–160].

It is well-known that the severity of fouling increases as the hydrophobicity of the membrane increases, because organic molecules have a higher affinity for hydrophobic materials [47,159,161–165]. Most commercial MF and UF membranes are made from relatively hydrophobic polymers (e.g., polysulfone, polyethersulfone, polypropylene, polyethylene, and polyvinylidene-fluoride (PVDF)), due to their excellent chemical resistance, and thermal and mechanical properties [83,166–170]. In some cases, the membranes are modified by additives to confer increased hydrophilicity for water applications. The charge on the membrane surface also plays a role in either exacerbating or mitigating fouling. Attractive electrostatic forces between a charged surface and the co-ions in the feed solution increase fouling [161], while repulsive electrostatic forces mitigate fouling. In general, a low surface charge or electrically neutral surfaces tend to show better anti-fouling properties during the initial stages of membrane fouling [163,171]. Since most colloidal particles, such as those in natural organic matter (NOM), that tend to deposit are negatively charged, membrane surfaces are usually negatively charged [161,168,169], although positively charged membrane surfaces are also used to repel positively charged colloids and cations [172]. Compared to having just a positively or negatively charged membrane surface, the addition of a zwitterionic charged material, one composed of neutral molecules that have a positive and a negative electrical charge, has been shown to confer better anti-fouling properties [173]. Inorganic nanoparticles (such as SiO_2 [174,175], Al_2O_3 [176,177], clay [178,179], ZrO_2 [180–182], TiO_2 [183,184], and ZnO [185,186]) also have been used to improve polymeric composite membranes, conferring improvements in the mechanical properties, thermal stability, hydrophilicity, permeation, and antifouling performance of membranes [165].

5.1.2. Fiber/Module Arrangement

Experimental efforts have consistently established that a lower HF packing density, either by having fewer fibers [108], widening the HF module [187,188], or by varying the module configuration [97,189], is linked to a lower fouling tendency. Other than the HF packing density, modules that are designed to enhance either lateral flow [190] or lateral movement of the fibers [188] are known to further improve the performance of submerged HF systems. In a carefully controlled arrangement with nine fibers in a matrix, a detailed analysis has shown that the performance of individual fibers varied with their position in the module, such that the fibers at the edge performed best, while those in the center surrounded by other fibers performed the worst [108]. The underlying reason for this was tied to the lower cross-flow velocity at the center of the module and the 'flux competition' for the surrounded fiber. Interestingly, whereas Yeo and Fane [108] found that a single fiber outperformed a multi-fiber HF module in a single-phase (no bubbling) system due to module blocking, Berube and Lei [191] found that a multi-fiber module performed better than a single fiber in the presence of bubbles (i.e., two-phase flow) due to inter-fiber interactions leading to mechanical erosion of the foulant layer.

Simulation results by Liu et al. [192] revealed that MBRs fitted with hollow fibers in a vertical orientation experienced 25% more membrane surface shear in the filtration zone than horizontally oriented fibers at the same aeration intensity. They also found that the addition of baffles in the membrane modules is a feasible way to promote turbulence and shear in the upper section of the membrane module.

5.1.3. Fiber Looseness

Fiber movement plays an important role in determining the extent of particle deposition on submerged HFs in bubbling cross-flow operation. Greater fiber movement, which can be achieved by using looser fibers, enhances the back-transport of the foulants from the membrane and also results in physical contact between the fibers, both of which reduce fouling [16]. Tight fibers studied by Wicaksana et al. [15] showed a 40% faster rate of TMP rise compared to fibers with a 4% looseness, where the looseness percent is based on the difference between the fiber length and the linear distance between the fixed ends of the fiber relative to the fiber length; this implies that approximately 40% of the fouling mitigation brought about by the bubbles was due to movement of the loose fibers. Simulations by Liu et al. [193] showed that fiber displacement and membrane surface shear are highly variable at different locations along the fiber and with time. In addition, increasing the fiber looseness from 0.5% to 1% increased the average surface shear by 50.4% (0.56–1.13 Pa) for fibers with the same diameter. Yeo et al. [194] found that increasing the looseness from 0% to 1% decreased the fouling tendency regardless of the bubble size, whereas increasing the looseness from 1% to 2% increased the fouling tendency. This suggests an optimum fiber looseness, whereby too much displacement may move the fiber away from the influence of bubbles. This effect will depend on the module geometry relative to the bubbling zone. It is also known that too much looseness can lead to fiber breakage.

5.1.4. Fiber Diameter

A smaller HF diameter has been experimentally shown to improve the performance of submerged HF systems [15,16,187] due to the attendant greater fiber mobility [15], which can be further enhanced by bubbling [16] or turbulence in general [187]. However, theoretical analysis shows that a smaller HF diameter leads to a higher pressure drop along the fiber lumen, such that a higher suction pressure is required for a given average imposed flux [187]. On the other hand, simulations show that the average surface shear was 67% higher for fibers with a diameter of 1.3 mm relative to fibers with a diameter of 1.0 mm that had an identical Young's modulus and looseness [193]. The interplay of these two effects suggests an optimal fiber diameter that balances flux enhancement with flux-distribution and suction-pressure considerations [187].

5.1.5. Fiber Length

Simulation studies have indicated that longer fibers lead to a higher lumen-side axial pressure drop and consequently an uneven permeation flux distribution along the fibers, which can result in the particle deposition being more severe near the suction ends relative to the closed end of the fibers due to a higher local flux [99]. Experimental evidence is in agreement that this non-uniform pattern of particle deposition along the fibers is more apparent for longer fibers (lengths tested were in the range of 0.3–1 m) [100,195–197]. On the other hand, fouling control is facilitated by the greater mobility of longer fibers [98], which can be enhanced by bubbling [15].

5.2. Hydrodynamics in Submerged HF Membranes

Enhanced surface shear, for example caused by an increased cross-flow velocity, is a common strategy to control concentration polarization and fouling in submerged HF systems [138,198,199]. Unsteady-state shear, such as two-phase flow (i.e., gas bubbles or fluidized particles) and vibration, is more energy-efficient than steady-state shear [123]. Air bubbling is particularly attractive in MBRs for aeration, mixing, and augmenting liquid flows [14,82,200,201].

5.2.1. Role of Air Bubbles

The use of bubbly flow has been reviewed by Cui et al. [14] and more recently by Wibisono et al. [82]. The major benefit of rising bubbles is the unsteady or transient shear stress at the membrane surface that causes particle back-transport away from this surface [14,15,22]. Figure 2, based on Cui et al. [14], illustrates the possible interactions of bubbles with the surface of hollow fibers and shows three different effects on the submerged hollow fibers: (i) a shear stress on the surface of the hollow fiber induced by the wake generated by the rising bubbles, (ii) fluctuating liquid flows transverse to the fibers induced by the bubbles, and (iii) lateral fiber movement induced by the bubbles that depends on the looseness of the fiber. Wibisono et al. [82] indicated that aeration intensity not only enhances the hydrodynamics, but also can affect the biomass properties in aerobic membrane bioreactors. Yeo et al. [202] [Yeo, 2017 #321] showed that aeration also influenced the biofilm growth. In addition, Cabassud et al. [203] reported that bubbles seem to alter the structure of the cake or fouling layer such that the specific resistance is reduced. They based this conclusion on the observation that gas sparging applied to the MF of particles increased the fluxes significantly after a period of flux decline at the higher feed concentrations. Wang et al. [204] correlated the bubble hydrodynamics with the critical flux and found that bubble momentum and the bubble-membrane contact area had the most positive correlation with the local critical flux. However, Du et al. [201] found that the shear stress associated with the bubbles was insufficient to mitigate the deposition of fine (1.75 μm) particles.

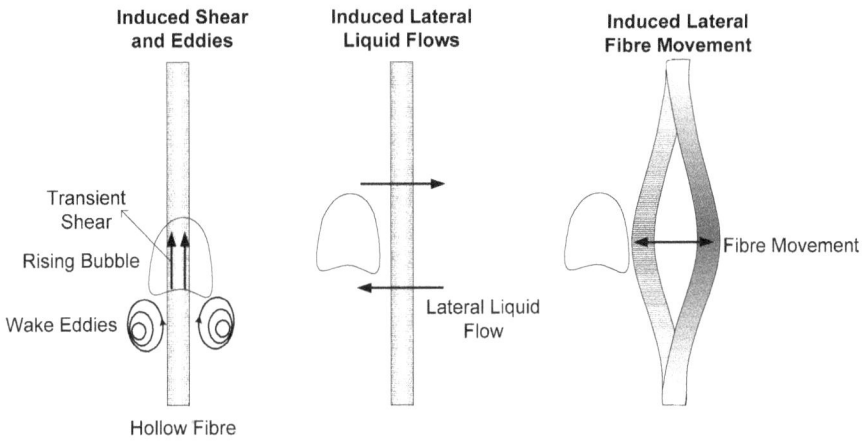

Figure 2. Fouling and cake control mechanisms by bubbles outside fibers [14]. Reproduced with permission from [14], Copyright Journal of Membrane Science, 2003.

5.2.2. Bubble Characteristics

Several techniques have been used to characterize the hydrodynamic conditions in submerged HF systems. Particle-image velocimetry (PIV) [205] has shown that the bubble size increases with height along the membrane module due to the reduced hydrostatic pressure as the air bubbles move upward, and varies over a wide range of 0.2–50 mm, with a predominant size range of 3–5 mm. A strong sheltering effect attributed to the hydrodynamics was observed within the hollow fiber module that resulted in a 10-fold reduction in the axial velocity relative to the velocity outside the fiber bundle [110]. Nguyen Cong Duc et al. [206] used a bi-optical probe to characterize the bubble velocity, distribution, and size throughout submerged full-scale HF modules. The shear stress was observed to be an important parameter in controlling particle back-transport from the membrane surfaces. Fulton et al. [207] studied the sparged bubble characteristics and the induced shear forces at the surface of submerged hollow fiber membranes using an electrochemical method. The shear stress was observed to be highly unpredictable over time and heterogeneously distributed within the module, ranging from 0.1 to over 10 Pa. Also, no correlation was observed between the shear stress and the bubble frequency or rise velocity. However, this does not corroborate with the general observation of better fouling control with increased aeration intensity (see Section 5.2.3) and also with the results of Yeo et al. [194], who observed an increase in the shear stress with increasing bubble frequency for all bubble types. These disparate observations highlight the challenge in achieving well-distributed two-phase flow in the submerged HF module. The importance of this was shown by Buetenholm et al. [208], who used X-ray computer tomography to detect the instantaneous displacement of fibers in an aerated HF bundle. The data were then incorporated into a computational fluid dynamics (CFD) simulation to allow optimization of the module design and aeration. More recently, Wang et al. [204] used a high speed video camera to characterize the bubble characteristics and direct observation through the membrane (DOTM) to determine the corresponding critical flux of micron-sized polystyrene particles. They found that the local bubble momentum and bubble size had the most positive correlation with the local critical flux.

5.2.3. Effect of Gas Flowrate

It has been reported in many studies that the filtration performance can be improved by controlling the bubbling rate [15,16,209,210]. On the one hand, the critical flux improves in an approximately linear fashion with respect to gas flow in submerged HF systems [100–103,204]. On the other hand, for different sizes of fibers, it has been shown that a modest gas flow can increase the final flux by a factor of 3–6 in a yeast suspension, but the enhancement quickly plateaus at higher gas flowrates such that further increases in the gas flowrate achieve negligible enhancement [187]. This disparate behavior stems from the slugging phenomenon, whereby large changes in the gas flow have a negligible effect on the velocity of the film that is formed adjacent to the surface of the fiber [211]. Similar plateaus have been observed in submerged HF systems for sewage treatment and drinking-water treatment [191,210]. It has been observed that at high gas flowrates (namely, 40 L/h), a maximum flux was observed, after which the flux started to decline [212]. This phenomenon was explained by the relationship between the bubble size and air flowrate; since bubble size tends to increase with the air flowrate, the bubbles become so large after an optimum air flowrate that they start to prevent the liquid from reaching the membrane surface, a phenomenon that is also linked to slugging flow.

5.2.4. Aeration Modes

Armed with the knowledge that the shear stress induced by bubbles is the dominant mechanism in controlling particle back-transport from membrane surfaces, the bubbling mode (e.g., continuous, alternating, pulsed) can be optimized to minimize the energy cost while achieving an optimum shear to improve system performance [213,214]. Yeom et al. [215] carried out a study of the frequency or duration of cycling between filtration and bubbling phases and showed that intermittent aeration is effective for fouling control in a denitrification MBR. Guibert et al. [216] studied the positioning of aeration ports and reported that the injection of air in different zones around the fiber bundles greatly improved the overall system performance. Fulton and Berube [188] studied the effectiveness of continuous, alternating, and pulsed bubbling modes and found that, even though the volume of gas used by pulse sparging was half of that used by the other sparging conditions, relatively similar induced shear stress was observed for all three bubbling modes. Similarly, Tung et al. [217] observed that semi-continuous aeration could suppress the membrane fouling at the same level as at continuous aeration.

5.3. Shear Stress on Membrane Surface by Non-Bubbling Techniques

Submerged HFs also are amenable to fouling control by methods that do not involve bubbling, as discussed below.

5.3.1. Vibrations

Vibrations have been proven to be an effective way to induce shear on a membrane surface and consequently reduce concentration polarization (CP) and fouling [123,218,219]. Various modes of vibration are applicable for different membrane systems. A submerged HF system can be vibrated longitudinally or axially (Figure 3a), transversely (Figure 3b), and rotationally (Figure 3c). Although anaerobic systems are gaining momentum in the wastewater industry due to their potential for energy production, bubbling by recycled biogas has some challenges; hence, the vibration approach is attractive for fouling mitigation in anaerobic MBR (AnMBR) systems. In particular, transverse vibration has been proven to be an effective way to mitigate the fouling in AnMBR applications [220,221].

Many studies have been carried out to probe the effectiveness of the different modes of vibration (i.e., longitudinally, transversely, or rotational). Low et al. [222] found that the use of vibrations slowed the flux decline for the submerged HF system that they investigated, and that longitudinal oscillation outperforms rotational oscillation. Li et al. [223] showed that vibration was more effective for a bentonite suspension compared to a washed yeast solution that may cause internal fouling; hence, vibration is more effective primarily for cake removal but not for the mitigation of internal fouling, which was corroborated by Kola et al. [220]. It was also shown that transverse vibration decreases the fouling rate much more effectively than longitudinal vibration [223]. Genkin et al. [224] reported that adding transverse vibration to longitudinal vibration (by using chess-patterned vanes) resulted in an almost doubling of the critical fluxes at the same frequency; adding coagulants further elevated the critical flux of the vibrating system, although floc breakup at higher frequencies (namely, 10 Hz) tended to reduce the critical flux. Beier and Jonsson [225] also found that vibration facilitates the separation of macromolecules (e.g., BSA) and larger components (e.g., yeast cells) at sub-critical fluxes by loosening and removing the built-up cake in the filtration of a mixed suspension for a membrane with pores larger than the macromolecular components. Fiber spacing and looseness were found to be important parameters to improve the benefits of vibration with respect to the turbulence kinetic energy and eddy length scale [226]. A recent study investigated rotating instead of vibrating the HF membranes and found membrane rotation to be more effective than gas scouring [227].

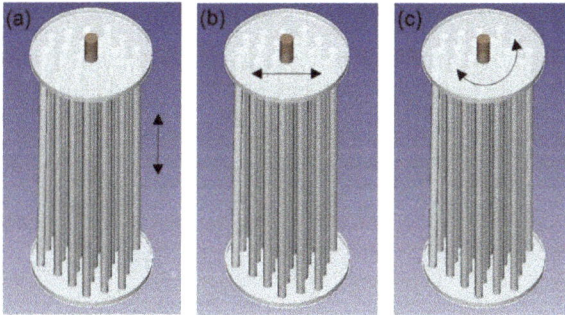

Figure 3. Schematic of an HF (hollow fiber) module with different modes of vibration: (**a**) longitudinal vibration; (**b**) transverse vibration; and (**c**) rotational vibration.

Figure 4 shows that both higher amplitudes and higher frequencies for longitudinal vibration contribute towards fouling mitigation, with an observed reduction in the fouling rate by as much as 90% [23]. Genkin et al. [224] found that the critical flux has a stronger dependency on frequency at higher frequencies but a weaker dependency at lower frequencies, presumably due to a change of the flow regime in the vibrating system. Chatzikonstantinou et al. [228] used high-frequency vibration in a pilot-scale submerged MBR and found it promising with respect to energy savings compared to conventional air-cleaning systems.

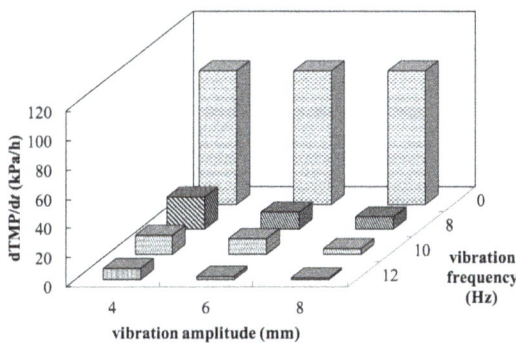

Figure 4. Effect of vibration amplitude and frequency on the fouling rate expressed as the time rate-of-change of the TMP (transmembrane pressure) for filtration of a 4 g/L bentonite suspension [23] for a constant flux of 30 LMH. Reproduced with permission from [23], Copyright Journal of membrane science, 2013.

The unsteady-state shear induced by vibration can be related to performance enhancement to provide a quantitative assessment of the beneficiation [123]. Beier and coworkers [229–231] applied longitudinal vibration, with frequencies between 0–30 Hz and amplitudes of 0.2, 0.7, and 1.175 mm, on HF bundles with nominal pore diameters of 0.45 μm, and found a general correlation between the critical flux (J_{crit}) and average shear rate (γ_{ave}) induced by vibration on the membrane surface as follows:

$$J_{crit} = a\gamma_{ave}^{n} \tag{1}$$

The correlations based on Equation (1) using different values of a and n for three aqueous suspensions are shown in Figure 5.

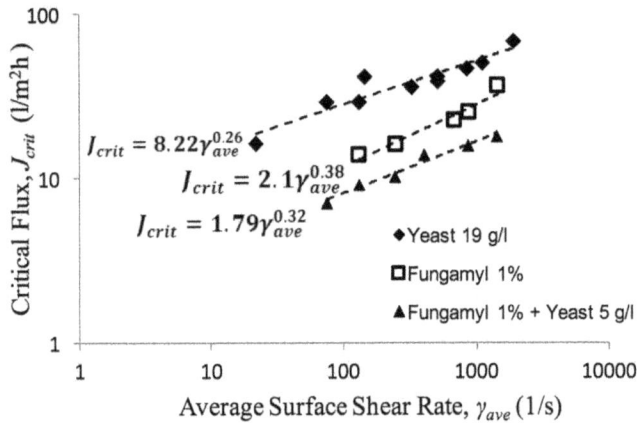

Figure 5. Correlation between the critical flux (J_{crit}) and average shear rate (γ_{ave}) for yeast suspensions with concentration of 19 g/L; 1% Fungymal solutions; and 1% Fungymal + 5 g/L yeast [231]. Reproduced with permission from [231], Copyright Separation and Purification Technology, 2007.

To calculate the shear rate (γ) on the surface of a vibrating fiber, Beier et al. [229] proposed the following equation:

$$\gamma = A(2\pi f)^{1.5} v^{-0.5} \cos\left(\omega t - \frac{3\pi}{4}\right) \tag{2}$$

where A is the amplitude, f is the frequency of the vibration, v is kinematic viscosity of the fluid, and t is the time. As expected, the shear rate displays a periodic behavior as a function of time since the membrane vibrates with a velocity of:

$$u = 2\pi \, A \, f \, \cos(\omega t) \tag{3}$$

Equation (2) does not depend on the fiber diameter because the Navier-Stokes equation was solved under the assumption that the curvature of the fiber near the membrane surface is negligible [229]. However, recently, Zamani et al. [232] showed that the shear rate of a vibrating fiber is also a function of the fiber diameter. They showed that Equation (2) can have a relative error of 40% for a fiber with a diamter of 1 mm vibrating with an amplitude and frequency of 10 mm and 1 Hz, repectively.

Krantz et al. [233] applied longitudinal vibrations to a bundle of silicon hollow tube membranes to enhance the mass transfer to the liquid on the lumen side of the membrane, and found that the mass-transfer coefficient was increased by a factor of 2.65 relative to that without vibrations. An analytical solution was developed for the velocity profile of the laminar flow within the vibrating tube. However, this solution is not applicable for commonly used submerged HF applications, since the external shear on the fiber is of more interest. However, it could be useful for novel submerged HF processes, such as forward osmosis (FO) and membrane distillation (MD) applications.

5.3.2. Particle Scouring

Particle scouring is one of several unsteady-state shear techniques useful in membrane processes for fouling mitigation and improving the mass-transfer coefficient [123]. In practice, the solid particles are brought into close contact with the membrane surface via fluidization, which is the process whereby the particles are dispersed and suspended by the liquid such that they behave like a fluid [234,235]; thereby, the fouling layer on the membrane surface is mechanically scoured by the particles. A recent review assessed the mechanical cleaning concepts in membrane filtration [27]. Wang et al. [204] concluded that particle fluidization is similar to cleaning via bubbling in terms of (i) the momentum of both bubbles and fluidized granular activated carbon (GAC) correlates more strongly with the critical flux, rather than to the velocities or concentrations; and (ii) an increase in energy input increases the critical flux. In contrast to bubbling, particle fluidization is different with respect to (i) the local critical flux decreasing instead of increasing with height; (ii) its optimization involves a complex interplay of particle size, concentration, and liquid flowrate, instead of simply involving increasing the gas flowrate.

As early as the 1970s, the beneficial impact of the fluidization of particles for membrane processes was recognized as being attributable to both the mixing action of the particles to reduce the solute concentration gradient, and the mechanical action of the particles to both vibrate and clean the membrane surface [236–239]. These studies predate the submerged HF. However, various types of inert solids subsequently have been shown to be beneficial for the mitigation of membrane fouling in submerged HF applications via the scouring mechanism [240–244], although negative effects such as the break-up of sludge flocs [240] and poor filterability of the activated sludge suspension [241] also have been noted. Another potential effect is membrane damage [238,242], which necessitates careful selection of scouring conditions.

The use of powdered activated carbon (PAC) is primarily targeted for mitigating organic accumulation, biological degradation, and reducing the cake-resistance (e.g., [245–253]), all of which contribute towards improving the permeate flux. Almost as an afterthought, PAC also was recognized as being beneficial for inducing fouling-mitigating shear on the membrane [254–257]. Because the increased inertia associated with larger-sized particles can lead to more effective scouring, granular activated carbon (GAC), whose mean diameter is an order-of-magnitude larger than that of PAC, has recently gained interest [24,130,257–264]. Although PAC is more effective than GAC in terms of adsorption capability [257], it has been claimed that GAC is more effective at the higher concentrations encountered in practice and in the longer term [24]. Several reports of the apparent success in the use of GAC in submerged HF membrane systems (namely, the fluidized-bed membrane reactor) for low-cost, sustainable operation have appeared in recent years; hence, a closer look is warranted. The first report was on the use of a two-stage AFBR-AFMBR (i.e., anaerobic fluidized-bed bioreactor-anaerobic fluidized-bed membrane bioreactor) for sustainable control of membrane fouling [24]. Extensive tests subsequently have been carried out [24,27,128,130,244,258–278], especially in view of the potentially lower energy cost than that of bubbling [24,204] and suitability for the anaerobic MBR. Effects of treating different types of wastewater [258,259,261,262,276] (e.g., using municipal versus synthetic wastewater [24,258]), trace organics [128,278], membrane type [273] (including effects on membrane integrity [271,272,275]), screen size [259], fluidized media [236,244,279,280] (including size and packing amount [244,267,268,271,275]), operating conditions [260,271] such as temperature [260–262], scale [261], design [130,261,264,265,276] (e.g., single (AFMBR) versus two-stage (AFBR-AFMBR) systems [130]), which collectively proved the efficacy of GAC in scouring the membranes. Different embodiments of the AFMBR include single (AFMBR) versus two-stage (AFBR-AFMBR) systems [130], as well as simplifications of the two-stage AFBR-AFMBR system termed an IAFMBR (i.e., integrated anaerobic fluidized-bed membrane bioreactor) [277], and hybrids such as the MFC (i.e., microbial fuel cell)-AFMBR [264] and the fluidized bed membrane bioelectrochemical reactor (MBER) [263]. A study on the extent of fouling mitigation by fluidized GAC in an HF module found that larger-sized GAC particles, higher packing densities, and a ratio of hollow fiber spacing to fluidized particle size of approximately 3–5 are beneficial for fouling control [271]. Collectively, these efforts prove the efficacy of particle fluidization in mitigating fouling in submerged HF systems.

The benefits conferred by particle fluidization include low energy cost [24,123,242,258,261,264,281], and amenability for scale-up and continuous operation [24,235], all of which make it an attractive means to improve membrane operations. In particular, the AFMBR energy requirement was only 0.028 kWh/m^3, which is much less than that reported for AMBRs using gas sparging [24]. The momentum, velocity, and concentration of the fluidized GAC particles have been found to play significant roles in membrane fouling mitigation via both experiments and simulation [266–269]. Note that membrane-particle interactions have to be managed to avoid membrane damage. Also, the module geometry must allow movement of the particles to avoid blockages, as highlighted in Section 4.

6. Techniques for Fouling Control in Dead-End Submerged Membrane Systems

As noted earlier, the submerged HF concept is widely used in the dead-end mode in the water-treatment industry. Strategies to mitigate fouling are required to avoid decline in the membrane permeability in dead-end submerged HF systems [6,282,283]. It is possible to minimize fouling both by choosing a suitable membrane material with a reduced tendency to adsorb substances in the feed and by optimizing the operating conditions in the system [5,161,284,285]. The application of relaxation (intermittent cessation of permeation), backwashing (reversal of permeate flow through the pores), and air backwashing with or without air scouring are common physical approaches to remove fouling in submerged systems. It has been shown in many studies that relaxation and backwashing provide an effective removal of the fouling layer, thereby prolonging the filtration process in submerged membrane systems, especially at high imposed fluxes [88,90,131,286,287]. A significant challenge in the application of relaxation and backwashing is that only partial recovery of the permeability is achieved at the end of a filtration cycle, which implies a gradual loss of the effective filtration area due to fouling. Subsequently, in the next filtration cycle the less fouled areas will have to experience increased local fluxes to maintain the overall average flux, which in turn results in a higher fouling rate [94]. However, Figure 6, which shows a plot of the TMP versus time for both continuous and periodic backwashing and filtration, indicates that even with a partial recovery during each cycle, filtration with intermittent backwashing and relaxation outperforms continuous operation. A further discussion of backwashing and relaxation is provided in the following sections.

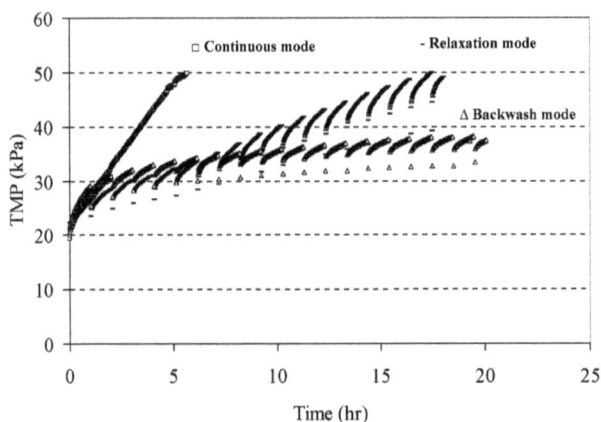

Figure 6. TMP versus time for continuous filtration relative to periodic relaxation and backwashing of real seawater [26]. Reproduced with permission from [26], Copyright Journal of Membrane Science, 2010.

6.1. Backwashing

Backwashing is commonly practiced in most HF filtration systems to limit fouling in both dead-end and cross-flow applications. Typical TMP profiles for dead-end filtration with intermittent backwashing for cake removal are shown in Figure 7. Two modes of operation are usually practiced, either (i) using a fixed cycle time (t_c), whereby backwashing is implemented after a designated filtration time, thereby causing the maximum TMP to increase with each cycle if residual fouling occurs (Figure 7a), or (ii) operating to achieve a fixed TMP_{max}, whereby backwashing is implemented whenever the TMP reaches a predetermined value, thereby requiring the frequency of backwashing to increase with each cycle if residual fouling occurs (Figure 7b). Backwashing is usually effective in reducing the TMP, but some deposits tend to remain attached and contribute an additional residual resistance to the filtration in subsequent cycles. Therefore, other than backwashing alone, cycling between backwashing and chemical cleaning (Section 8) is also a common practice to reduce the minimum TMP (TMP_{min}) attainable. Figure 8 illustrates a typical TMP profile with intermittent backwashing and chemical cleaning with a fixed cycle time.

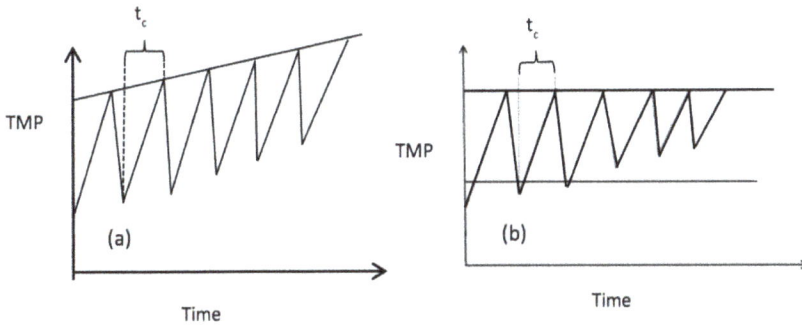

Figure 7. Typical TMP profiles with intermittent backwashing; (**a**) operation to achieve a fixed TMP_{max}; (**b**) operation for a fixed cycle time.

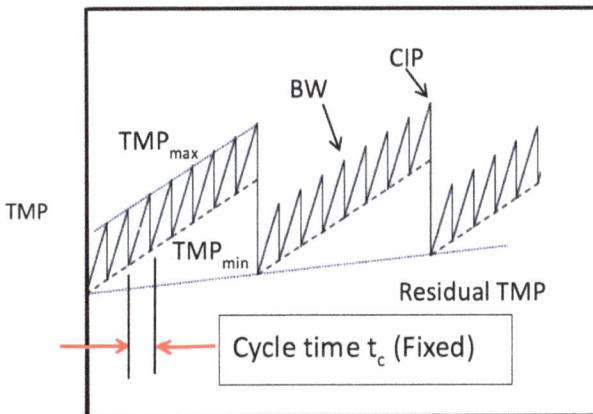

Figure 8. Typical profile of the TMP as a function of time showing the effect of intermittent backwashing and chemical cleaning for a fixed cycle time.

Although backwashing loosens and detaches the fouling cake from the membrane surface so that the foulants can be removed easily by cross-flow or air bubbles [6,25,90,93,288,289], some drawbacks also exist. In cases for which the cake layer serves as a secondary layer to protect the membrane from internal fouling by macromolecular components, over-frequent backwashing can provide more opportunity for macromolecules to enter the membrane pores [290] or change the chemical composition and/or structure of the fouling layer (e.g., from a mixed cake layer of particulates and macromolecules to a fouling structure dominated by the macromolecules after several filtration/cleaning cycles [291]) and consequently the fouling patterns [26]. Generally, the first few cycles of backwashing lead to more significant irreversible fouling, after which the percentage of irreversible fouling with respect to total fouling becomes constant. The reason for the augmented vulnerability to irreversible fouling of new membranes relative to a used membrane is the greater probability of blocking the larger pores in the pore-size distribution, which can be the dominant fouling mechanism in the first few cycles [26,90,91,292].

Overall, an increased backwashing flux was found to be slightly more effective than increased backwash duration when the same amount of backwash volume was used [26,91,93]. Similarly, Akhondi et al. [90] reported that excessive backwash duration and strength resulted in permeate loss, severe pore blocking, and high specific energy consumption. Ye et al. [26] investigated the effect of filtration duration (from 1200 to 5400 s per cycle) on membrane fouling during real seawater filtration while the other operating parameters were kept constant. It was found that the final TMP after 16 h of filtration and the percentage of reversible fouling that can be removed by backwashing did not increase when the filtration duration increased from 1200 to 3600 s, while a further increase in the filtration duration from 3600 to 5400 s promoted membrane fouling due to a more compact cake layer that was more irreversible.

Chua et al. [41] reported that, for a pilot-scale pressurized HF module, prolonging the duration of backwashing was found to be more effective than air scouring in controlling membrane plugging. Studies by Ye et al. [26] showed that increasing the backwash duration from 10 to 30 s led to the final TMP and fouling rate decreasing by more than 50% as well as a slight increase in the percentage of fouling removed by backwashing. However, a further increase in the duration beyond 30 s did not result in any additional improvement, but instead slightly reduced the percentage of fouling removed; this indicates that an excess backwash volume may cause membrane blockage or change the structure of the fouling cake due to impurities in the backwash flux.

Akhondi et al. [160] studied the effect of backwashing on the pore size of hollow fiber ultrafiltration membranes by using the evapoporometry [91,293,294] technique. They reported the following: (i) backwashing can enlarge the pores of a membrane with a greater effect on the larger pores for operation at the same TMP; (ii) pore enlargement due to backwashing was larger for amorphous (PVDF fibers) relative to glassy polymers (PAN fibers) due to the lower modulus-of-elasticity of the former; (iii) cyclic filtration and backwashing at constant flux could more effectively remove foulants both on and within the larger membrane pores compared to the small pores; and (iv) increasing the backwashing flux could remove foulants from smaller pores.

Results for seawater showed that the lowest final TMP and the maximum percentage of fouling removed by backwashing after 16 h of filtration was for the case for which the backwash flux was 1.5 times the filtration flux [26], which suggests the existence of an optimum backwash flux for fouling mitigation. The observation that a further increase in the backwash flux to twice that of the filtration flux led to an increase in the final TMP and a reduction of foulant removal implies that backwashing changes the fouling rate during the filtration cycle. Similar to excessive backwash duration, it seems that an excessive backwash flux also causes convection of impurities to the membrane pores or a residual fouling layer that results in less reversible fouling and a higher fouling rate. The existence of an optimum backwash flux for fouling mitigation was also reported by Chua et al. [41], who found that an increase in the backwash flowrate up to twice that of the permeate flowrate resulted in a process improvement, but no further benefits were observed for a further increase in the backwash flowrate. Compared to the duration or interval of backwashing, the effect of backwashing flux was found to be more significant for fouling mitigation [288].

It has been reported that air scouring during backwashing can assist fouling removal and improve backwash efficiency [289,295]. While the backwashing is expected to detach the cake layer from the fibers, air scouring loosens the deposits and carries them from the membrane surface into the bulk fluid [289,295]. The impact of aeration during backwashing on membrane fouling during seawater filtration was investigated by Ye at al. [26]. Their results showed that backwashing with a moderate air flowrate had a lower final TMP and also slowed down the fouling rate during the filtration. However, a high air flowrate limited the benefits of air scouring and did not improve the reversibility.

6.2. Relaxation

Relaxation, the intermittent cessation of permeation, has been incorporated in many membrane bioreactor (MBR) designs and some other submerged HF systems as a standard operating protocol to control membrane fouling [93,94,288]. For example, the investigation of an MBR found that relaxation was still beneficial even when the relaxation necessitated periods of higher flux to give the same production of permeate [288]. Different relaxation conditions resulted in distinctly different temporal TMP profiles, but all the runs that incorporated relaxation displayed a lower final TMP than the continuous mode. Relaxation was found to be more favorable than backwashing for this MBR application, because, while performances were similar, backwashing may have resulted in membrane pore clogging [288].

The protocol for intermittent filtration/relaxation can be optimized in terms of the ratio of the durations in each cycle (ratios between 0.5–50 were tested) to be more beneficial for fouling removal [94,296], but were not necessarily beneficial for retarding the TMP increase as filtration progressed. This suggests that the relaxation duration and interval should be carefully managed to achieve the best outcome in terms of reducing the fouling resistance during relaxation and retarding the TMP increase during filtration.

Ye et al. [26] investigated the effect of relaxation on the performance of dead-end (i.e., without bubbling) filtration using membranes with two different porosities and a seawater feed. Relaxation was confirmed to limit membrane fouling compared to continuous filtration, but was more effective for the membrane with a higher porosity. The difference in results due to porosity was hypothesized to be that, although the relaxation removed part of the foulant cake for membranes with lower porosities, when the filtration flux resumed the cake was reorganized to a more compact structure. Table 2 lists fouling control methods for submerged HF systems.

Table 2. Fouling control methods in submerged HF systems [18,297–299].

Operation Mode	Fouling Control Technique	Important Parameters	Applications	Benefits
Cross-flow	Bubbling	Bubble characteristics, gas flowrate, bubbling modes (intermittent or continuous)	MBR, AnMBR	Unsteady or transient shear stress; changes biomass properties
	Vibration	Vibration amplitude and frequency	AnMBR, MF, UF, MD, FO	Low energy cost; surface shear; effective cake removal; facilitates separation of macromolecules
	Particle Scouring	Size, fluidization rate	AFBR-AFMBR, IAFMBR	Reduced fouling; low energy cost; amenability for scale-up (disadvantages: membrane damage, blockage)
Dead-end	Backwashing	Backwash flux, backwash duration, backwash frequency	All HF systems	Internal fouling control; can be applied with air scouring
	Relaxation	Relaxation duration, relaxation frequency	All HF systems, especially MBR	-

AnMBR: anaerobic membrane bioreactor; MF: microfiltration; UF: ultrafiltration; MD: membrane distillation; FO: forward osmosis; AFBR-AFMBR: anaerobic fluidized-bed bioreactor-anaerobic fluidized-bed membrane bioreactor; IAFMBR: integrated anaerobic fluidized-bed membrane bioreactor.

7. Chemical Cleaning in Submerged HF Membranes—Procedure, Effect on Membrane Performance

Other than the physical cleaning means discussed in the previous section, chemical cleaning, which involves the use of acids, bases, oxidants, and surfactants, also aids in mitigating membrane fouling. Typically, physical cleaning is followed by chemical cleaning in membrane applications to effectively mitigate fouling [283]. Chemical cleaning was classified by Lin et al. [283] into four categories: (i) clean-in-place (CIP), which involves directly adding chemicals to the submerged HF system; (ii) clean-out-off-place (COP), which involves cleaning the membrane in a separate tank with a higher concentration of chemicals; (iii) chemical washing (CW), which involves adding chemicals to the feed stream; and (iv) chemically enhanced backwashing (CEB), which involves combining chemical and physical cleaning means.

For chemical cleaning, the key factors affecting efficiency in mitigating fouling are the type of chemical agents, cleaning duration and interval, concentration of chemicals, cleaning temperature, and flux [300,301]. The type of chemical used depends mainly on the application, feed characteristics (e.g., pH, ionic strength, and temperature), and membrane materials (e.g., compatibility of the membrane with the chemicals) [302]. Sodium hypochlorite (NaOCl) and citric acid are the most common chemical cleaning agents provided by the main MBR suppliers [18], although they are reported to be ineffective for removing iron species [303], and less effective than the coupling of NaOCl and caustic soda for removing natural organic matter (NOM) [304].

While the robust nature of most submerged HF membranes allows the use of relatively aggressive cleaning, some changes may occur. Kweon et al. [304] evaluated the effectiveness and changes in the membrane surface properties by acidic and alkali cleaning of PVDF HF membranes during the microfiltration of two feed waters. The results indicated that the feed-water quality played an important role in the cleaning efficiency; hence, experiments with the actual feed are necessary for the selection of cleaning procedures. In addition, chemical cleaning leads to changes in the surface properties of the membranes, which may lead to a gradual decrease in the recoverable flux.

8. Submerged HF Membrane Integrity and Failure

Given the chemical and physical stresses experienced by submerged HF membranes during operation, the lifespan of the membrane fibers tends to be significantly shortened. The prorated warranty provided by membrane manufacturers can range from 3 to 10 years [305]; however, experience teaches that the effective membrane life can either exceed or fall short of the manufacturer's expectations. In one particular case, a UF plant treating wastewater from a manufacturer of cosmetics experienced rapid membrane failure resulting in an average membrane lifespan of less than 6 months (note that the membranes were cleaned once a week with an alkaline bleach product, and backwashed monthly) [306], and the main cause was found to be high local shear forces due to fibrous material in the wastewater. In another study performed by De Wilde et al., the lifespan of the membranes was determined to be 13 years by extrapolating data based on 3 years of operation [307].

Given that the paramount operating objective is to avoid any failure that could compromise quality and restrict capacity, the development of a strategy that relies exclusively on the manufacturer's warranty to estimate the membrane lifespan and replacement schedules is fraught with uncertainty. With the prevalent variability in the integrity and productivity of membrane modules, operators of full-scale plants would need to manage an inventory of several thousand membranes; thus, anticipating and scheduling activities for the replacement of these membranes in service becomes a unique challenge for drinking-water plants utilizing membrane technology.

Even though membrane ageing and failure are closely related, a distinction should be made between these two factors. Membrane degradation is the result of ageing and the onset of its adverse effects, which in turn leads to membrane failure. Membrane failure, on the other hand, results in a loss of process removal efficiency, and a reduction in product-water throughput as well as product-water non-compliancy.

8.1. Ageing

Membrane ageing of commonly used composite membranes is defined as the deterioration of the surface layer and sub-layers of membranes due to the irreversible deposition of foulants or by frequent exposure to chemical cleaning agents, which leads to the deterioration of the membrane performance [308,309]. While the active layer of the membrane has been found to be chemically modified, pore reduction has been found in the intermediate sublayer (i.e., between the active layer and the porous support) [308]. Membrane material also affects ageing; for example, polyethylsulfone (PES) membranes were found to be more resistant to acid than alkali [310].

To control membrane fouling caused by the retention of dissolved salts, organics, microorganisms, and suspended solids after extended operation, the industry employs routine chemical cleaning protocols involving specific concentrations, temperatures, and extended cleaning times; in some cases, with submerged HFs, strong oxidants such as sodium hypochlorite (NaOCl) are used to control fouling, causing membrane ageing to be exacerbated after repeated cleanings. Prolonged filtration and cleaning cycles not only have an adverse effect on membrane integrity, but can also lead to the internal fouling of membranes, which is irreversible, detrimental to membrane performance, and also reduces the lifespan of the membrane by increasing the likelihood of membrane failure. For submerged HFs, physical cleaning by backwashing (Section 6.1) can also cause changes in the membrane properties. For example, it has been shown that backwashing can cause a change in the pore-size distribution by increasing the diameters of the largest pores [311]. This could make the membranes more susceptible to fouling as these larger pores become blocked.

8.2. Failure

Membrane failure is defined as the loss of mechanical integrity leading to the inability to achieve the rated log-removal values (LRV) of pathogens [312–314]. Membrane failure can occur during two phases of the operational lifespan of a membrane, namely, damage during the manufacturing and installation process, and during membrane filtration. Inconsistent manufacturing and fabrication techniques as well as handling error during installation often cause failure in the former phase. This issue is kept in check via the implementation of rigorous product quality-control methods and integrity testing of membrane modules before commissioning. On the other hand, unlike failure during the manufacturing and installation process, membrane failure during the filtration operation can mainly be attributed to operating parameters and maintenance protocols [315]. During operation, the likelihood of damage to the membrane is high given the stringent nature of operating protocols such as vigorous mechanical cleaning, chemical cleaning using strong oxidants, and high-pressure backwashing. Although these measures ensure that the membrane performance is maintained, they indirectly put a strain on the membrane integrity, leading to membrane ageing and failure.

According to Childress et al. [313], fiber failure can occur via four different mechanisms: chemical attack; damage during operation due to improper installation; faulty membrane module design; and punctures and scores due to the presence of foreign bodies. Furthermore, membrane ageing, coupled with excessive fiber movement due to external loads, can also cause submerged HF membranes to fail. This is discussed further in Sections 8.4 and 8.5.

8.3. Chemical Oxidation

Owing to membrane fouling being an inevitable phenomenon, membrane maintenance protocols using chemical cleaning to control fouling and restore the membrane flux are employed. There is a wide variety of chemical cleaning agents utilized by the industry, with the most common being sodium hypochlorite (NaOCl) because of its ready availability, relatively low price, and high cleaning efficiency. Unfortunately, such oxidants are the main causes of deterioration in the membrane integrity [316], whereby prolonged exposure causes oxidative damage to the membrane [317], which accelerates membrane ageing and degradation that in turn not only leads to discoloration of the membrane fibers, but also embrittlement of the fibers that subsequently increases the likelihood of membrane-fiber fracture [316,318,319]. The embrittlement rate for hollow fibers has been found to be four times that of flat-sheet membranes [309].

8.4. Module Design

The optimization of membrane-module designs in terms of the potting of the membrane fibers and the design of the membrane housing is usually performed to reduce membrane fouling and to maintain membrane integrity. In submerged HF modules, the fibers are located in a constrained geometry. For example, the GE-Zenon system has membranes assembled into cartridges and held in a supporting frame that connects to the aeration and permeate suction header. The ends of the fibers are potted into the permeate carrier. These features are typical of most submerged HF modules. More details can be found elsewhere [3,58].

The potting efficacy of the fibers can significantly affect the performance and integrity of the module. Current membrane modules consist of up to 20,000 hollow fibers held together with either an epoxy or urethane resin; depending on the manufacturing process, the resin can be cured under static or dynamic conditions. In the slower static curing method the resin is allowed to cure without heat or external forces acting on it. Membranes that are potted statically would have the resin wick up the edge of the fiber due to capillary forces, which leads to the development of a sharp edge that potentially can cause fiber breakage (see Figure 9a). As seen in Figure 9b, with this method an elastomer overlay is usually added on top of the potting material to minimize the sharp edges. On the other hand, for the dynamic curing method, centrifugal forces are used under elevated temperatures to cure the potting resin, thereby preventing resin wicking and avoiding the development of sharp edges [313] (see Figure 9c). Notably, with the implementation of air-scouring and external loads, the probability of fiber breakage at the potting site increases; thus, proper selection of the potting method and material is important. Moreover, an optimized arrangement of the fibers within the module has been reported to significantly improve process performance by as much as 200% [119,190].

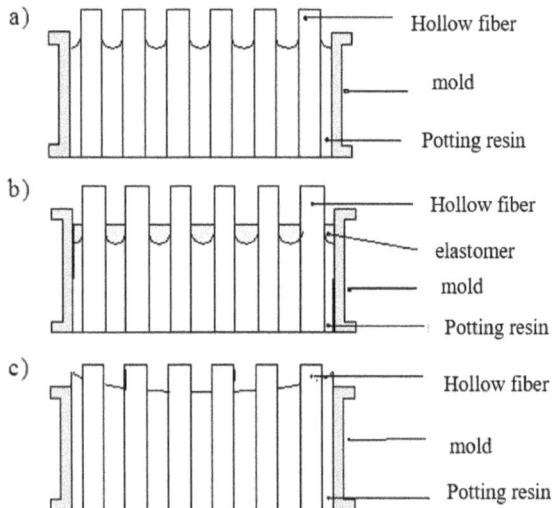

Figure 9. Membrane potting methods employing different conditions: (**a**) Static Conditions, (**b**) Static with Elastomer Overlay, (**c**) Dynamic Conditions.

The design of the membrane module and housing plays an equally important role in maintaining membrane integrity. Although there are many membrane housing designs that try to minimize excessive fiber movement, routine membrane filtration and backwashing carried out at pressures higher than the manufacturer's recommendations can lead to membrane failure via the rupturing of membrane fibers, damage to the membrane module housing, and degradation of the membrane module seals. Consequently, the tolerable stress load on a membrane module depends on the membrane material and structure, and the packing density of the HF membranes in the module. Pilot plant testing indicated that membrane symmetry, which affects the stress at the juncture between the potting material and the fiber, was more important than the potting technique for hollow fiber integrity [313]. At present there is a lack of fundamental data on the stresses experienced by fibers during the filtration and cleaning cycles in the presence of air scouring [320]. This, coupled with the difficulty in accurately measuring and calculating stresses in a multiple fiber system, has significantly limited the development of improved membrane modules. Therefore, for improved membrane-module, design a need exists for both a better understanding of membrane potting methods and the stress-strain forces acting on the membrane fibers.

8.5. Excessive Fiber Movement

The advantage of submerged membranes is that the hydrostatic pressure generated eliminates the need for the membrane modules to be pressurized. For such configurations, air-scouring or bubbling is employed to provide a shear force along the membrane surface to help alleviate the fouling phenomenon [14]. A higher shear force on the membrane surface results in a more efficient removal of foulants. In addition, specifically for bubbly flow around hollow fibers, another mechanism at play is the back and forth movement of the fibers induced by the bubbles, which causes a transverse vibration for loose fibers that leads to enhanced secondary mixing [15]. However, although the bubble-induced shear and fiber movements were able to reduce the fouling rate by up to 10-fold less, the excessive membrane movement due to a higher shear force can also lead to fiber breakage [15]. This phenomenon, coupled with the degradation of the membrane fibers due to ageing, could lead to a higher occurrence of fiber failure. Excessive fiber movement is also constrained if the fiber looseness is limited, for which the practical limits are typically 1–5% [3].

8.6. Foreign Bodies

Membrane damage and integrity compromise also can be caused by unexpected water-quality fluctuations together with the failure of the pretreatment processes, leading to the inadequate removal of foreign material [314]. These foreign bodies, coupled with the effects of strong aeration, can score or puncture the membrane fibers. A membrane autopsy performed by Zappia et al. concluded that the unexpected presence of silicon dioxide spicules (needle-like structures) resulted in multiple membrane occlusions and punctures, leading to a loss in membrane integrity [321]. The erosive effect of fluidized particles [322–325] in the feed stream is also known to compromise the membrane integrity by impacting the membrane surface [123,238,242,326–329]. Patterns of particle scraping was clearly observed [238,279], as well as a decrease in the membrane rejection for the larger fluidized glass beads (3 mm) [242]. As a result, care is needed for submerged HF processes that deliberately introduce scouring by suspended or fluidized media (Section 5.3.2) to control fouling.

8.7. Future Trends for Integrity Assessment

The current state-of-the-art tools for monitoring membrane integrity are limited to detecting compromises via a variety of in situ and ex situ techniques and tools. Despite extensive research performed on membrane-failure mechanisms and their resultant effects, these studies often are based on ex situ or offline analytical techniques, which can only provide information when a serious breach in membrane integrity is detected. Therefore, this underscores the need for the development of non-destructive, computer-aided modeling techniques to predict membrane failure and optimize module design. One possible approach is the prediction of failure in membrane systems via the use of finite element analysis (FEA). FEA is a modeling technique that is widely used by structural, mechanical, and biomedical engineers to perform mechanical analyses on complex structures to determine displacements from applied loads [330,331]. Through FEA, high stress locations along the membrane fiber can be determined that will help identify areas where failure is most likely to occur. Such analyses would aid in membrane and membrane-module design as well as the optimization of operating strategies. FEA can also be used as a diagnostic tool to provide supportive interpretations at similar operating conditions when performing autopsies on failed membranes and modules.

9. Conclusions and Research Opportunities

In a little over two decades, submerged HFs have gone from a curiosity to the mainstream of membrane technology. The major applications are for low-pressure membranes (MF and UF) in the water industry. The submerged HF concept is ideally suited for the dead-end filtration of dilute feeds (surface waters, pretreatment for RO) because effective cake removal can be achieved by backwash flow from the lumen side of the fiber. The submerged HF concept is also well-suited for more concentrated feeds, such as in aerobic MBR, where bubble-induced fiber movement helps to control fouling. The development of the submerged HF system has seen some advances in membranes per se (e.g., improved strength, flexibility, etc.) but the major efforts have been in module design and process optimization, such as fiber geometry, looseness, packing density, bubbling characteristics, backwash protocol, and modifying feed properties (e.g., by adjusting the bioprocess parameters in the MBR). The current generation of submerged HFs is clearly very effective, but there are opportunities for further development that could improve the concept and its applications. These are briefly discussed in the following section.

9.1. Hydrodynamics and Bubbling

Although hydrodynamics and bubbling in submerged HF systems have been actively studied, there may be opportunities for further incremental improvement. Areas for research include identifying the optimal bubble size (somewhere between relatively few large bubbles and many micro-bubbles) and the means for generating these bubble sizes. Also, the potential energy benefits of intermittent bubbling should be further evaluated.

9.2. Non-Bubbled Hydrodynamics

This review has provided an overview of the research activities in vibrations applied to submerged HFs. Further development is required to optimize this strategy in terms of module arrangement (alignment, packing density, use of flow promoters, etc.), vibration frequency and amplitude, etc. An important driver for development is likely to be the anaerobic MBR (AnMBR), where bubbling by recycled biogas has its challenges. Fluidized media to control hydrodynamics may also find application in the AnMBR. Further studies are required to evaluate and compare energy demand for these alternative hydrodynamic control methods.

9.3. Backwashing and Relaxation

Backwashing and relaxation are effective methods for fouling mitigation. However, the chosen protocols are likely to be conservative and suboptimal. Improved performance could involve an adjustable protocol (backwash frequency, duration, flux, etc.) that responds to changes to feed conditions and required production rate. Such a system could use an online monitor for 'fouling propensity' linked to a neural-network-based control system.

9.4. Identifying Sustainable Flux

The concept of critical flux is discussed in Section 5. Due to the limited practical applicability of critical flux, threshold and sustainable flux have been proposed [158,332]. While the threshold flux demarcates a low fouling from a high fouling region, the sustainable flux is one at which moderate fouling is tolerated based on balancing capital and operating costs; some guidelines are available that link measured threshold flux to sustainable flux [158]. However, in practice, it is difficult to measure the threshold flux in an operating plant as it involves flux-stepping involving conditions of high fouling rate that could be detrimental. This could be overcome by the development of a small-scale threshold-flux monitor that simulates the operating plant that could be flux-stepped. The challenge is to design a sufficiently accurate and reliable simulator based on the submerged HF concept.

9.5. Potential Non-Filtration Applications

The submerged HF concept is widely used in membrane filtration (MF and UF) applications. The concept is also amenable to other membrane separations, such as membrane distillation (MD) and forward osmosis (FO). Indeed 'high-retention' MBRs based on MD and FO with submerged HFs have been developed [333,334]. Further work is required to optimize these systems, including development of externally skinned FO hollow fiber membranes to minimize fouling. An interesting extension of FO is pressure-retarded osmosis (PRO), used for harnessing the salinity gradient for the generation of electricity, that can use hollow fibers with pressurized 'draw solute' on the lumen side [335,336]. It would be feasible to use the submerged HF arrangement for PRO with the low salinity feed in the tank.

9.6. Membrane Integrity

The strategies to improve submerged HF membrane integrity have been addressed in Section 9. An additional need is an effective online monitor to detect the loss of integrity. One example of such a device is given in Krantz et al. [337]. The key requirements are reliability, easy implementation, and modest cost for industrial applications.

Acknowledgments: We gratefully acknowledge funding from the Singapore Ministry of Education Academic Research Funds Tier 2 (MOE2014-T2-2-074; ARC16/15) and Tier 1 (2015-T1-001-023; RG7/15), and the Joint Singapore-Germany Research Project Fund (SGP-PROG3-019). We acknowledge support from the Singapore Economic Development Board to the Singapore Membrane Technology Centre.

Conflicts of Interest: The authors declare no conflict of interest.

References

1. Laîné, J.M.; Vial, D.; Moulart, P. Status after 10 years of operation—Overview of uf technology today. *Desalination* **2000**, *131*, 17–25. [CrossRef]
2. Choi, Y.-J.; Oh, H.; Lee, S.; Nam, S.-H.; Hwang, T.-M. Investigation of the filtration characteristics of pilot-scale hollow fiber submerged mf system using cake formation model and artificial neural networks model. *Desalination* **2012**, *297*, 20–29. [CrossRef]
3. Fane, A.G. Submerged membranes. In *Advanced Membrane Technology and Applications*; Li, N.N., Fane, A.G., Ho, W.S.W., Matsuura, T., Eds.; John Wiley & Sons: Hoboken, NJ, USA, 2008; Chapter 10.
4. Ye, Y.; Le Clech, P.; Chen, V.; Fane, A.G.; Jefferson, B. Fouling mechanisms of alginate solutions as model extracellular polymeric substances. *Desalination* **2005**, *175*, 7–20. [CrossRef]
5. Nakatsuka, S.; Nakate, I.; Miyano, T. Drinking water treatment by using ultrafiltration hollow fiber membranes. *Desalination* **1996**, *106*, 55–61. [CrossRef]
6. Günther, J.; Schmitz, P.; Albasi, C.; Lafforgue, C. A numerical approach to study the impact of packing density on fluid flow distribution in hollow fiber module. *J. Membr. Sci.* **2010**, *348*, 277–286. [CrossRef]
7. Di Profio, G.; Ji, X.; Curcio, E.; Drioli, E. Submerged hollow fiber ultrafiltration as seawater pretreatment in the logic of integrated membrane desalination systems. *Desalination* **2011**, *269*, 128–135. [CrossRef]
8. Chellam, S.; Jacangelo, J.G.; Bonacquisti, T.P. Modeling and experimental verification of pilot-scale hollow fiber, direct flow microfiltration with periodic backwashing. *Environ. Sci. Technol.* **1998**, *32*, 75–81. [CrossRef]
9. Raffin, M.; Germain, E.; Judd, S.J. Influence of backwashing, flux and temperature on microfiltration for wastewater reuse. *Sep. Purif. Technol.* **2012**, *96*, 147–153. [CrossRef]
10. Aimar, P. Slow colloidal aggregation and membrane fouling. *J. Membr. Sci.* **2010**, *360*, 70–76. [CrossRef]

11. Tang, C.Y.; Chong, T.H.; Fane, A.G. Colloidal interactions and fouling of nf and ro membranes: A review. *Adv. Colloid Interface Sci.* **2011**, *164*, 126–143. [CrossRef] [PubMed]

12. Fane, A.G.; Chong, T.H.; Le-Clech, P. Fouling in membrane processes. In *Membrane Operations, Innovative Separations and Transformations*; Drioli, E., Giorno, L., Eds.; Wiley-VCH: Weinheim, Germany, 2009; Chapter 6.

13. Vigneswaran, S.; Kwon, D.Y.; Ngo, H.H.; Hu, J.Y. Improvement of microfiltration performance in water treatment: Is critical flux a viable solution? *Water Sci. Technol.* **2000**, *41*, 309–315.

14. Cui, Z.F.; Chang, S.; Fane, A.G. The use of gas bubbling to enhance membrane processes. *J. Membr. Sci.* **2003**, *221*, 1–35. [CrossRef]

15. Wicaksana, F.; Fane, A.G.; Chen, V. Fibre movement induced by bubbling using submerged hollow fibre membranes. *J. Membr. Sci.* **2006**, *271*, 186–195. [CrossRef]

16. Chang, S.; Fane, A.G. Filtration of biomass with laboratory-scale submerged hollow fibre modules—Effect of operating conditions and module configuration. *J. Chem. Technol. Biotechnol.* **2002**, *77*, 1030–1038. [CrossRef]

17. Gomaa, H.G.; Rao, S. Analysis of flux enhancement at oscillating flat surface membranes. *J. Membr. Sci.* **2011**, *374*, 59–66. [CrossRef]

18. Le-Clech, P.; Chen, V.; Fane, T.A.G. Fouling in membrane bioreactors used in wastewater treatment. *J. Membr. Sci.* **2006**, *284*, 17–53. [CrossRef]

19. Radjenović, J.; Matošić, M.; Mijatović, I.; Petrović, M.; Barceló, D. Membrane bioreactor (MBR) as an advanced wastewater treatment technology. In *Emerging Contaminants from Industrial and Municipal Waste*; Springer: Berlin/Heidelberg, Germany, 2008; pp. 37–101.

20. Le-Clech, P. Membrane bioreactors and their uses in wastewater treatments. *Appl. Microbiol. Biotechnol.* **2010**, *88*, 1253–1260. [CrossRef] [PubMed]

21. Lu, Y.; Ding, Z.; Liu, L.; Wang, Z.; Ma, R. The influence of bubble characteristics on the performance of submerged hollow fiber membrane module used in microfiltration. *Sep. Purif. Technol.* **2008**, *61*, 89–95. [CrossRef]

22. Tian, J.-Y.; Xu, Y.-P.; Chen, Z.-L.; Nan, J.; Li, G.-B. Air bubbling for alleviating membrane fouling of immersed hollow-fiber membrane for ultrafiltration of river water. *Desalination* **2010**, *260*, 225–230. [CrossRef]

23. Li, T.; Law, A.W.-K.; Cetin, M.; Fane, A.G. Fouling control of submerged hollow fibre membranes by vibrations. *J. Membr. Sci.* **2013**, *427*, 230–239. [CrossRef]

24. Kim, J.; Kim, K.; Ye, H.; Lee, E.; Shin, C.; McCarty, P.L.; Bae, J. Anaerobic fluidized bed membrane bioreactor for wastewater treatment. *Environ. Sci. Technol.* **2011**, *45*, 576–581. [CrossRef] [PubMed]

25. Hilal, N.; Ogunbiyi, O.O.; Miles, N.J.; Nigmatullin, R. Methods employed for control of fouling in mf and uf membranes: A comprehensive review. *Sep. Sci. Technol.* **2005**, *40*, 1957–2005. [CrossRef]

26. Ye, Y.; Sim, L.N.; Herulah, B.; Chen, V.; Fane, A.G. Effects of operating conditions on submerged hollow fibre membrane systems used as pre-treatment for seawater reverse osmosis. *J. Membr. Sci.* **2010**, *365*, 78–88. [CrossRef]

27. Aslam, M.; Charfi, A.; Lesage, G.; Heran, M.; Kim, J. Membrane bioreactors for wastewater treatment: A review of mechanical cleaning by scouring agents to control membrane fouling. *Chem. Eng. J.* **2017**, *307*, 897–913. [CrossRef]

28. Krzeminski, P.; Leverette, L.; Malamis, S.; Katsou, E. Membrane bioreactors—A review on recent developments in energy reduction, fouling control, novel configurations, lca and market prospects. *J. Membr. Sci.* **2017**, *527*, 207–227. [CrossRef]

29. Arnal, J.M.; Garcia-Fayos, B.; Verdu, G.; Lora, J. Ultrafiltration as an alternative membrane technology to obtain safe drinking water from surface water: 10 years of experience on the scope of the aquapot project. *Desalination* **2009**, *248*, 34–41. [CrossRef]

30. Madaeni, S.S. The application of membrane technology for water disinfection. *Water Res.* **1999**, *33*, 301–308. [CrossRef]

31. Ray, C.; Jain, R. *Low Cost Emergency Water Purification Technologies: Integrated Water Security Series*; Butterworth-Heinemann: Oxford, UK, 2014; pp. 1–205.

32. Loo, S.L.; Fane, A.G.; Krantz, W.B.; Lim, T.T. Emergency water supply: A review of potential technologies and selection criteria. *Water Res.* **2012**, *46*, 3125–3151. [CrossRef] [PubMed]

33. Iannelli, R.; Ripari, S.; Casini, B.; Buzzigoli, A.; Privitera, G.; Verani, M.; Carducci, A. Feasibility assessment of surface water disinfection by ultrafiltration. *Water Sci. Technol. Water Supply* **2014**, *14*, 522–531. [CrossRef]

34. Carvajal, G.; Branch, A.; Sisson, S.A.; Roser, D.J.; van den Akker, B.; Monis, P.; Reeve, P.; Keegan, A.; Regel, R.; Khan, S.J. Virus removal by ultrafiltration: Understanding long-term performance change by application of bayesian analysis. *Water Res.* **2017**, *122*, 269–279. [CrossRef] [PubMed]

35. Zularisam, A.W.; Ismail, A.F.; Sakinah, M. Application and challenges of membrane in surface water treatment. *J. Appl. Sci.* **2010**, *10*, 380–390. [CrossRef]

36. Zularisam, A.W.; Ismail, A.F.; Salim, R. Behaviours of natural organic matter in membrane filtration for surface water treatment—A review. *Desalination* **2006**, *194*, 211–231. [CrossRef]

37. Lin, T.; Shen, B.; Chen, W.; Zhang, X. Interaction mechanisms associated with organic colloid fouling of ultrafiltration membrane in a drinking water treatment system. *Desalination* **2014**, *332*, 100–108. [CrossRef]

38. Guastalli, A.R.; Simon, F.X.; Penru, Y.; de Kerchove, A.; Llorens, J.; Baig, S. Comparison of dmf and uf pre-treatments for particulate material and dissolved organic matter removal in swro desalination. *Desalination* **2013**, *322*, 144–150. [CrossRef]

39. Jamaly, S.; Darwish, N.N.; Ahmed, I.; Hasan, S.W. A short review on reverse osmosis pretreatment technologies. *Desalination* **2014**, *354*, 30–38. [CrossRef]

40. Jeong, S.; Park, Y.; Lee, S.; Kim, J.; Lee, K.; Lee, J.; Chon, H.T. Pre-treatment of swro pilot plant for desalination using submerged mf membrane process: Trouble shooting and optimization. *Desalination* **2011**, *279*, 86–95. [CrossRef]

41. Chua, K.T.; Hawlader, M.N.A.; Malek, A. Pretreatment of seawater: Results of pilot trials in singapore. *Desalination* **2003**, *159*, 225–243. [CrossRef]

42. Wu, B.; Fane, A.G. Microbial relevant fouling in membrane bioreactors: Influencing factors, characterization, and fouling control. *Membranes* **2012**, *2*, 565–584. [CrossRef] [PubMed]

43. Mo, L.; Huanga, X. Fouling characteristics and cleaning strategies in a coagulation-microfiltration combination process for water purification. *Desalination* **2003**, *159*, 1–9. [CrossRef]

44. Huang, H.; Schwab, K.; Jacangelo, J.G. Pretreatment for low pressure membranes in water treatment: A review. *Environ. Sci. Technol.* **2009**, *43*, 3011–3019. [CrossRef] [PubMed]

45. Huang, H.; O'Melia, C.R. Direct-flow microfiltration of aquasols: II. On the role of colloidal natural organic matter. *J. Membr. Sci.* **2008**, *325*, 903–913. [CrossRef]

46. Zhang, M.; Li, C.; Benjamin, M.M.; Chang, Y. Fouling and natural organic matter removal in adsorbent/membrane systems for drinking water treatment. *Environ. Sci. Technol.* **2003**, *37*, 1663–1669. [CrossRef] [PubMed]

47. Jones, K.L.; O'Melia, C.R. Protein and humic acid adsorption onto hydrophilic membrane surfaces: Effects of ph and ionic strength. *J. Membr. Sci.* **2000**, *165*, 31–46. [CrossRef]

48. Lorain, O.; Hersant, B.; Persin, F.; Grasmick, A.; Brunard, N.; Espenan, J.M. Ultrafiltration membrane pre-treatment benefits for reverse osmosis process in seawater desalting. Quantification in terms of capital investment cost and operating cost reduction. *Desalination* **2007**, *203*, 277–285. [CrossRef]

49. Pearce, G.; Talo, S.; Chida, K.; Basha, A.; Gulamhusein, A. Pretreatment options for large scale swro plants: Case studies of of trials at kindasa, saudi arabia, and conventional pretreatment in spain. *Desalination* **2004**, *167*, 175–189. [CrossRef]

50. Pearce, G.K. The case for UF/MF pretreatment to RO in seawater applications. *Desalination* **2007**, *203*, 286–295. [CrossRef]

51. Pearce, G.K. UF/MF pre-treatment to RO in seawater and wastewater reuse applications: A comparison of energy costs. *Desalination* **2008**, *222*, 66–73. [CrossRef]

52. Brehant, A.; Bonnelye, V.; Perez, M. Comparison of MF/UF pretreatment with conventional filtration prior to RO membranes for surface seawater desalination. *Desalination* **2002**, *144*, 353–360. [CrossRef]

53. Bonnélye, V.; Guey, L.; Del Castillo, J. UF/MF as RO pre-treatment: The real benefit. *Desalination* **2008**, *222*, 59–65. [CrossRef]

54. Teng, C.K.; Hawlader, M.N.A.; Malek, A. An experiment with different pretreatment methods. *Desalination* **2003**, *156*, 51–58. [CrossRef]

55. Sohn, J.; Valavala, R.; Han, J.; Her, N.; Yoon, Y. Pretreatment in reverse osmosis seawater desalination: A short review. *Environ. Eng. Res.* **2011**, *16*, 205–212.

56. Akhondi, E.; Wu, B.; Sun, S.; Marxer, B.; Lim, W.; Gu, J.; Liu, L.; Burkhardt, M.; McDougald, D.; Pronk, W.; et al. Gravity-driven membrane filtration as pretreatment for seawater reverse osmosis: Linking biofouling layer morphology with flux stabilization. *Water Res.* **2015**, *70*, 158–173. [CrossRef] [PubMed]

57. Wu, B.; Hochstrasser, F.; Akhondi, E.; Ambauen, N.; Tschirren, L.; Burkhardt, M.; Fane, A.G.; Pronk, W. Optimization of gravity-driven membrane (GDM) filtration process for seawater pretreatment. *Water Res.* **2016**, *93*, 133–140. [CrossRef] [PubMed]

58. Judd, S.; Judd, C. *The MBR Book: Principles and Applications of Membrane Bioreactors for Water and Wastewater Treatment*, 1st ed.; Elsevier: Amsterdam, The Netherlands; Boston, MA, USA; London, UK, 2006; pp. 207–272.

59. Santos, A.; Ma, W.; Judd, S.J. Membrane bioreactors: Two decades of research and implementation. *Desalination* **2011**, *273*, 148–154. [CrossRef]

60. Zhang, Q.; Singh, S.; Stuckey, D.C. Fouling reduction using adsorbents/flocculants in a submerged anaerobic membrane bioreactor. *Bioresour. Technol.* **2017**, *239*, 226–235. [CrossRef] [PubMed]

61. Svojitka, J.; Dvořák, L.; Studer, M.; Straub, J.O.; Frömelt, H.; Wintgens, T. Performance of an anaerobic membrane bioreactor for pharmaceutical wastewater treatment. *Bioresour. Technol.* **2017**, *229*, 180–189. [CrossRef] [PubMed]

62. Chen, C.; Guo, W.; Ngo, H.H.; Chang, S.W.; Duc Nguyen, D.; Dan Nguyen, P.; Bui, X.T.; Wu, Y. Impact of reactor configurations on the performance of a granular anaerobic membrane bioreactor for municipal wastewater treatment. *Int. Biodeterior. Biodegrad.* **2017**, *121*, 131–138. [CrossRef]

63. Ozgun, H.; Dereli, R.K.; Ersahin, M.E.; Kinaci, C.; Spanjers, H.; van Lier, J.B. A review of anaerobic membrane bioreactors for municipal wastewater treatment: Integration options, limitations and expectations. *Sep. Purif. Technol.* **2013**, *118*, 89–104. [CrossRef]

64. Johir, M.A.H.; George, J.; Vigneswaran, S.; Kandasamy, J.; Sathasivan, A.; Grasmick, A. Effect of imposed flux on fouling behavior in high rate membrane bioreactor. *Bioresour. Technol.* **2012**, *122*, 42–49. [CrossRef] [PubMed]

65. Monsalvo, V.M.; McDonald, J.A.; Khan, S.J.; Le-Clech, P. Removal of trace organics by anaerobic membrane bioreactors. *Water Res.* **2014**, *49*, 103–112. [CrossRef] [PubMed]

66. Liao, B.-Q.; Kraemer, J.T.; Bagley, D.M. Anaerobic membrane bioreactors: Applications and research directions. *Crit. Rev. Environ. Sci. Technol.* **2006**, *36*, 489–530. [CrossRef]

67. Ng, H.Y.; Hermanowicz, S.W. Membrane bioreactor operation at short solids retention times: Performance and biomass characteristics. *Water Res.* **2005**, *39*, 981–992. [CrossRef] [PubMed]

68. Chang, I.S.; Kim, S.N. Wastewater treatment using membrane filtration—Effect of biosolids concentration on cake resistance. *Process Biochem.* **2005**, *40*, 1307–1314. [CrossRef]

69. Sablani, S.; Goosen, M.; Al-Belushi, R.; Wilf, M. Concentration polarization in ultrafiltration and reverse osmosis: A critical review. *Desalination* **2001**, *141*, 269–289. [CrossRef]

70. Schäfer, A.I.; Fane, A.G.; Waite, T.D. Fouling effects on rejection in the membrane filtration of natural waters. *Desalination* **2000**, *131*, 215–224. [CrossRef]

71. Zydney, A.L.; Colton, C.K. A concentration polarization model for the filtrate flow in cross-flow microfiltration of particulate suspensions. *Chem. Eng. Commun.* **1986**, *47*, 1–21. [CrossRef]

72. Goosen, M.F.A.; Sablani, S.S.; Al-Hinai, H.; Al-Obeidani, S.; Al-Belushi, R.; Jackson, D. Fouling of reverse osmosis and ultrafiltration membranes: A critical review. *Sep. Sci. Technol.* **2004**, *39*, 2261–2297. [CrossRef]

73. Aoustin, E.; Schäfer, A.I.; Fane, A.G.; Waite, T.D. Ultrafiltration of natural organic matter. *Sep. Purif. Technol.* **2001**, *22*, 63–78. [CrossRef]

74. Peinemann, K.V.; Pereira, N.S. *Membrane Technology: Volume 4: Membranes for Water Treatmen*; Wiley-VCH: Weinheim, Germany, 2010.

75. Huang, H.; Young, T.A.; Jacangelo, J.G. Unified membrane fouling index for low pressure membrane filtration of natural waters: Principles and methodology. *Environ. Sci. Technol.* **2008**, *42*, 714–720. [CrossRef] [PubMed]

76. Cogan, N.G.; Chellam, S. Incorporating pore blocking, cake filtration, and eps production in a model for constant pressure bacterial fouling during dead-end microfiltration. *J. Membr. Sci.* **2009**, *345*, 81–89. [CrossRef]

77. Ho, C.-C.; Zydney, A.L. A combined pore blockage and cake filtration model for protein fouling during microfiltration. *J. Colloid Interface Sci.* **2000**, *232*, 389–399. [CrossRef] [PubMed]

78. Iritani, E.; Katagiri, N.; Takenaka, T.; Yamashita, Y. Membrane pore blocking during cake formation in constant pressure and constant flux dead-end microfiltration of very dilute colloids. *Chem. Eng. Sci.* **2015**, *122*, 465–473. [CrossRef]

79. Abdelrasoul, A.; Doan, H.; Lohi, A.; Cheng, C.H. Modeling of fouling and foulant attachments on heterogeneous membranes in ultrafiltration of latex solution. *Sep. Purif. Technol.* **2014**, *135*, 199–210. [CrossRef]

80. Hong, S.; Elimelech, M. Chemical and physical aspects of natural organic matter (NOM) fouling of nanofiltration membranes. *J. Membr. Sci.* **1997**, *132*, 159–181. [CrossRef]
81. Lee, E.K.; Chen, V.; Fane, A.G. Natural organic matter (NOM) fouling in low pressure membrane filtration —Effect of membranes and operation modes. *Desalination* **2008**, *218*, 257–270. [CrossRef]
82. Wibisono, Y.; Cornelissen, E.R.; Kemperman, A.J.B.; Van Der Meer, W.G.J.; Nijmeijer, K. Two-phase flow in membrane processes: A technology with a future. *J. Membr. Sci.* **2014**, *453*, 566–602. [CrossRef]
83. Fane, A.G.; Fell, C.J.D. A review of fouling and fouling control in ultrafiltration. *Desalination* **1987**, *62*, 117–136. [CrossRef]
84. Pimentel, G.A.; Dalmau, M.; Vargas, A.; Comas, J.; Rodriguez-Roda, I.; Rapaport, A.; Vande Wouwer, A. Validation of a simple fouling model for a submerged membrane bioreactor. *IFAC-PapersOnLine* **2015**, *48*, 737–742. [CrossRef]
85. Tang, C.Y.; Fu, Q.S.; Criddle, C.S.; Leckie, J.O. Effect of flux (transmembrane pressure) and membrane properties on fouling and rejection of reverse osmosis and nanofiltration membranes treating perfluorooctane sulfonate containing wastewater. *Environ. Sci. Technol.* **2007**, *41*, 2008–2014. [CrossRef] [PubMed]
86. Vrijenhoek, E.M.; Hong, S.; Elimelech, M. Influence of membrane surface properties on initial rate of colloidal fouling of reverse osmosis and nanofiltration membranes. *J. Membr. Sci.* **2001**, *188*, 115–128. [CrossRef]
87. Iritani, E. A review on modeling of pore-blocking behaviors of membranes during pressurized membrane filtration. *Dry. Technol.* **2013**, *31*, 146–162. [CrossRef]
88. Vera, L.; González, E.; Díaz, O.; Delgado, S. Application of a backwashing strategy based on transmembrane pressure set-point in a tertiary submerged membrane bioreactor. *J. Membr. Sci.* **2014**, *470*, 504–512. [CrossRef]
89. De Souza, N.P.; Basu, O.D. Relaxation: A beneficial operational step for the reduction of fouling in hollow fiber membranes for drinking water treatment. In Proceedings of the Water Quality Technology Conference and Exposition 2012, Toronto, ON, Canada, 4–8 Novemmmber 2012.
90. Akhondi, E.; Wicaksana, F.; Fane, A.G. Evaluation of fouling deposition, fouling reversibility and energy consumption of submerged hollow fiber membrane systems with periodic backwash. *J. Membr. Sci.* **2014**, *452*, 319–331. [CrossRef]
91. Akhondi, E.; Wicaksana, F.; Krantz, W.B.; Fane, A.G. Influence of dissolved air on the effectiveness of cyclic backwashing in submerged membrane systems. *J. Membr. Sci.* **2014**, *456*, 77–84. [CrossRef]
92. Khirani, S.; Smith, P.J.; Manéro, M.H.; Aim, R.B.; Vigneswaran, S. Effect of periodic backwash in the submerged membrane adsorption hybrid system (SMAHS) for wastewater treatment. *Desalination* **2006**, *191*, 27–34. [CrossRef]
93. Zsirai, T.; Buzatu, P.; Aerts, P.; Judd, S. Efficacy of relaxation, backflushing, chemical cleaning and clogging removal for an immersed hollow fibre membrane bioreactor. *Water Res.* **2012**, *46*, 4499–4507. [CrossRef] [PubMed]
94. Hong, S.P.; Bae, T.H.; Tak, T.M.; Hong, S.; Randall, A. Fouling control in activated sludge submerged hollow fiber membrane bioreactors. *Desalination* **2002**, *143*, 219–228. [CrossRef]
95. Davis, R.H. Modeling of fouling of cross-flow microfiltration membranes. *Sep. Purif. Methods* **1992**, *21*, 75–126. [CrossRef]
96. Braak, E.; Alliet, M.; Schetrite, S.; Albasi, C. Aeration and hydrodynamics in submerged membrane bioreactors. *J. Membr. Sci.* **2011**, *379*, 1–18. [CrossRef]
97. Abdullah, S.Z.; Wray, H.E.; Bérubé, P.R.; Andrews, R.C. Distribution of surface shear stress for a densely packed submerged hollow fiber membrane system. *Desalination* **2015**, *357*, 117–120. [CrossRef]
98. Wicaksana, F.; Fane, A.G.; Chen, V. The relationship between critical flux and fibre movement induced by bubbling in a submerged hollow fibre system. *Water Sci. Technol.* **2005**, *51*, 115–122. [PubMed]
99. Chang, S.; Fane, A.G.; Vigneswaran, S. Modeling and optimizing submerged hollow fiber membrane modules. *AIChE J.* **2002**, *48*, 2203–2212. [CrossRef]
100. Kim, J.; DiGiano, F.A. Defining critical flux in submerged membranes: Influence of length-distributed flux. *J. Membr. Sci.* **2006**, *280*, 752–761. [CrossRef]
101. Fane, A.G.; Chang, S.; Chardon, E. Submerged hollow fibre membrane module—Design options and operational considerations. *Desalination* **2002**, *146*, 231–236. [CrossRef]
102. Madec, A.; Buisson, H.; Ben Aim, R. Aeration to enhance membrane critical flux. In Proceedings of the World Filtration Congress, Brighton, UK, 3–7 April 2000; pp. 199–202.

103. Wu, Z.; Wang, Z.; Huang, S.; Mai, S.; Yang, C.; Wang, X.; Zhou, Z. Effects of various factors on critical flux in submerged membrane bioreactors for municipal wastewater treatment. *Sep. Purif. Technol.* **2008**, *62*, 56–63. [CrossRef]

104. Judd, S.; Judd, C. Design. In *The MBR Book: Principles and Applications of Membrane Bioreactors for Water and Wastewater Treatment*, 1st ed.; Elsevier: Amsterdam, The Netherlands; Boston, MA, USA; London, UK, 2006; Chapter 3, pp. 123–162.

105. Cho, B.D.; Fane, A.G. Fouling transients in nominally sub-critical flux operation of a membrane bioreactor. *J. Membr. Sci.* **2002**, *209*, 391–403. [CrossRef]

106. Zhang, J.; Chua, H.C.; Zhou, J.; Fane, A.G. Factors affecting the membrane performance in submerged membrane bioreactors. *J. Membr. Sci.* **2006**, *284*, 54–66. [CrossRef]

107. Judd, S. The status of membrane bioreactor technology. *Trends Biotechnol.* **2008**, *26*, 109–116. [CrossRef] [PubMed]

108. Yeo, A.; Fane, A.G. Performance of individual fibers in a submerged hollow fiber bundle. *Water Sci. Technol.* **2005**, *51*, 165–172. [PubMed]

109. Judd, S.; Judd, C. Fundamentals. In *The Mbr Book: Principles and Applications of Membrane Bioreactors for Water and Wastewater Treatment*, 1st ed.; Elsevier: Amsterdam, The Netherlands; Boston, MA, USA; London, UK, 2006; Chapter 2, pp. 22–121.

110. Yeo, A.P.S.; Law, A.W.K.; Fane, A.G. Factors affecting the performance of a submerged hollow fiber bundle. *J. Membr. Sci.* **2006**, *280*, 969–982. [CrossRef]

111. Costello, M.J.; Fane, A.G.; Hogan, P.A.; Schofield, R.W. The effect of shell side hydrodynamics on the performance of axial-flow hollow-fiber modules. *J. Membr. Sci.* **1993**, *80*, 1–11. [CrossRef]

112. Zheng, J.M.; Xu, Y.Y.; Xu, Z.K. Flow distribution in a randomly packed hollow fiber membrane module. *J. Membr. Sci.* **2003**, *211*, 263–269. [CrossRef]

113. Wu, J.; Chen, V. Shell-side mass transfer performance of randomly packed hollow fiber modules. *J. Membr. Sci.* **2000**, *172*, 59–74. [CrossRef]

114. Wickramasinghe, S.R.; Semmens, M.J.; Cussler, E.L. Mass-transfer in various hollow fiber geometries. *J. Membr. Sci.* **1992**, *69*, 235–250. [CrossRef]

115. Lipnizki, F.; Field, R.W. Mass transfer performance for hollow fibre modules with shell-side axial feed flow: Using an engineering approach to develop a framework. *J. Membr. Sci.* **2001**, *193*, 195–208. [CrossRef]

116. Busch, J.; Cruse, A.; Marquardt, W. Modeling submerged hollow-fiber membrane filtration for wastewater treatment. *J. Membr. Sci.* **2007**, *288*, 94–111. [CrossRef]

117. Ding, Z.W.; Liu, L.Y.; Ma, R.Y. Study on the effect of flow maldistribution on the performance of the hollow fiber modules used in membrane distillation. *J. Membr. Sci.* **2003**, *215*, 11–23.

118. Li, Z.X.; Zhang, L.Z. Flow maldistribution and performance deteriorations in a counter flow hollow fiber membrane module for air humidification/dehumidification. *Int. J. Heat Mass Transf.* **2014**, *74*, 421–430. [CrossRef]

119. Yang, X.; Wang, R.; Fane, A.G. Novel designs for improving the performance of hollow fiber membrane distillation modules. *J. Membr. Sci.* **2011**, *384*, 52–62. [CrossRef]

120. Chang, S.; Fane, A.G.; Waite, T.D.; Yeo, A. Unstable filtration behavior with submerged hollow fiber membranes. *J. Membr. Sci.* **2008**, *308*, 107–114. [CrossRef]

121. Chen, H.; Cao, C.; Xu, L.L.; Xiao, T.H.; Jiang, G.L. Experimental velocity measurements and effect of flow maldistribution on predicted permeator performances. *J. Membr. Sci.* **1998**, *139*, 259–268. [CrossRef]

122. Wicaksana, F.; Fane, A.G.; Law, A.W.K. The use of constant temperature anemometry for permeate flow distribution measurement in a submerged hollow fibre system. *J. Membr. Sci.* **2009**, *339*, 195–203. [CrossRef]

123. Zamani, F.; Chew, J.W.; Akhondi, E.; Krantz, W.B.; Fane, A.G. Unsteady-state shear strategies to enhance mass-transfer for the implementation of ultrapermeable membranes in reverse osmosis: A review. *Desalination* **2015**, *356*, 328–348. [CrossRef]

124. Gabarron, S.; Gomez, M.; Dvorak, L.; Ruzickova, I.; Rodriguez-Roda, I.; Comas, J. Ragging in mbr: Effects of operational conditions, chemical cleaning, and pre-treatment improvements. *Sep. Sci. Technol.* **2014**, *49*, 2115–2123. [CrossRef]

125. Sanchez, A.; Buntner, D.; Garrido, J.M. Impact of methanogenic pre-treatment on the performance of an aerobic mbr system. *Water Res.* **2013**, *47*, 1229–1236. [CrossRef] [PubMed]

126. Yu, W.Z.; Xu, L.; Graham, N.; Qu, J.H. Pre-treatment for ultrafiltration: Effect of pre-chlorination on membrane fouling. *Sci. Rep.* **2014**, *4*, 6513. [CrossRef] [PubMed]

127. Nguyen, T.T.; Ngo, H.H.; Guo, W.S. Pilot scale study on a new membrane bioreactor hybrid system in municipal wastewater treatment. *Bioresour. Technol.* **2013**, *141*, 8–12. [CrossRef] [PubMed]

128. Dutta, K.; Lee, M.Y.; Lai, W.W.P.; Lee, C.H.; Lin, A.Y.C.; Lin, C.F.; Lin, J.G. Removal of pharmaceuticals and organic matter from municipal wastewater using two-stage anaerobic fluidized membrane bioreactor. *Bioresour. Technol.* **2014**, *165*, 42–49. [CrossRef] [PubMed]

129. Sheldon, M.S.; Zeelie, P.J.; Edwards, W. Treatment of paper mill effluent using an anaerobic/aerobic hybrid side-stream membrane bioreactor. *Water Sci. Technol.* **2012**, *65*, 1265–1272. [CrossRef] [PubMed]

130. Bae, J.; Shin, C.; Lee, E.; Kim, J.; Mccarty, P.L. Anaerobic treatment of low-strength wastewater: A comparison between single and staged anaerobic fluidized bed membrane bioreactors. *Bioresour. Technol.* **2014**, *165*, 75–80. [CrossRef] [PubMed]

131. Wang, Z.; Ma, J.; Tang, C.Y.; Kimura, K.; Wang, Q.; Han, X. Membrane cleaning in membrane bioreactors: A review. *J. Membr. Sci.* **2014**, *468*, 276–307. [CrossRef]

132. Yang, W.B.; Cicek, N.; Ilg, J. State-of-the-art of membrane bioreactors: Worldwide research and commercial applications in north america. *J. Membr. Sci.* **2006**, *270*, 201–211. [CrossRef]

133. Choo, K.H.; Kang, I.J.; Yoon, S.H.; Park, H.; Kim, J.H.; Adiya, S.; Lee, C.H. Approaches to membrane fouling control in anaerobic membrane bioreactors. *Water Sci. Technol.* **2000**, *41*, 363–371.

134. Kang, I.J.; Yoon, S.H.; Lee, C.H. Comparison of the filtration characteristics of organic and inorganic membranes in a membrane-coupled anaerobic bioreactor. *Water Res.* **2002**, *36*, 1803–1813. [CrossRef]

135. Lee, S.M.; Jung, J.Y.; Chung, Y.C. Novel method for enhancing permeate flux of submerged membrane system in two-phase anaerobic reactor. *Water Res.* **2001**, *35*, 471–477. [CrossRef]

136. Brepols, C.; Drensla, K.; Janot, A.; Trimborn, M.; Engelhardt, N. Strategies for chemical cleaning in large scale membrane bioreactors. *Water Sci. Technol.* **2008**, *57*, 457–463. [CrossRef] [PubMed]

137. Brannock, M.W.D.; De Wever, H.; Wang, Y.; Leslie, G. Computational fluid dynamics simulations of mbrs: Inside submerged versus outside submerged membranes. *Desalination* **2009**, *236*, 244–251. [CrossRef]

138. Liu, R.; Huang, X.; Wang, C.; Chen, L.; Qian, Y. Study on hydraulic characteristics in a submerged membrane bioreactor process. *Process Biochem.* **2000**, *36*, 249–254. [CrossRef]

139. Brannock, M.; Wang, Y.; Leslie, G. Mixing characterisation of full-scale membrane bioreactors: CFD modelling with experimental validation. *Water Res.* **2010**, *44*, 3181–3191. [CrossRef] [PubMed]

140. Ji, J.; Li, J.F.; Qiu, J.P.; Li, X.D. Polyacrylamide-starch composite flocculant as a membrane fouling reducer: Key factors of fouling reduction. *Sep. Purif. Technol.* **2014**, *131*, 1–7. [CrossRef]

141. Nouri, N.; Mehrnia, M.R.; Sarrafzadeh, M.H.; Nabizadeh, R. Performance of membrane bioreactor in presence of flocculants. *Desalination Water Treat.* **2014**, *52*, 2933–2938. [CrossRef]

142. Zhang, H.F.; Gao, Z.Y.; Zhang, L.H.; Song, L.F. Performance enhancement and fouling mitigation by organic flocculant addition in membrane bioreactor at high salt shock. *Bioresour. Technol.* **2014**, *164*, 34–40. [CrossRef] [PubMed]

143. Melo-Guimaraes, A.; Torner-Morales, F.J.; Duran-Alvarez, J.C.; Jimenez-Cisneros, B.E. Removal and fate of emerging contaminants combining biological, flocculation and membrane treatments. *Water Sci. Technol.* **2013**, *67*, 877–885. [CrossRef] [PubMed]

144. Nguyen, T.T.; Guo, W.S.; Ngo, H.H.; Vigneswaran, S. A new combined inorganic-organic flocculant (CIOF) as a performance enhancer for aerated submerged membrane bioreactor. *Sep. Purif. Technol.* **2010**, *75*, 204–209. [CrossRef]

145. Johir, M.A.; Shanmuganathan, S.; Vigneswaran, S.; Kandasamy, J. Performance of submerged membrane bioreactor (SMBR) with and without the addition of the different particle sizes of gac as suspended medium. *Bioresour. Technol.* **2013**, *141*, 13–18. [CrossRef] [PubMed]

146. Nguyen, L.N.; Hai, F.I.; Kang, J.G.; Price, W.E.; Nghiem, L.D. Removal of trace organic contaminants by a membrane bioreactor-granular activated carbon (MBR-GAC) system. *Bioresour. Technol.* **2012**, *113*, 169–173. [CrossRef] [PubMed]

147. Gur-Reznik, S.; Katz, I.; Dosoretz, C.G. Removal of dissolved organic matter by granular-activated carbon adsorption as a pretreatment to reverse osmosis of membrane bioreactor effluents. *Water Res.* **2008**, *42*, 1595–1605. [CrossRef] [PubMed]

148. Abegglen, C.; Joss, A.; Boehler, M.; Buetzer, S.; Siegrist, H. Reducing the natural color of membrane bioreactor permeate with activated carbon or ozone. *Water Sci. Technol.* **2009**, *60*, 155–165. [CrossRef] [PubMed]

149. Guo, J.F.; Xia, S.Q.; Lu, Y.J. Characteristics of combined submerged membrane bioreactor with granular activated carbon (GAC) in treating lineal alkylbenzene sulphonates (LAS) wastewater. *AIP Conf. Proc.* **2010**, *1251*, 65–68.

150. Hai, F.I.; Yamamoto, K.; Nakajima, F.; Fukushi, K. Bioaugmented membrane bioreactor (MBR) with a gac-packed zone for high rate textile wastewater treatment. *Water Res.* **2011**, *45*, 2199–2206. [CrossRef] [PubMed]

151. Pham, T.T.; Nguyen, V.A.; Van der Bruggen, B. Pilot-scale evaluation of gac adsorption using low-cost, high-performance materials for removal of pesticides and organic matter in drinking water production. *J. Environ. Eng.* **2013**, *139*, 958–965. [CrossRef]

152. Wu, B.; Yi, S.; Fane, A.G. Microbial behaviors involved in cake fouling in membrane bioreactors under different solids retention times. *Bioresour. Technol.* **2011**, *102*, 2511–2516. [CrossRef] [PubMed]

153. Judd, S.; Judd, C. Commercial technologies. In *The Mbr Book: Principles and Applications of Membrane Bioreactors for Water and Wastewater Treatment*, 1st ed.; Elsevier: Amsterdam, The Netherlands; Boston, MA, USA; London, UK, 2006; Chapter 4, pp. 163–205.

154. Zamani, F.; Ullah, A.; Akhondi, E.; Tanudjaja, H.J.; Cornelissen, E.R.; Honciuc, A.; Fane, A.G.; Chew, J.W. Impact of the surface energy of particulate foulants on membrane fouling. *J. Membr. Sci.* **2016**, *510*, 101–111. [CrossRef]

155. Botton, S.; Verliefde, A.R.D.; Quach, N.T.; Cornelissen, E.R. Influence of biofouling on pharmaceuticals rejection in nf membrane filtration. *Water Res.* **2012**, *46*, 5848–5860. [CrossRef] [PubMed]

156. Cornelissen, E.R.; van den Boomgaard, T.; Strathmann, H. Physicochemical aspects of polymer selection for ultrafiltration and microfiltration membranes. *Colloids Surf. A Physicochem. Eng. Asp.* **1998**, *138*, 283–289. [CrossRef]

157. Van Oss, C.J. Introduction. In *Interfacial Forces in Aqueous Media*, 2nd ed.; CRC Press: Boca Raton, FL, USA, 2006; Chapter 1.

158. Field, R.W.; Pearce, G.K. Critical, sustainable and threshold fluxes for membrane filtration with water industry applications. *Adv. Colloid Interface Sci.* **2011**, *164*, 38–44. [CrossRef] [PubMed]

159. Bildyukevich, A.V.; Plisko, T.V.; Liubimova, A.S.; Volkov, V.V.; Usosky, V.V. Hydrophilization of polysulfone hollow fiber membranes via addition of polyvinylpyrrolidone to the bore fluid. *J. Membr. Sci.* **2017**, *524*, 537–549. [CrossRef]

160. Akhondi, E.; Zamani, F.; Law, A.W.K.; Krantz, W.B.; Fane, A.G.; Chew, J.W. Influence of backwashing on the pore size of hollow fiber ultrafiltration membranes. *J. Membr. Sci.* **2017**, *521*, 33–42. [CrossRef]

161. Rana, D.; Matsuura, T. Surface modifications for antifouling membranes. *Chem. Rev.* **2010**, *110*, 2448–2471. [CrossRef] [PubMed]

162. Marshall, A.D.; Munro, P.A.; Trägårdh, G. The effect of protein fouling in microfiltration and ultrafiltration on permeate flux, protein retention and selectivity: A literature review. *Desalination* **1993**, *91*, 65–108. [CrossRef]

163. Pasmore, M.; Todd, P.; Smith, S.; Baker, D.; Silverstein, J.; Coons, D.; Bowman, C.N. Effects of ultrafiltration membrane surface properties on pseudomonas aeruginosa biofilm initiation for the purpose of reducing biofouling. *J. Membr. Sci.* **2001**, *194*, 15–32. [CrossRef]

164. Brant, J.A.; Childress, A.E. Colloidal adhesion to hydrophilic membrane surfaces. *J. Membr. Sci.* **2004**, *241*, 235–248. [CrossRef]

165. Kochkodan, V.; Hilal, N. A comprehensive review on surface modified polymer membranes for biofouling mitigation. *Desalination* **2015**, *356*, 187–207. [CrossRef]

166. Reddy, A.V.R.; Mohan, D.J.; Bhattacharya, A.; Shah, V.J.; Ghosh, P.K. Surface modification of ultrafiltration membranes by preadsorption of a negatively charged polymer: I. Permeation of water soluble polymers and inorganic salt solutions and fouling resistance properties. *J. Membr. Sci.* **2003**, *214*, 211–221. [CrossRef]

167. Wang, D.; Li, K.; Teo, W.K. Preparation and characterization of polyvinylidene fluoride (PVDF) hollow fiber membranes. *J. Membr. Sci.* **1999**, *163*, 211–220. [CrossRef]

168. Kochkodan, V.; Johnson, D.J.; Hilal, N. Polymeric membranes: Surface modification for minimizing (bio) colloidal fouling. *Adv. Colloid Interface Sci.* **2014**, *206*, 116–140. [CrossRef] [PubMed]

169. Bernardes, P.C.; De Andrade, N.J.; Da Silva, L.H.M.; De Carvalho, A.F.; Fernandes, P.E.; Araújo, E.A.; Lelis, C.A.; Mol, P.C.G.; De Sá, J.P.N. Modification of polysulfone membrane used in the water filtration process to reduce biofouling. *J. Nanosci. Nanotechnol.* **2014**, *14*, 6355–6367. [CrossRef] [PubMed]

170. Zhu, X.; Loo, H.-E.; Bai, R. A novel membrane showing both hydrophilic and oleophobic surface properties and its non-fouling performances for potential water treatment applications. *J. Membr. Sci.* **2013**, *436*, 47–56. [CrossRef]

171. Lee, W.; Ahn, C.H.; Hong, S.; Kim, S.; Lee, S.; Baek, Y.; Yoon, J. Evaluation of surface properties of reverse osmosis membranes on the initial biofouling stages under no filtration condition. *J. Membr. Sci.* **2010**, *351*, 112–122. [CrossRef]

172. Ulbricht, M.; Richau, K.; Kamusewitz, H. Chemically and morphologically defined ultrafiltration membrane surfaces prepared by heterogeneous photo-initiated graft polymerization. *Colloids Surf. A Physicochem. Eng. Asp.* **1998**, *138*, 353–366. [CrossRef]

173. Susanto, H.; Ulbricht, M. Photografted thin polymer hydrogel layers on pes ultrafiltration membranes: Characterization, stability, and influence on separation performance. *Langmuir* **2007**, *23*, 7818–7830. [CrossRef] [PubMed]

174. Yu, L.-Y.; Xu, Z.-L.; Shen, H.-M.; Yang, H. Preparation and characterization of PVDF–SiO$_2$ composite hollow fiber uf membrane by sol–gel method. *J. Membr. Sci.* **2009**, *337*, 257–265. [CrossRef]

175. Cui, A.; Liu, Z.; Xiao, C.; Zhang, Y. Effect of micro-sized SiO$_2$-particle on the performance of PVDF blend membranes via tips. *J. Membr. Sci.* **2010**, *360*, 259–264. [CrossRef]

176. Yan, L.; Li, Y.S.; Xiang, C.B. Preparation of poly (vinylidene fluoride)(PVDF) ultrafiltration membrane modified by nano-sized alumina (Al$_2$O$_3$) and its antifouling research. *Polymer* **2005**, *46*, 7701–7706. [CrossRef]

177. Yan, L.; Li, Y.S.; Xiang, C.B.; Xianda, S. Effect of nano-sized al2o3-particle addition on PVDF ultrafiltration membrane performance. *J. Membr. Sci.* **2006**, *276*, 162–167. [CrossRef]

178. Koh, M.J.; Hwang, H.Y.; Kim, D.J.; Kim, H.J.; Hong, Y.T.; Nam, S.Y. Preparation and characterization of porous PVDF-hfp/clay nanocomposite membranes. *J. Mater. Sci. Technol.* **2010**, *26*, 633–638. [CrossRef]

179. Rajabi, H.; Ghaemi, N.; Madaeni, S.S.; Daraei, P.; Khadivi, M.A.; Falsafi, M. Nanoclay embedded mixed matrix PVDF nanocomposite membrane: Preparation, characterization and biofouling resistance. *Appl. Surf. Sci.* **2014**, *313*, 207–214. [CrossRef]

180. Shen, X.; Xie, T.; Wang, J.; Liu, P.; Wang, F. An anti-fouling poly (vinylidene fluoride) hybrid membrane blended with functionalized ZrO$_2$ nanoparticles for efficient oil/water separation. *RSC Adv.* **2017**, *7*, 5262–5271. [CrossRef]

181. Yang, X.; He, Y.; Zeng, G.; Zhan, Y.; Pan, Y.; Shi, H.; Chen, Q. Novel hydrophilic PVDF ultrafiltration membranes based on a ZrO$_2$. *J. Mater. Sci.* **2016**, *51*, 8965–8976. [CrossRef]

182. Yang, H.C.; Hou, J.; Chen, V.; Xu, Z.K. Surface and interface engineering for organic-inorganic composite membranes. *J. Mater. Chem. A* **2016**, *4*, 9716–9729. [CrossRef]

183. Oh, S.J.; Kim, N.; Lee, Y.T. Preparation and characterization of PVDF/TiO$_2$ organic–inorganic composite membranes for fouling resistance improvement. *J. Membr. Sci.* **2009**, *345*, 13–20. [CrossRef]

184. Wu, H.; Liu, Y.; Mao, L.; Jiang, C.; Ang, J.; Lu, X. Doping polysulfone ultrafiltration membrane with TiO$_2$-PDA nanohybrid for simultaneous self-cleaning and self-protection. *J. Membr. Sci.* **2017**, *532*, 20–29. [CrossRef]

185. Hong, J.; He, Y. Effects of nano sized zinc oxide on the performance of PVDF microfiltration membranes. *Desalination* **2012**, *302*, 71–79. [CrossRef]

186. Liang, S.; Xiao, K.; Mo, Y.; Huang, X. A novel zno nanoparticle blended polyvinylidene fluoride membrane for anti-irreversible fouling. *J. Membr. Sci.* **2012**, *394*, 184–192. [CrossRef]

187. Chang, S.; Fane, A.G. The effect of fibre diameter on filtration and flux distribution—Relevance to submerged hollow fibre modules. *J. Membr. Sci.* **2001**, *184*, 221–231. [CrossRef]

188. Fulton, B.G.; Bérubé, P.R. Optimizing the sparging condition and membrane module spacing for a ZW500 submerged hollow fiber membrane system. *Desalination Water Treat.* **2012**, *42*, 8–16. [CrossRef]

189. Shimizu, Y.; Okuno, Y.-I.; Uryu, K.; Ohtsubo, S.; Watanabe, A. Filtration characteristics of hollow fiber microfiltration membranes used in membrane bioreactor for domestic wastewater treatment. *Water Res.* **1996**, *30*, 2385–2392. [CrossRef]

190. Yang, X.; Fridjonsson, E.O.; Johns, M.L.; Wang, R.; Fane, A.G. A non-invasive study of flow dynamics in membrane distillation hollow fiber modules using low-field nuclear magnetic resonance imaging (MRI). *J. Membr. Sci.* **2014**, *451*, 46–54. [CrossRef]

191. Bérubé, P.R.; Lei, E. The effect of hydrodynamic conditions and system configurations on the permeate flux in a submerged hollow fiber membrane system. *J. Membr. Sci.* **2006**, *271*, 29–37. [CrossRef]

192. Liu, X.; Wang, Y.; Waite, T.D.; Leslie, G. Numerical simulations of impact of membrane module design variables on aeration patterns in membrane bioreactors. *J. Membr. Sci.* **2016**, *520*, 201–213. [CrossRef]

193. Liu, X.; Wang, Y.; Waite, T.D.; Leslie, G. Fluid structure interaction analysis of lateral fibre movement in submerged membrane reactors. *J. Membr. Sci.* **2016**, *504*, 240–250. [CrossRef]

194. Yeo, A.P.S.; Law, A.W.K.; Fane, A.G. The relationship between performance of submerged hollow fibers and bubble-induced phenomena examined by particle image velocimetry. *J. Membr. Sci.* **2007**, *304*, 125–137. [CrossRef]

195. Kim, J.; DiGiano, F.A. Particle fouling in submerged microfiltration membranes: Effects of hollow-fiber length and aeration rate. *J. Water Supply Res. Technol.-Aqua* **2006**, *55*, 535–547. [CrossRef]

196. Lee, M.; Kim, J. Analysis of local fouling in a pilot-scale submerged hollow-fiber membrane system for drinking water treatment by membrane autopsy. *Sep. Purif. Technol.* **2012**, *95*, 227–234. [CrossRef]

197. Li, X.; Li, J.; Cui, Z.; Yao, Y. Modeling of filtration characteristics during submerged hollow fiber membrane microfiltration of yeast suspension under aeration condition. *J. Membr. Sci.* **2016**, *510*, 455–465. [CrossRef]

198. Lin, C.-J.; Rao, P.; Shirazi, S. Effect of operating parameters on permeate flux decline caused by cake formation—A model study. *Desalination* **2005**, *171*, 95–105. [CrossRef]

199. Pradhan, M.; Aryal, R.; Vigneswaran, S.; Kandasamy, J. Application of air flow for mitigation of particle deposition in submerged membrane microfiltration. *Desalination Water Treat.* **2011**, *32*, 201–207. [CrossRef]

200. Visvanathan, C.; Ben Aim, R.; Parameshwaran, K. Membrane separation bioreactors for wastewater treatment. *Crit. Rev. Environ. Sci. Technol.* **2000**, *30*, 1–48. [CrossRef]

201. Du, X.; Qu, F.-S.; Liang, H.; Li, K.; Bai, L.-M.; Li, G.-B. Control of submerged hollow fiber membrane fouling caused by fine particles in photocatalytic membrane reactors using bubbly flow: Shear stress and particle forces analysis. *Sep. Purif. Technol.* **2017**, *172*, 130–139. [CrossRef]

202. Yeo, B.J.L.; Goh, S.; Livingston, A.G.; Fane, A.G. Controlling biofilm development in the extractive membrane bioreactor. *Sep. Sci. Technol.* **2017**, *52*, 113–121. [CrossRef]

203. Cabassud, C.; Laborie, S.; Durand-Bourlier, L.; Lainé, J.M. Air sparging in ultrafiltration hollow fibers: Relationship between flux enhancement, cake characteristics and hydrodynamic parameters. *J. Membr. Sci.* **2001**, *181*, 57–69. [CrossRef]

204. Wang, J.; Fane, A.G.; Chew, J.W. Effect of bubble characteristics on critical flux in the microfiltration of particulate foulants. *J. Membr. Sci.* **2017**, *535*, 279–293. [CrossRef]

205. Liu, N.; Zhang, Q.; Chin, G.-L.; Ong, E.-H.; Lou, J.; Kang, C.-W.; Liu, W.; Jordan, E. Experimental investigation of hydrodynamic behavior in a real membrane bio-reactor unit. *J. Membr. Sci.* **2010**, *353*, 122–134. [CrossRef]

206. Nguyen Cong Duc, E.; Fournier, L.; Levecq, C.; Lesjean, B.; Grelier, P.; Tazi-Pain, A. Local hydrodynamic investigation of the aeration in a submerged hollow fibre membranes cassette. *J. Membr. Sci.* **2008**, *321*, 264–271. [CrossRef]

207. Fulton, B.G.; Redwood, J.; Tourais, M.; Bérubé, P.R. Distribution of surface shear forces and bubble characteristics in full-scale gas sparged submerged hollow fiber membrane modules. *Desalination* **2011**, *281*, 128–141. [CrossRef]

208. Buetehorn, S.; Volmering, D.; Vossenkaul, K.; Wintgens, T.; Wessling, M.; Melin, T. CFD simulation of single- and multi-phase flows through submerged membrane units with irregular fiber arrangement. *J. Membr. Sci.* **2011**, *384*, 184–197. [CrossRef]

209. Bouhabila, E.H.; Ben Aïm, R.; Buisson, H. Microfiltration of activated sludge using submerged membrane with air bubbling (application to wastewater treatment). *Desalination* **1998**, *118*, 315–322. [CrossRef]

210. Ueda, T.; Hata, K.; Kikuoka, Y.; Seino, O. Effects of aeration on suction pressure in a submerged membrane bioreactor. *Water Res.* **1997**, *31*, 489–494. [CrossRef]

211. Chang, S.; Fane, A.G. Filtration of biomass with axial inter-fibre upward slug flow: Performance and mechanisms. *J. Membr. Sci.* **2000**, *180*, 57–68. [CrossRef]

212. Qaisrani, T.M.; Samhaber, W.M. Impact of gas bubbling and backflushing on fouling control and membrane cleaning. *Desalination* **2011**, *266*, 154–161. [CrossRef]

213. Chan, C.C.V.; Bérubé, P.R.; Hall, E.R. Relationship between types of surface shear stress profiles and membrane fouling. *Water Res.* **2011**, *45*, 6403–6416. [CrossRef] [PubMed]

214. Xia, L.; Law, A.W.-K.; Fane, A.G. Hydrodynamic effects of air sparging on hollow fiber membranes in a bubble column reactor. *Water Res.* **2013**, *47*, 3762–3772. [CrossRef] [PubMed]

215. Yeom, I.-T.; Nah, Y.-M.; Ahn, K.-H. Treatment of household wastewater using an intermittently aerated membrane bioreactor. *Desalination* **1999**, *124*, 193–203. [CrossRef]

216. Guibert, D.; Aim, R.B.; Rabie, H.; Côté, P. Aeration performance of immersed hollow-fiber membranes in a bentonite suspension. *Desalination* **2002**, *148*, 395–400. [CrossRef]

217. Tung, K.-L.; Damodar, H.-R.; Damodar, R.-A.; Tsai, J.-H.; Chen, C.-H.; You, S.-J.; Huang, M.-S. Imaging the effect of aeration on particle fouling mitigation in a submerged membrane filtration using a photointerrupt sensor array. *Sep. Sci. Technol.* **2017**, *52*, 228–239. [CrossRef]

218. Jaffrin, M.Y. Dynamic shear-enhanced membrane filtration: A review of rotating disks, rotating membranes and vibrating systems. *J. Membr. Sci.* **2008**, *324*, 7–25. [CrossRef]

219. Pourbozorg, M.; Li, T.; Law, A.W.K. Effect of turbulence on fouling control of submerged hollow fibre membrane filtration. *Water Res.* **2016**, *99*, 101–111. [CrossRef] [PubMed]

220. Kola, A.; Ye, Y.; Ho, A.; Le-Clech, P.; Chen, V. Application of low frequency transverse vibration on fouling limitation in submerged hollow fibre membranes. *J. Membr. Sci.* **2012**, *409–410*, 54–65. [CrossRef]

221. Kola, A.; Ye, Y.; Le-Clech, P.; Chen, V. Transverse vibration as novel membrane fouling mitigation strategy in anaerobic membrane bioreactor applications. *J. Membr. Sci.* **2014**, *455*, 320–329. [CrossRef]

222. Low, S.C.; Juan, H.H.; Siong, L.K. A combined VSEP and membrane bioreactor system. *Desalination* **2005**, *183*, 353–362. [CrossRef]

223. Li, T.; Law, A.W.-K.; Fane, A.G. Submerged hollow fibre membrane filtration with transverse and longitudinal vibrations. *J. Membr. Sci.* **2014**, *455*, 83–91. [CrossRef]

224. Genkin, G.; Waite, T.D.; Fane, A.G.; Chang, S. The effect of vibration and coagulant addition on the filtration performance of submerged hollow fibre membranes. *J. Membr. Sci.* **2006**, *281*, 726–734. [CrossRef]

225. Prip Beier, S.; Jonsson, G. A vibrating membrane bioreactor (VMBR): Macromolecular transmission—Influence of extracellular polymeric substances. *Chem. Eng. Sci.* **2009**, *64*, 1436–1444. [CrossRef]

226. Pourbozorg, M.; Li, T.; Law, A.W.K. Fouling of submerged hollow fiber membrane filtration in turbulence: Statistical dependence and cost-benefit analysis. *J. Membr. Sci.* **2017**, *521*, 43–52. [CrossRef]

227. Ruigomez, I.; Gonzalez, E.; Guerra, S.; Rodriguez-Gomez, L.E.; Vera, L. Evaluation of a novel physical cleaning strategy based on hf membrane rotation during the backwashing/relaxation phases for anaerobic submerged mbr. *J. Membr. Sci.* **2017**, *526*, 181–190. [CrossRef]

228. Chatzikonstantinou, K.; Tzamtzis, N.; Aretakis, N.; Pappa, A. The effect of various high-frequency powerful vibration (HFPV) types on fouling control of hollow fiber membrane elements in a small pilot-scale smbr system. *Desalination Water Treat.* **2016**, *57*, 27905–27913. [CrossRef]

229. Beier, S.P.; Guerra, M.; Garde, A.; Jonsson, G. Dynamic microfiltration with a vibrating hollow fiber membrane module: Filtration of yeast suspensions. *J. Membr. Sci.* **2006**, *281*, 281–287. [CrossRef]

230. Beier, S.P.; Jonsson, G. Dynamic microfiltration with a vibrating hollow fiber membrane module. *Desalination* **2006**, *199*, 499–500. [CrossRef]

231. Beier, S.P.; Jonsson, G. Separation of enzymes and yeast cells with a vibrating hollow fiber membrane module. *Sep. Purif. Technol.* **2007**, *53*, 111–118. [CrossRef]

232. Zamani, F.; Law, A.W.K.; Fane, A.G. Hydrodynamic analysis of vibrating hollow fibre membranes. *J. Membr. Sci.* **2013**, *429*, 304–312. [CrossRef]

233. Krantz, W.B.; Bilodeau, R.R.; Voorhees, M.E.; Elgas, R.J. Use of axial membrane vibrations to enhance mass transfer in a hollow tube oxygenator. *J. Membr. Sci.* **1997**, *124*, 283–299. [CrossRef]

234. Epstein, N. Liquid–solids fluidization. In *Handbook of Fluidization and Fluid-Particle Systems*; Yang, W.-C., Ed.; Marcel Dekker: New York, NY, USA, 2003.

235. Kunii, D.; Levenspiel, O. *Fluidization Engineering*; Butterworth-Heinemann: Newton, MA, USA, 1991.

236. Bixler, H.J.; Rappe, G.C. Ultrafiltration Process. U.S. Patent 3,541,006, 17 November 1970.

237. Lowe, E.; Durkee, E.L. Dynamic turbulence promotion in reverse osmosis processing of liquid foods. *J. Food Sci.* **1971**, *36*, 31–32. [CrossRef]

238. Van der Waal, M.J.; van der Velden, P.M.; Koning, J.; Smolders, C.A.; Vanswaay, W.P.M. Use of fluidized-beds as turbulence promotors in tubular membrane systems. *Desalination* **1977**, *22*, 465–483. [CrossRef]

239. Hamer, E.A.G. Semipermeable Membrane Cleaning Means. U.S. Patent 3,425,562, 4 February 1969.

240. Wei, C.H.; Huang, X.; Wang, C.W.; Wen, X.H. Effect of a suspended carrier on membrane fouling in a submerged membrane bioreactor. *Water Sci. Technol.* **2006**, *53*, 211–220. [CrossRef] [PubMed]

241. Yang, Q.Y.; Chen, J.H.; Zhang, F. Membrane fouling control in a submerged membrane bioreactor with porous, flexible suspended carriers. *Desalination* **2006**, *189*, 292–302. [CrossRef]

242. De Boer, R.; Zomerman, J.J.; Hiddink, J.; Aufderheyde, J.; Vanswaay, W.P.M.; Smolders, C.A. Fluidized-beds as turbulence promoters in the concentration of food liquids by reverse-osmosis. *J. Food Sci.* **1980**, *45*, 1522–1528. [CrossRef]

243. Zhong, Z.X.; Liu, X.; Chen, R.Z.; Xing, W.H.; Xu, N.P. Adding microsized silica particles to the catalysis/ultrafiltration system: Catalyst dissolution inhibition and flux enhancement. *Ind. Eng. Chem. Res.* **2009**, *48*, 4933–4938. [CrossRef]

244. Aslam, M.; McCarty, P.L.; Bae, J.; Kim, J. The effect of fluidized media characteristics on membrane fouling and energy consumption in anaerobic fluidized membrane bioreactors. *Sep. Purif. Technol.* **2014**, *132*, 10–15. [CrossRef]

245. Urbain, V.; Benoit, R.; Manem, J. Membrane bioreactor: A new treatment tool. *J. Am. Water Works Assoc.* **1996**, *88*, 75–86.

246. Williams, M.D.; Pirbazari, M. Membrane bioreactor process for removing biodegradable organic matter from water. *Water Res.* **2007**, *41*, 3880–3893. [CrossRef] [PubMed]

247. Li, Y.Z.; He, Y.L.; Liu, Y.H.; Yang, S.C.; Zhang, G.J. Comparison of the filtration characteristics between biological powdered activated carbon sludge and activated sludge in submerged membrane bioreactors. *Desalination* **2005**, *174*, 305–314. [CrossRef]

248. Akram, A.; Stuckey, D.C. Flux and performance improvement in a submerged anaerobic membrane bioreactor (SAMBR) using powdered activated carbon (PAC). *Process Biochem.* **2008**, *43*, 93–102. [CrossRef]

249. Satyawali, Y.; Balakrishnan, M. Effect of pac addition on sludge properties in an mbr treating high strength wastewater. *Water Res.* **2009**, *43*, 1577–1588. [CrossRef] [PubMed]

250. Ng, C.A.; Sun, D.; Fane, A.G. Operation of membrane bioreactor with powdered activated carbon addition. *Sep. Sci. Technol.* **2006**, *41*, 1447–1466.

251. Kim, J.S.; Lee, C.H. Effect of powdered activated carbon on the performance of an aerobic membrane bioreactor: Comparison between cross-flow and submerged membrane systems. *Water Environ. Res.* **2003**, *75*, 300–307. [CrossRef] [PubMed]

252. Munz, G.; Gori, R.; Mori, G.; Lubello, C. Powdered activated carbon and membrane bioreactors (MBRPAC) for tannery wastewater treatment: Long term effect on biological and filtration process performances. *Desalination* **2007**, *207*, 349–360. [CrossRef]

253. Tian, J.Y.; Liang, H.; Yang, Y.L.; Tian, S.; Li, G.B. Membrane adsorption bioreactor (MABR) for treating slightly polluted surface water supplies: As compared to membrane bioreactor (MBR). *J. Membr. Sci.* **2008**, *325*, 262–270. [CrossRef]

254. Park, H.; Choo, K.H.; Lee, C.H. Flux enhancement with powdered activated carbon addition in the membrane anaerobic bioreactor. *Sep. Sci. Technol.* **1999**, *34*, 2781–2792. [CrossRef]

255. Remy, M.; Potier, V.; Temmink, H.; Rulkens, W. Why low powdered activated carbon addition reduces membrane fouling in mbrs. *Water Res.* **2010**, *44*, 861–867. [CrossRef] [PubMed]

256. Ng, C.A.; Sun, D.; Zhang, J.S.; Wu, B.; Fane, A.G. Mechanisms of fouling control in membrane bioreactors by the addition of powdered activated carbon. *Sep. Sci. Technol.* **2010**, *45*, 873–889. [CrossRef]

257. Hu, A.Y.; Stuckey, D.C. Activated carbon addition to a submerged anaerobic membrane bioreactor: Effect on performance, transmembrane pressure, and flux. *J. Environ. Eng.* **2007**, *133*, 73–80. [CrossRef]

258. Yoo, R.; Kim, J.; McCarty, P.L.; Bae, J. Anaerobic treatment of municipal wastewater with a staged anaerobic fluidized membrane bioreactor (SAF-MBR) system. *Bioresour. Technol.* **2012**, *120*, 133–139. [CrossRef] [PubMed]

259. Bae, J.; Yoo, R.; Lee, E.; McCarty, P.L. Two-stage anaerobic fluidized-bed membrane bioreactor treatment of settled domestic wastewater. *Water Sci. Technol.* **2013**, *68*, 394–399. [CrossRef] [PubMed]

260. Yoo, R.H.; Kim, J.H.; McCarty, P.L.; Bae, J.H. Effect of temperature on the treatment of domestic wastewater with a staged anaerobic fluidized membrane bioreactor. *Water Sci. Technol.* **2014**, *69*, 1145–1150. [CrossRef] [PubMed]

261. Shin, C.; McCarty, P.L.; Kim, J.; Bae, J. Pilot-scale temperate-climate treatment of domestic wastewater with a staged anaerobic fluidized membrane bioreactor (SAF-MBR). *Bioresour. Technol.* **2014**, *159*, 95–103. [CrossRef] [PubMed]

262. Gao, D.W.; Hu, Q.; Yao, C.; Ren, N.Q.; Wu, W.M. Integrated anaerobic fluidized-bed membrane bioreactor for domestic wastewater treatment. *Chem. Eng. J.* **2014**, *240*, 362–368. [CrossRef]

263. Li, J.; Ge, Z.; He, Z. A fluidized bed membrane bioelectrochemical reactor for energy-efficient wastewater treatment. *Bioresour. Technol.* **2014**, *167*, 310–315. [CrossRef] [PubMed]

264. Ren, L.J.; Ahn, Y.; Logan, B.E. A two-stage microbial fuel cell and anaerobic fluidized bed membrane bioreactor (MFC-AFMBR) system for effective domestic wastewater treatment. *Environ. Sci. Technol.* **2014**, *48*, 4199–4206. [CrossRef] [PubMed]

265. Li, J.; Luo, S.; He, Z. Cathodic fluidized granular activated carbon assisted-membrane bioelectrochemical reactor for wastewater treatment. *Sep. Purif. Technol.* **2016**, *169*, 241–246. [CrossRef]

266. Cahyadi, A.; Yang, S.; Chew, J.W. CFD study on the hydrodynamics of fluidized granular activated carbon in anfmbr applications. *Sep. Purif. Technol.* **2017**, *178*, 75–89. [CrossRef]

267. Wang, J.; Wu, B.; Yang, S.; Liu, Y.; Fane, A.G.; Chew, J.W. Characterizing the scouring efficiency of granular activated carbon (GAC) particles in membrane fouling mitigation via wavelet decomposition of accelerometer signals. *J. Membr. Sci.* **2016**, *498*, 105–115. [CrossRef]

268. Wang, J.; Zamani, F.; Cahyadi, A.; Toh, J.Y.; Yang, S.; Wu, B.; Liu, Y.; Fane, A.G.; Chew, J.W. Correlating the hydrodynamics of fluidized granular activated carbon (GAC) with membrane-fouling mitigation. *J. Membr. Sci.* **2016**, *510*, 38–49. [CrossRef]

269. Wang, J.; Wu, B.; Liu, Y.; Fane, A.G.; Chew, J.W. Effect of fluidized granular activated carbon (GAC) on critical flux in the microfiltration of particulate foulants. *J. Membr. Sci.* **2017**, *523*, 409–417. [CrossRef]

270. Wu, B.; Zamani, F.; Lim, W.; Liao, D.; Wang, Y.; Liu, Y.; Chew, J.W.; Fane, A.G. Effect of mechanical scouring by granular activated carbon (GAC) on membrane fouling mitigation. *Desalination* **2017**, *403*, 80–87. [CrossRef]

271. Wu, B.; Wang, Y.; Lim, W.; Chew, J.W.; Fane, A.G.; Liu, Y. Enhanced performance of submerged hollow fibre microfiltration by fluidized granular activated carbon. *J. Membr. Sci.* **2016**, *499*, 47–55. [CrossRef]

272. Shin, C.; Kim, K.; McCarty, P.L.; Kim, J.; Bae, J. Integrity of hollow-fiber membranes in a pilot-scale anaerobic fluidized membrane bioreactor (AFMBR) after two-years of operation. *Sep. Purif. Technol.* **2016**, *162*, 101–105. [CrossRef]

273. Aslam, M.; McCarty, P.L.; Shin, C.; Bae, J.; Kim, J. Low energy single-staged anaerobic fluidized bed ceramic membrane bioreactor (AFCMBR) for wastewater treatment. *Bioresour. Technol.* **2017**, *240*, 33–41. [CrossRef] [PubMed]

274. Charfi, A.; Aslam, M.; Lesage, G.; Heran, M.; Kim, J. Macroscopic approach to develop fouling model under gac fluidization in anaerobic fluidized bed membrane bioreactor. *J. Ind. Eng. Chem.* **2017**, *49*, 219–229. [CrossRef]

275. Wu, B.; Wong, P.C.Y.; Fane, A.G. The potential roles of granular activated carbon in anaerobic fluidized membrane bioreactors: Effect on membrane fouling and membrane integrity. *Desalination Water Treat.* **2015**, *53*, 1450–1459. [CrossRef]

276. Wu, B.; Li, Y.; Lim, W.; Lee, S.L.; Guo, Q.; Fane, A.G.; Liu, Y. Single-stage versus two-stage anaerobic fluidized bed bioreactors in treating municipal wastewater: Performance, foulant characteristics, and microbial community. *Chemosphere* **2017**, *171*, 158–167. [CrossRef] [PubMed]

277. Gao, D.W.; Hu, Q.; Yao, C.; Ren, N.Q. Treatment of domestic wastewater by an integrated anaerobic fluidized-bed membrane bioreactor under moderate to low temperature conditions. *Bioresour. Technol.* **2014**, *159*, 193–198. [CrossRef] [PubMed]

278. McCurry, D.L.; Bear, S.E.; Bae, J.; Sedlak, D.L.; McCarty, P.L.; Mitch, W.A. Superior removal of disinfection byproduct precursors and pharmaceuticals from wastewater in a staged anaerobic fluidized membrane bioreactor compared to activated sludge. *Environ. Sci. Technol. Lett.* **2014**, *1*, 459–464. [CrossRef]

279. Düppenbecker, B.; Engelhart, M.; Cornel, P. Fouling mitigation in anaerobic membrane bioreactor using fluidized glass beads: Evaluation fitness for purpose of ceramic membranes. *J. Membr. Sci.* **2017**, *537*, 69–82. [CrossRef]

280. Aslam, M.; Charfi, A.; Kim, J. Membrane scouring to control fouling under fluidization of non-adsorbing media for wastewater treatment. *Environ. Sci. Pollut. Res.* **2017**, 1–11. [CrossRef] [PubMed]

281. McCarty, P.L.; Bae, J.; Kim, J. Domestic wastewater treatment as a net energy producer—Can this be achieved? *Environ. Sci. Technol.* **2011**, *45*, 7100–7106. [CrossRef] [PubMed]

282. Ohn, T.; Jami, M.S.; Iritani, E.; Mukai, Y.; Katagiri, N. Filtration behaviors in constant rate microfiltration with cyclic backwashing of coagulated sewage secondary effluent. *Sep. Sci. Technol.* **2003**, *38*, 951–966. [CrossRef]

283. Lin, J.C.T.; Lee, D.J.; Huang, C. Membrane fouling mitigation: Membrane cleaning. *Sep. Sci. Technol.* **2010**, *45*, 858–872. [CrossRef]

284. Ma, H.; Bowman, C.N.; Davis, R.H. Membrane fouling reduction by backpulsing and surface modification. *J. Membr. Sci.* **2000**, *173*, 191–200. [CrossRef]

285. Ma, H.; Hakim, L.F.; Bowman, C.N.; Davis, R.H. Factors affecting membrane fouling reduction by surface modification and backpulsing. *J. Membr. Sci.* **2001**, *189*, 255–270. [CrossRef]

286. De Souza, N.P.; Basu, O.D. Comparative analysis of physical cleaning operations for fouling control of hollow fiber membranes in drinking water treatment. *J. Membr. Sci.* **2013**, *436*, 28–35. [CrossRef]

287. Ferrer, O.; Lefèvre, B.; Prats, G.; Bernat, X.; Gibert, O.; Paraira, M. Reversibility of fouling on ultrafiltration membrane by backwashing and chemical cleaning: Differences in organic fractions behaviour. *Desalination Water Treat.* **2016**, *57*, 8593–8607. [CrossRef]

288. Wu, J.; Le-Clech, P.; Stuetz, R.M.; Fane, A.G.; Chen, V. Effects of relaxation and backwashing conditions on fouling in membrane bioreactor. *J. Membr. Sci.* **2008**, *324*, 26–32. [CrossRef]

289. Bessiere, Y.; Guigui, C.; Remize, P.J.; Cabassud, C. Coupling air-assisted backwash and rinsing steps: A new way to improve ultrafiltration process operation for inside-out hollow fibre modules. *Desalination* **2009**, *240*, 71–77. [CrossRef]

290. Wu, J.; Le-Clech, P.; Stuetz, R.M.; Fane, A.G.; Chen, V. Novel filtration mode for fouling limitation in membrane bioreactors. *Water Res.* **2008**, *42*, 3677–3684. [CrossRef] [PubMed]

291. Ye, Y.; Chen, V.; Le-Clech, P. Evolution of fouling deposition and removal on hollow fibre membrane during filtration with periodical backwash. *Desalination* **2011**, *283*, 198–205. [CrossRef]

292. Howe, K.J.; Marwah, A.; Chiu, K.-P.; Adham, S.S. Effect of membrane configuration on bench-scale mf and uf fouling experiments. *Water Res.* **2007**, *41*, 3842–3849. [CrossRef] [PubMed]

293. Akhondi, E.; Zamani, F.; Chew, J.W.; Krantz, W.B.; Fane, A.G. Improved design and protocol for evapoporometry determination of the pore-size distribution. *J. Membr. Sci.* **2015**, *496*, 334–343. [CrossRef]

294. Krantz, W.B.; Greenberg, A.R.; Kujundzic, E.; Yeo, A.; Hosseini, S.S. Evapoporometry: A novel technique for determining the pore-size distribution of membranes. *J. Membr. Sci.* **2013**, *438*, 153–166. [CrossRef]

295. Serra, C.; Durand-Bourlier, L.; Clifton, M.J.; Moulin, P.; Rouch, J.-C.; Aptel, P. Use of air sparging to improve backwash efficiency in hollow-fiber modules. *J. Membr. Sci.* **1999**, *161*, 95–113. [CrossRef]

296. Christensen, M.L.; Bugge, T.V.; Hede, B.H.; Nierychlo, M.; Larsen, P.; Jørgensen, M.K. Effects of relaxation time on fouling propensity in membrane bioreactors. *J. Membr. Sci.* **2016**, *504*, 176–184. [CrossRef]

297. Chang, I.S.; Le Clech, P.; Jefferson, B.; Judd, S. Membrane fouling in membrane bioreactors for wastewater treatment. *J. Environ. Eng.* **2002**, *128*, 1018–1029. [CrossRef]

298. Drews, A. Membrane fouling in membrane bioreactors-characterisation, contradictions, cause and cures. *J. Membr. Sci.* **2010**, *363*, 1–28. [CrossRef]

299. Meng, F.G.; Chae, S.R.; Drews, A.; Kraume, M.; Shin, H.S.; Yang, F.L. Recent advances in membrane bioreactors (MBRs): Membrane fouling and membrane material. *Water Res.* **2009**, *43*, 1489–1512. [CrossRef] [PubMed]

300. Chen, J.P.; Kim, S.L.; Ting, Y.P. Optimization of membrane physical and chemical cleaning by a statistically designed approach. *J. Membr. Sci.* **2003**, *219*, 27–45. [CrossRef]

301. Porcelli, N.; Judd, S. Effect of cleaning protocol on membrane permeability recovery: A sensitivity analysis. *J. Am. Water Works Assoc.* **2010**, *102*, 78–86.

302. Trägårdh, G. Membrane cleaning. *Desalination* **1989**, *71*, 325–335. [CrossRef]

303. Zhang, Z.; Bligh, M.W.; Wang, Y.; Leslie, G.L.; Bustamante, H.; Waite, T.D. Cleaning strategies for iron-fouled membranes from submerged membrane bioreactor treatment of wastewaters. *J. Membr. Sci.* **2015**, *475*, 9–21. [CrossRef]

304. Kweon, J.H.; Jung, J.H.; Lee, S.R.; Hur, H.W.; Shin, Y.; Choi, Y.H. Effects of consecutive chemical cleaning on membrane performance and surface properties of microfiltration. *Desalination* **2012**, *286*, 324–331. [CrossRef]

305. Cote, P.; Alam, Z.; Penny, J. Hollow fiber membrane life in membrane bioreactors (MBR). *Desalination* **2012**, *288*, 145–151. [CrossRef]

306. Huisman, I.H.; Williams, K. Autopsy and failure analysis of ultrafiltration membranes from a waste-water treatment system. *Desalination* **2004**, *165*, 161–164. [CrossRef]

307. De Wilde, W.; Thoeye, C.; De Gueldre, G. Membrane life expectancy assessment after 3 years of MBR operation at WWTP schilde. In Proceedings of the 4th International Water Association Conference on Membranes for Water and Wastewater Treatment, Harrogate, UK, 15–17 May 2007.

308. Benavente, J.; Vázquez, M. Effect of age and chemical treatments on characteristic parameters for active and porous sublayers of polymeric composite membranes. *J. Colloid Interface Sci.* **2004**, *273*, 547–555. [CrossRef] [PubMed]

309. Thominette, F.; Farnault, O.; Gaudichet-Maurin, E.; Machinal, C.; Schrotter, J.-C. Ageing of polyethersulfone ultrafiltration membranes in hypochlorite treatment. *Desalination* **2006**, *200*, 7–8. [CrossRef]

310. Anton, E.; Alvarez, J.R.; Palacio, L.; Pradanos, P.; Hernandez, A.; Pihlajamaki, A.; Luque, S. Ageing of polyethersulfone ultrafiltration membranes under long-term exposures to alkaline and acidic cleaning solutions. *Chem. Eng. Sci.* **2015**, *134*, 178–195. [CrossRef]

311. Akhondi, E. Submerged Hollow Fibre Membrane Fouling: Characterization and Control. Ph.D. Thesis, Nanyang Technological University, Singapore, 2014.

312. Mallevialle, J.; Odendaal, P.E.; Foundation, A.R.; Wiesner, M.R.; eaux-Dumez, L.D.; Commission, S.A.W.R. *Water Treatment Membrane Processes*; McGraw-Hill: New York, NY, USA, 1996.

313. Childress, A.; Le-Clech, P.; Daugherty, J.; Chen, C.; Leslie, G. Mechanical analysis of hollow fiber membrane integrity in water reuse applications. *Desalination* **2005**, *180*, 5–14. [CrossRef]

314. Gijsbertsen-Abrahamse, A.J.; Cornelissen, E.R.; Hofman, J.A.M.H. Fiber failure frequency and causes of hollow fiber integrity loss. *Desalination* **2006**, *194*, 251–258. [CrossRef]

315. Johnson, W.; MacCormick, T. Issues of operational integrity in membrane drinking water plants. *Water Sci. Technol. Water Supply* **2003**, *3*, 73–80.

316. Arkhangelsky, E.; Kuzmenko, D.; Gitis, N.; Vinogradov, M.; Kuiry, S.; Gitis, V. Hypochlorite cleaning causes degradation of polymer membranes. *Tribol. Lett.* **2007**, *28*, 109–116. [CrossRef]

317. Antony, A.; Leslie, G. Degradation of polymeric membranes in water and wastewater treatment. In *Advanced Membrane Science and Technology for Sustainable Energy and Environmental Applications*; Woodhead Publishing Limited: Cambridge, UK, 2011; pp. 718–745.

318. Gaudichet-Maurin, E.; Thominette, F. Ageing of polysulfone ultrafiltration membranes in contact with bleach solutions. *J. Membr. Sci.* **2006**, *282*, 198–204. [CrossRef]

319. Wang, P.; Wang, Z.; Wu, Z.; Zhou, Q.; Yang, D. Effect of hypochlorite cleaning on the physiochemical characteristics of polyvinylidene fluoride membranes. *Chem. Eng. J.* **2010**, *162*, 1050–1056. [CrossRef]

320. Regula, C.; Carretier, E.; Wyart, Y.; Gésan-Guiziou, G.; Vincent, A.; Boudot, D.; Moulin, P. Chemical cleaning/disinfection and ageing of organic uf membranes: A review. *Water Res.* **2014**, *56*, 325–365. [CrossRef] [PubMed]

321. Zappia, L.R.; Hayes, D.; Nolan, P. Source water characterisation and implications for ultrafiltration in the east kimberly, western australia. In Proceedings of the AWA Membranes and Desalination Conference, Brisbane, Australia, 1 July 2013.

322. Chew, J.W.; Hays, R.; Findlay, J.G.; Knowlton, T.M.; Karri, S.B.R.; Cocco, R.A.; Hrenya, C.M. Reverse core-annular flow of geldart group b particles in risers. *Powder Technol.* **2012**, *221*, 1–12. [CrossRef]

323. King, D.H.; Smith, J.W. Wall mass transfer in liquid-fluidized beds. *Can. J. Chem. Eng.* **1967**, *45*, 329–333. [CrossRef]

324. Zhu, J.; Grace, J.R.; Lim, C.J. Tube wear in gas-fluidized beds. 1. Experimental findings. *Chem. Eng. Sci.* **1990**, *45*, 1003–1015. [CrossRef]

325. Zhu, J.; Lim, C.J.; Grace, J.R.; Lund, J.A. Tube wear in gas-fluidized beds. 2. Low velocity impact erosion and semiempirical model for bubbling and slugging fluidized-beds. *Chem. Eng. Sci.* **1991**, *46*, 1151–1156. [CrossRef]

326. Bethune, B. Surface cracking of glassy polymers under a sliding spherical indenter. *J. Mater. Sci.* **1976**, *11*, 199–205. [CrossRef]

327. Cicek, N.; Dionysiou, D.; Suidan, M.T.; Ginestet, P.; Audic, J.M. Performance deterioration and structural changes of a ceramic membrane bioreactor due to inorganic abrasion. *J. Membr. Sci.* **1999**, *163*, 19–28. [CrossRef]

328. Doll, T.E.; Frimmel, F.H. Cross-flow microfiltration with periodical back-washing for photocatalytic degradation of pharmaceutical and diagnostic residues-evaluation of the long-term stability of the photocatalytic activity of TiO_2. *Water Res.* **2005**, *39*, 847–854. [CrossRef] [PubMed]

329. Lawn, B.; Wilshaw, R. Indentation fracture—Principles and applications. *J. Membr. Sci.* **1975**, *10*, 1049–1081. [CrossRef]

330. Nicholson, D.W. *Finite Element Analysis: Thermomechanics of Solids*; CRC Press: Boca Raton, FL, USA, 2003.

331. Xie, Z.; Swain, M.V.; Hoffman, M.J. Structural integrity of enamel: Experimental and modeling. *J. Dent. Res.* **2009**, *88*, 529–533. [CrossRef] [PubMed]

332. Bacchin, P.; Aimar, P.; Field, R.W. Critical and sustainable fluxes: Theory, experiments and applications. *J. Membr. Sci.* **2006**, *281*, 42–69. [CrossRef]

333. Phattaranawik, J.; Fane, A.G.; Pasquier, A.C.S.; Bing, W. A novel membrane bioreactor based on membrane distillation. *Desalination* **2008**, *223*, 386–395. [CrossRef]

334. Yap, W.J.; Zhang, J.S.; Lay, W.C.L.; Cao, B.; Fane, A.G.; Liu, Y. State of the art of osmotic membrane bioreactors for water reclamation. *Bioresour. Technol.* **2012**, *122*, 217–222. [CrossRef] [PubMed]

335. Achilli, A.; Cath, T.Y.; Childress, A.E. Power generation with pressure retarded osmosis: An experimental and theoretical investigation. *J. Membr. Sci.* **2009**, *343*, 42–52. [CrossRef]

336. Cath, T.Y.; Childress, A.E.; Elimelech, M. Forward osmosis: Principles, applications, and recent developments. *J. Membr. Sci.* **2006**, *281*, 70–87. [CrossRef]

337. Krantz, W.B.; Lin, C.S.; Sin, P.C.Y.; Yeo, A.; Fane, A.G. An integrity sensor for assessing the performance of low pressure membrane modules in the water industry. *Desalination* **2011**, *283*, 117–122. [CrossRef]

MDPI

St. Alban-Anlage 66

4052 Basel

Switzerland

Tel. +41 61 683 77 34

Fax +41 61 302 89 18

www.mdpi.com

Applied Sciences Editorial Office

E-mail: applsci@mdpi.com

www.mdpi.com/journal/applsci